Sustainability and Innovation

Technological, institutional and social innovation promotes economic development and international competitiveness, and can do much to reduce environmental burdens. Thus, innovation is an essential factor in the realisation of the principles of sustainable development.

The series 'Sustainability and Innovation' aims to positively contribute to a better understanding of the connection between innovation and sustainability. It is interdisciplinary in coverage, concentrating primarily on the links between environmental and innovation research. Individual titles in the series focus on topical issues and cover both theoretical and empirical aspects: concepts of major practical relevance are also discussed. The series is addressed to researchers and policy makers, as well as to specialists and leaders in business and industry.

More information about this series at http://www.springer.com/series/6891

Jens Horbach • Christiane Reif
Editors

New Developments in Eco-Innovation Research

Springer

Editors
Jens Horbach
Faculty of Business
Augsburg University of Applied Sciences
Augsburg, Germany

Christiane Reif
Centre for European Economic Research (ZEW)
Mannheim, Germany

ISSN 1860-1030 ISSN 2197-926X (electronic)
Sustainability and Innovation
ISBN 978-3-030-06570-6 ISBN 978-3-319-93019-0 (eBook)
https://doi.org/10.1007/978-3-319-93019-0

This Springer imprint is published by the registered company Springer Nature Switzerland AG
The registered company address is: Gewerbestrasse 11, 6330 Cham, Switzerland

To the memory of Klaus Rennings

Contents

1 **New Developments in Eco-Innovation Research: Aim of the Book and Overview of the Different Chapters** 1
Jens Horbach and Christiane Reif

2 **Changing Patterns in Eco-Innovation Research: A Bibliometric Analysis** 13
Serdar Türkeli and René Kemp

3 **On the Economic Returns of Eco-Innovation: Where Do We Stand?** 55
Claudia Ghisetti

4 **Shaping System Innovation: Transformative Environmental Policies** 81
Klaus Jacob

5 **Outlook: Can Environmental Product Standards Enable Eco-Innovation?** 95
Albert Roger

6 **Disentangling Technological Innovations: A Micro-Econometric Analysis of their Determinants** 123
Andreas Ziegler

7 **The Impact of Resource Efficiency Measures on the Performance of Small and Medium-Sized Enterprises** 147
Jens Horbach

8 **Good Enough! Are Socially Responsible Companies the More Successful Environmental Innovators?** 163
Christiane Reif and Sascha Rexhäuser

9 Environmental Innovation and Corporate Sustainability: A 15-Year Comparison Based on Survey Data 193
Hüseyin Doluca, Benedikt Holzner, and Marcus Wagner

10 Effects of Innovation and Domestic Market Factors on OECD Countries' Exports of Wind Power Technologies 219
Joachim Schleich and Rainer Walz

11 Exploring the Role of Instrument Design and Instrument Interaction for Eco-Innovation: A Survey-Based Analysis of Renewable Energy Innovation in Germany 233
Karoline S. Rogge and Joachim Schleich

12 Corporate Social Responsibility in the Fashion Industry: How Eco-Innovations Can Lead to a (More) Sustainable Business Model in the Fashion Industry 257
Thomas Cleff, Gwen van Driel, Lisa-Marie Mildner, and Nadine Walter

13 Towards a Dynamic Understanding of Innovation Systems: An Integrated TIS-MLP Approach for Wind Turbines 277
Rainer Walz

List of Contributors

Thomas Cleff Pforzheim University, Pforzheim, Germany

Hüseyin Doluca University of Augsburg, Augsburg, Germany

Claudia Ghisetti European Commission, DG Joint Research Centre, Unit I.1, Modelling, Indicators & Impact Evaluation, Ispra, VA, Italy

Benedikt Holzner University of Augsburg, Augsburg, Germany

Jens Horbach Faculty of Business, Augsburg University of Applied Sciences, Augsburg, Germany

Klaus Jacob Freie Universität Berlin, Forschungszentrum für Umweltpolitik (ffu), Berlin, Germany

René Kemp United Nations University-MERIT, Maastricht University, Maastricht, The Netherlands

Lisa-Marie Mildner Pforzheim University, Pforzheim, Germany

Christiane Reif Centre for European Economic Research (ZEW), Mannheim, Germany

Sascha Rexhäuser Centre for European Economic Research (ZEW), Mannheim, Germany

Albert Roger Centre for European Economic Research (ZEW), Mannheim, Germany

Karoline S. Rogge SPRU – Science Policy Research Unit, University of Sussex, Brighton, UK

Fraunhofer Institute for Systems and Innovation Research ISI, Karlsruhe, Germany

Joachim Schleich Fraunhofer Institute for Systems and Innovation Research ISI, Karlsruhe, Germany

Grenoble Ecole de Management, Grenoble, France

Serdar Türkeli United Nations University-MERIT, Maastricht University, Maastricht, The Netherlands

Gwen van Driel Tias School for Business and Society, Tilburg University, Utrecht, The Netherlands

Marcus Wagner Chair of Management, Innovation and International Business, University of Augsburg, Augsburg, Germany

Nadine Walter Pforzheim University, Pforzheim, Germany

Rainer Walz Competence Center Sustainability & Infrastructure Systems, Fraunhofer Institute for Systems and Innovation Research ISI, Karlsruhe, Germany

Andreas Ziegler Department of Economics, University of Kassel, Kassel, Germany

List of Figures

Fig. 2.1	Cumulative CO_2 emissions, 1850–2011 (% of world total)	14
Fig. 2.2	Co-occurrence analysis: temporal network of the authors' keywords used in eco-innovation reviews (n: 24 reviews)	37
Fig. 2.3	Co-occurrences analysis: received citation-scored network of the authors' keywords in eco-innovation reviews (n: 24 reviews)	38
Fig. 2.4	Co-occurrence analysis/no of keywords: 178 (two or more occurrences) authors' keywords, legend: average normalized citations (n: 350)	39
Fig. 2.5	Co-occurrence analysis/no of keywords: 178 (two or more occurrences) authors' keywords, legend: average publication year (n: 350)	40
Fig. 2.6	Authors (Co-citation analysis, unit of analysis: cited first author, weight: total citations received per author)	43
Fig. 2.7	Organizations: bibliographic coupling, legend: avg. publication year	43
Fig. 2.8	Journals: citation analysis: legend: avg. normalized citations, size of nodes: number of publications	44
Fig. 2.9	Countries: bibliographic coupling of countries, normalized citations received by the country	44
Fig. 2.10	Eco-innovation research and publications	48
Fig. 2.11	Co-authorship network: unit of analysis: country level, weights: citations received	49
Fig. 3.1	Eco-innovation adoption for end users, by sector	64
Fig. 3.2	Eco-innovation adoption by firms, by sector	65
Fig. 3.3	Eco-innovation adoption for end users, by country	65
Fig. 3.4	Eco-innovation adoption by firms, by country	66
Fig. 3.5	Eco-innovation and employment, scatter plot and fit	67
Fig. 3.6	Eco-innovation and value added, scatter plot and fit	68

Fig. 5.1 Classification of different information transmission approaches in EPS 96
Fig. 5.2 Types of EPS by Sector 100
Fig. 5.3 Comparing market objectives of different types of EPS 102

Fig. 8.1 Unbalanced panel data sample (6737 observations) 181
Fig. 8.2 Overview panel data and innovation and CSR categories by continents 181
Fig. 8.3 Overview panel data and innovation and CSR categories 182

Fig. 9.1 Company size by year and country 196
Fig. 9.2 Adoption of EMS 201
Fig. 9.3 Boxplots of operational activities vs. EMS 202
Fig. 9.4 Boxplots of managerial activities vs. EMS 202
Fig. 9.5 Boxplot of operational activities by aggregate size category in Germany 206
Fig. 9.6 Boxplots of operational activities by aggregate industry category in Germany 209
Fig. 9.7 Boxplot of managerial activities by aggregate size category in Germany 211
Fig. 9.8 Boxplots of managerial activities by aggregate industry category in Germany 213

Fig. 10.1 Annual export volume of wind power technology for 12 OECD countries 222

Fig. 11.1 Analytical framework for exploring the determinants of eco-innovation 236

Fig. 13.1 Level of aggregation of technological innovation system within the multi-level perspective 284
Fig. 13.2 World shares and specialization of China's wind power industry 285
Fig. 13.3 World shares and specialization of China's wind power industry 288
Fig. 13.4 Feedback loops and actors involved in the first development phase of the Chinese wind energy innovation system 289
Fig. 13.5 Feedback loops and actors involved in the take-off phase of the Chinese wind energy innovation system 290
Fig. 13.6 Feedback loops and actors involved in the mature phase of the Chinese wind energy innovation system 292

List of Tables

Table 2.1	Varieties and selections of labels (n: number of publications in WoS core collection)	29
Table 2.2	Eco-innovation	31
Table 2.3	Environmental innovation	32
Table 2.4	Common set, Top 3 articles w.r.t. received citations	32
Table 2.5	Comparative content analysis through co-occurrences of certain keywords	33
Table 2.6	Comparative content analysis through co-occurrences of certain keywords	34
Table 2.7	A review of reviews	35
Table 2.8	Highly-cited keywords published in the last 5 years (Circular economy-related themes in bold)	41
Table 2.9	Multi-level funding structure in the Netherlands	45
Table 2.10	Multi-level funding structure in Spain	46
Table 2.11	Funding agencies of the most cited articles published in 2015	47
Table 2.12	Main journals for eco-innovation scientific knowledge dissemination (>1%)	50
Table 2.13	Citing countries (Top 25)	51
Table 3.1	Sectors included in the analysis	63
Table 3.2	Main EI variables	63
Table 3.3	Variables descriptive statistics	69
Table 3.4	Log-linear model estimations	70
Table 3.5	Log-linear model estimations augmented by environmental policy	72
Table 3.6	First difference estimation results	74
Table 5.1	Overview of behavioral biases	112
Table 5.2	Summary behavioral experiments on EPS part I	114
Table 5.3	Summary behavioral experiments on EPS part II	116

Table 6.1 Descriptive statistics 132
Table 6.2 Pairwise (Pearson) correlation coefficients, number of firms varies between 441 and 475 133
Table 6.3 Maximum likelihood estimates in bivariate probit models (382 observations in model 1, 386 observations in model 2) ... 135
Table 6.4 Simulated maximum likelihood estimates in the multivariate probit model (372 observations) 136

Table 7.1 Distribution of the sample by firm size and sectors 152
Table 7.2 Resource efficiency measures and green products 153
Table 7.3 Performance effects of resource efficiency measures—all sectors 155
Table 7.4 Performance effects of resource efficiency measures by different sectors 157

Table 8.1 Adoption decision and relative frequencies 173
Table 8.2 Estimation results base models 176
Table 8.3 Estimation Results GMM Model 4 178
Table 8.4 Overview industry sectors (6737 observations) 182
Table 8.5 Overview variables and descriptive statistics (6737 observations) 183
Table 8.6 Overview environmental R&D and CSR variables 183
Table 8.7 Estimation Results GMM Models 5 and 6 184
Table 8.8 Estimation results GMM models for manufacturing industries . . 185
Table 8.9 Estimation results GMM models for CSR variable sustainability external audit 187
Table 8.10 Estimation results Anderson-Hsiao estimator 189

Table 9.1 Operational activities by country and year 197
Table 9.2 Managerial activities by country and year 199
Table 9.3 Internal social sustainability activities 203
Table 9.4 External social sustainability activities 204
Table 9.5 Firm size distribution by aggregate industry category 206
Table 9.6 Adoption of operational activities by aggregate size category in Germany 208
Table 9.7 Adoption of operational activities by aggregate industry category in Germany 210
Table 9.8 Adoption of managerial activities by aggregate size category in Germany 212
Table 9.9 Adoption of managerial activities by aggregate industry category in Germany 214

Table 10.1 Definition of variables 223
Table 10.2 Descriptive statistics of dependent and explanatory variables (1991–2011) 225
Table 10.3 Results for fixed-effects and random-effects estimators (*p*-values in parentheses) 227

Table 11.1 Operationalization of variables for instrument design features of the EEG 241
Table 11.2 Descriptive statistics of dependent and explanatory variables ($N = 160$) 243
Table 11.3 Regression results 246

Table 13.1 History of the Chinese wind energy industry and policy 287

Chapter 1
New Developments in Eco-Innovation Research: Aim of the Book and Overview of the Different Chapters

Jens Horbach and Christiane Reif

1.1 Introduction

Eco-innovations are crucial for reducing the environmental damages arising from economic activities. They can be regarded as one of the main drivers of a successful transition towards sustainable development and for the solution of climate change problems. Companies are increasingly attaching great importance to eco-innovation due to the growing environmental concerns of consumers and governments, but also because of long-run benefits. During the last 15 years, the literature on eco-innovation has been growing fast. The European Commission considers eco-innovation to be key to addressing global environmental and economic challenges. For this purpose, the specially convened 'Eco-Innovation Observatory' provides information and analyses of the trends in eco-innovation.

The book is dedicated to Dr. Klaus Rennings, one of the leading researchers in eco-innovation, who unexpectedly passed away in September 2015. In his memory, we organized a workshop in order to present and discuss ongoing and future developments in eco-innovation research. The workshop on "New Developments in Eco-Innovation Research" took place at the Centre for European Economic Research (ZEW) in Mannheim/Germany in November 2016. Some of the contributions to this book were among the papers presented at the workshop. This chapter introduces the basic concept of eco-innovation and provides an outline of the consecutive chapters.

J. Horbach (✉)
Faculty of Business, Augsburg University of Applied Sciences, Augsburg, Germany
e-mail: jens.horbach@hs-augsburg.de

C. Reif
Centre for European Economic Research (ZEW), Mannheim, Germany
e-mail: reif@zew.de

J. Horbach, C. Reif (eds.), *New Developments in Eco-Innovation Research*, Sustainability and Innovation, https://doi.org/10.1007/978-3-319-93019-0_1

Why do we need specific analyses of eco-innovation compared with other innovations? An answer to this question firstly requires a definition of eco-innovation. In the literature, other terms for eco-innovation are green or environmental innovation (e.g. Kemp 2010; Ghisetti and Pontoni 2015; Horbach 2018). We use these terms interchangeably (like e.g. Horbach et al. 2012), although other authors differentiate between them according to whether they effect environmental performance only or also economic performance (e.g. Ghisetti and Pontoni 2015). The Oslo-Manual by the OECD (2005: 46) defines innovation as "[. . .] the implementation of a new or significantly improved product (good or service), or process, a new marketing method, or a new organizational method in business practices, workplace organization or external relations." As such, it subsumes product, process, organizational and marketing innovations. While the Oslo-Manual points out environmental factors and effects of innovation, it does not provide an explicit definition of eco-innovation. A definition of eco-innovation has been developed in the MEI (Measuring Eco-Innovation) project (Kemp and Pearson 2008: 7): "Eco-innovation is the production, assimilation or exploitation of a product, production process, service or management or business method that is novel to the organisation (developing or adopting it) and which results, throughout its life cycle, in a reduction of environmental risk, pollution and other negative impacts of resources use (including energy use) compared to relevant alternatives."

The crucial distinction between eco-innovations and other, conventional innovations is therefore the environmental performance, which incorporates a reduction of the natural resources used as well as of the harmful substances released, irrespective of whether the protection of the environment was initially intended or not (see e.g. Driessen and Hillebrand 2002; Horbach et al. 2012). Furthermore, eco-innovations have specific determinants. A fast growing empirical literature shows that incentives like regulation measures, organizational innovation activities and cost-savings are more important for eco-innovations than for other innovations (see Horbach 2018 for an overview of this literature).

After clarifying the term eco-innovation, it is necessary to recap the historical roots of eco-innovation research to obtain a better understanding of the current state of the art and future trends. In its 'Limits to Growth' report (Meadows et al. 1972), the Club of Rome already pointed out the crucial role of innovation to preserve the environment and discuss innovation in the context of sustainability. The sustainability notion was taken up by the 'International Union for Conservation of Nature (IUCN)', which introduced the term 'sustainable development'. This term was conceptualized and defined within the Brundtland report (WCED 1987: 16) as "Humanity has the ability to make development sustainable to ensure that it meets the needs of the present without compromising the ability of future generations to meet their own needs. The concept of sustainable development does imply limits—not absolute limits but limitations imposed by the present state of technology and social organization on environmental resources and by the ability of the biosphere to absorb the effects of human activities. But technology and social organization can be both managed and improved to make way for a new era of economic growth." While the Club of Rome (1972) only lists ecological and economic stability as sustainability, in the definition of the Brundtland report (WCED 1987) sustainable development includes economic, ecologic and social concerns, which are nowadays the common three dimensions of sustainability. Furthermore, the ‘Earth Summit’ in Rio in 1992 laid the foundation for global principles on sustainable development and further political ambitions like the UN sustainable development goals. Although eco-innovations emerge in the realm of sustainability concerns, they do not include social aspects. Therefore, the term ‘sustainable innovation’ has become established as a broader concept of innovation incorporating all three dimensions of sustainability, which distinguishes it from eco-innovations. However, there is no common definition of ‘sustainable innovation’ due to missing clear-cut definitions of sustainability and innovation (see e.g. Horbach 2005). Klewitz and Hansen (2013) review journal articles from 1987 to 2010 and conclude that the research focus is mainly related to eco-innovation rather than sustainable innovation including all three dimensions of sustainability. In fact, this is not surprising because of possible conflicts between different sustainability dimensions. For example, the introduction of electro-mobility might be accompanied by positive environmental effects and even market opportunities but might also cause a reduction in the number of jobs. An empirical assessment of the sustainability of this innovation would then require a normative weighting between different sustainability dimensions (Horbach 2018).

Despite the definition of the different terms in the context of sustainability, the crucial question is how the industry has responded to this politically driven shift towards sustainable development? One reaction of companies towards the introduction of new regulations or the threat of tougher regulations is innovation. The often criticized and tested (e.g. Rexhäuser and Rammer 2014) ‘Porter hypothesis’ by Porter and van der Linde (1995) goes even a step further by suggesting that the shift not only encourages innovations but also leads to a competitive advantage. Examples like the ‘Montreal Protocol’, ratified by all UN member states in 2009, show industries’ reaction, which in this case consisted in the creation of innovative products replacing ozone-depleting substances. Another example are ‘Environmental Management Systems (EMS)’, supporting companies in implementing eco-innovations by helping them overcome barriers. Empirical research has proven the importance of EMS especially for technological innovations that reduce costs (e.g. Rennings et al. 2006; Rehfeld et al. 2007; Wagner 2007, 2008; Khanna et al. 2009; Ziegler and Nogareda 2009).

This movement in politics and the industrial reactions to has also resulted in a change of eco-innovation research. Rennings (2000) raised the question if sustainable innovations—including eco-innovations—call for a particular theory and specific policies or if the same methodological and theoretical frameworks as for conventional innovations applies. Since then, the interest in eco-innovation research has been increasing. This gain in importance in the research field of innovation has also led to literature reviews focusing on different perspectives. Berkhout (2011) provides a review on eco-innovation identifying the four social science research trends: disaggregated empirical analysis, knowledge and technology flows, connection of economic and physical models, and attention to policy and governance of eco-innovation. In a recent overview applying a main path analysis, Barbieri et al.

(2016) have shown that eco-innovation research has so far mainly concentrated on the following fields: determinants of eco-innovation, economic and environmental effects and policy inducement.

Shifts in the research field are the main focus of the review by Türkeli and Kemp (Chap. 2 of this book). They base their review on different bibliometric analyses on Web of Science Core Collection data and can identify seven perspectives researchers take in their analyses of eco-innovation (e.g. supply or demand side). Also the analytical concepts used have been changing, e.g. industrial ecology, industrial symbiosis, and circular economy. The review by Türkeli and Kemp already gives as an impression of how the research in eco-innovation has developed and which future tendencies can be expected. In the following paragraphs, we take a closer look at the current state of empirical research in eco-innovation. We will first draw attention to present reviews provided in the book which cover different focus areas. Next, we discuss current empirical research findings on eco-innovation which set future trends in the research field.

1.2 Current State of the Art in Eco-Innovation

In the last decades, politics, industry and research especially wanted to know if eco-innovations benefit the environment and at the same time reduce costs or increase revenues. The economic effects of eco-innovations are of particular interest for policy-makers to avoid resentment of environmental regulation. Furthermore, evidence of a win-win situation could pave the way for companies voluntarily investing in eco-innovations to gain a competitive advantage. The related empirical research is mainly influenced and encouraged by the so-called Porter hypothesis (Porter and van der Linde 1995). The weak version of this hypothesis postulates that regulation triggers eco-innovation whereas the strong version says that regulation driven eco-innovation even leads to higher competitiveness. The empirical evidence for the weak version is often confirmed (e. g. Demirel and Kesidou 2011; Horbach et al. 2012) whereas there are only mixed results for the strong version (Lanoie et al. 2011).

Ambec and Lanoie (2008) provide an overview on empirical results concerning environmental and economic performance. They show that a positive link cannot always be made. Therefore, they identify seven aspects, distinguished into the increase in revenues and the reduction of costs, in which environmental and economic performance go hand in hand. Nevertheless, they also stress that even after decades of research efforts it is still not clear if environmental performance influences economic performance or vice versa, or if there is another factor influencing both. Also, Orsato (2009) emphasizes that only in some cases environmental performance also leads to an economic benefit. He claims that eco-investments are only beneficial for the company if stakeholders value this action. Ghisetti (Chap. 3) reviews the current state of the art in eco-innovation research by summarizing the main findings on how eco-innovations affect economic performance. She gives an extensive overview on different levels of economic performance measurement, taking short- and long-term performance into account. Furthermore, she provides new empirical evidence on the economic effects of eco-innovations, using panel data from the Community Innovation Surveys (CIS) on selected European countries and sectors. Chapter 3 also contains first results on the current CIS 2014 on eco-innovation.

The databases and their adaptation to the current political and research agenda already indicate the emergence of new political targets in the realm of eco-innovation and sustainability. As Chap. 2 (Türkeli and Kemp) focuses on the shifts in the research of eco-innovation, the following two Chaps. 4 and 5 provide the policy perspective.

At the latest the UN Agenda 2030, adopted at the United Nations Sustainable Development Summit in 2015, initiates a new era in the political agenda setting with the transition to sustainable development. It is also on the European and on national agendas. Such a global goal is only reachable with technological changes and innovation on different levels. Although the first steps have been made with the UN resolution "Transforming our world: the 2030 Agenda for Sustainable Development" (A/RES/70/1) (United Nations 2015), the very details determining at which level policies should be implanted are unclear. Especially innovations incorporating environmental and social concerns are important to address global challenges like climate change and to pave the way for long-term transformation. These interrelated innovations are also called socio-technical systems and are the focus of Chap. 4. Jacob (Chap. 4) disaggregates the elements of such systems based on a literature review. He classifies different challenges and obstacles and specifies options to counteract them. Finally, he summarizes the findings with a toolbox for transformative environmental policies.

Despite this general view on environmental policy making, the introduction of product standards provides a possibility to meet sustainable development goals and supports the transformation process. Environmental product standards (EPS) enable firms to signal environmentally friendly aspects of their products to overcome information asymmetry. Although firms face rather high costs, they voluntarily engage in the certification process to convince consumers of their product in order to acquire customers. Recently, more and more certificates and labels have popped up and with this increasing number of certificates it becomes more difficult for firms, consumers and policy makers to judge the quality of a label and therefore the labelled product itself. Moreover, firms have more difficulties choosing suitable labels for specific products. Roger (Chap. 5) provides a systematic overview on EPS and how they support eco-innovation. For this purpose, different types of labels are distinguished and illustrated by best-practice examples. Additionally, Roger also provides the drivers of a successful implementation of EPS.

1.3 New Empirical Findings and Ways Forward

Although research in eco-innovation can be still classified as a young research line, especially in the last decade the number of publications in this field has risen enormously. This can be traced back to the historical developments and political agenda setting but also to a change in consumers' behavior and preferences as well as managers' attitude towards environmental friendly products, e.g. firms incorporating economic as well as environmental targets in their business plans.

Thus, an understanding of the drivers of eco-innovations is important for all actors and therefore a main empirical research topic (see e.g. Barbieri et al. 2016 for an overview). These determinants are roughly differentiated into supply side, demand side, as well as institutional and political influences (Horbach 2008). The research conducted on this topic is based on firm level data (e.g. Demirel and Kesidou 2011; Horbach et al. 2012), which allow the inclusion of different explanatory variables depending on the database. Empirical studies have confirmed regulation as a key driver of innovation (e.g. Cleff and Rennings 1999; Horbach 2008; Barbieri et al. 2016).

Recent trends in eco-innovation research also take further determinants into account. Demirel and Kesidou (2011) bring together the three types of determinants—regulation, supply and demand side—in their study on over 1500 UK firms based on a 2006 survey. Peng and Liu (2016) focus on managerial awareness in their study on 144 firms in China. Horbach (2016) analyses in particular Eastern European countries in an analysis on 19 European countries. His study shows that especially regulation and subsidies trigger eco-innovation in these countries and moreover that firms in Eastern Europe depend on a technology transfer from Western countries. Another study by Horbach et al. (2012) shows that besides regulation also environmental management tools and R&D investments drive eco-innovations. This perspective is also picked up by Ziegler (Chap. 6). His analysis is based on the German manufacturing sector, includes over 300 firms, and focuses on technological innovation. Ziegler shows that R&D activities are important for all types of technological innovation. Furthermore, the analysis reveals environmental organizational measures—certified or not—as a crucial driver of environmental technological innovations.

The studies discussed above focus on large firms, whereas others particularly concentrate on SMEs. Klewitz et al. (2012) show the influential role of intermediaries like local authorities in triggering eco-innovation in SMEs. Triguero et al. (2013) provide a study on SMEs in 27 European countries, disentangling the effect of different drivers on specific types of eco-innovation. Chapter 7 contributes to this rather new research area of eco-innovation in SMEs. Horbach (Chap. 7) analyses the effect of pro environmental behavior on firm's economic performance. He confirms the general result that resource efficiency measures trigger economic performance also in SMEs. However, differentiating between measures, he also shows the positive effect of increasing renewables and the negative effect of the reduction of water consumption on the financial performance. Furthermore, firms' and employees' self-perceived identification with green values support eco-innovations. The research on SMEs is still in its early stage and will get more prominent in the future, as more data on SMEs will hopefully be available.

Another current development in eco-innovation research is the connection to sustainability. As introduced above, sustainability comprises economic, environmental and social dimensions. Firms' voluntary sustainable measures are commonly described as corporate social responsibility (CSR). Especially in the last years, research on the effect of CSR on financial performance of firms has gained importance and numerous studies indicate a positive relation (see meta-analyses by Orlitzky et al. 2003 and Margolis et al. 2007). However, environmental innovations are hardly taken into account in this research context. Exceptions are, e.g., McWilliams and Siegel (2000) and papers extending this study (e.g. Hull and Rothenberg 2008). Reif and Rexhäuser (Chap. 8) focus on the neglected link between CSR and environmental innovation. They specifically analyze the complementarity of CSR and environmental innovation, which would indicate a higher financial performance when both measures are introduced together. Based on a worldwide panel dataset, they confirm that environmental R&D and CSR measured by the participation in the Global Reporting Initiative (GRI) are complementary. However, the authors also stress that for other types of CSR measures the link to environmental innovations might have a different effect. The results in this context raise the question of the historical development of CSR activities and environmental innovation. Wagner and colleagues (Chap. 9) investigate this question by using a recent survey dataset of German and UK manufacturing firms including the years 2001–2016. On the aggregate level, they show an increase in the usage of environmental measures and EMS certification. However, by differentiating large and small firms, the analysis reveals that the former prefer ISO 14001 certification, while for the latter EMAS gains importance. Especially this result calls for a distinction between large and small enterprises in research but also informs policy makers and practitioners that firms should adapt their strategy depending on their size.

Focusing on specific industries to get more detailed insights constitutes another step forward in eco-innovation research. Regulation might differ not only from country to country but also by industry. Empirical research has mainly focused on the manufacturing sector. A further distinction within this sector is necessary to gain more knowledge on the specific circumstances of eco-innovation in sub-sectors. Smith and Crotty (2008) particularly investigate the impact of the EU 'End of Life Vehicles Directive (ELVD)' on the UK automotive sector. Their results show a rather low influence of this directive on product innovation. The authors call for more restrictive regulations to promote product innovation. Also Schleich and Walz (Chap. 10) contribute to this new research line by focusing on wind power. They observe how innovation and support policies influence the exports of wind power in a panel of twelve OECD countries. In their analysis, they differentiate between innovation input and output. They find a positive relation of wind power exports to both innovation measures and to supportive policies, but the effects of feed-in tariffs do not seem to differ from the effects of other support policies. Besides the importance to observe sub-sectors more specifically, this study also demonstrates the relevance of accounting for international interrelation.

In connection with that, another recent line in eco-innovation research is to account for policy mixes rather than a single policy [see Flanagan et al. (2011) and Rogge and Reichardt (2016) for a discussion on the term policy mix]. Normally, if a new policy is implemented other policies are already in place. These policies might interact with each other. Thus, focusing only on the effect of one specific policy could lead to wrong conclusions. This topic has been raised in the context of innovation policies for example by the OECD (2010). Especially innovation policy is characterized by different combined policy instruments (see e.g. Borrás and Edquist 2013; Magro and Wilson 2013) and becomes more pronounced within the attempts of achieving sustainable transitions, such as the transition to low-carbon energy systems (Rogge et al. 2017). For this purpose, further developments in existing datasets and setting up new data are necessary. Rogge and Schleich (Chap. 11) contribute to this young research field. The authors have specifically developed a policy mix module which they have incorporated into a standard company innovation survey based on the CIS to analyze renewable energy innovation in Germany. In their analysis the authors focus on the role of the interaction of multiple instruments and the role of instrument design features for innovation in renewable power generation technologies. This new attempt to observe the effects of policy mixes is informative for policy makers as well as researchers and therefore sets future research trends.

Another step in the direction of focusing on a specific industry and the generation of new data is the usage of case studies. Especially for those industries in which the supply chain across countries and several production steps makes it difficult to track eco-innovation and sustainability efforts, case studies are an important research tool. The fashion industry is an example for a sector with complex supply chains and additionally high usage of natural resources (see e.g. de Brito et al. 2008). Cleff et al. (Chap. 12) provide a case study, conducting a structured interview with ten participants. In general, the three participating fashion companies and the seven sustainability experts confirm that measures to increase resource efficiency are being implemented but only slowly. Moreover, they stress the importance of awareness on all levels—government, business and consumers—to make progress in sustainability issues. Cleff et al. conclude that especially eco-innovations are required to pave the way for a more sustainable fashion industry.

Incorporating dynamics has been a further expansion of eco-innovation research in recent years. Innovations in general follow paths. These path dependencies are especially interesting for sustainable transition and therefore also for the related eco- und sustainable innovations. There the consideration of interacting different levels is important for the conceptualization of research analyses. Walz (Chap. 13) provides such a concept by combining a technological innovation system with a multi-level perspective approach (see Markard and Truffer 2008 for a review on the usage of both concepts), which allows him to analyze the dynamics in the innovation process. Based on this methodology, he conducts a case study for the Chinese wind energy industry, which is an example for a sector characterized by different phases of innovation.

This short overview covers the different perspectives on 'New Developments in Eco-Innovation Research' taken by the following chapters. The historical view on eco-innovations shows us the development of the still young research field and particularly the boom in the last decade (Chap. 2). The reviews on the current state of the art in eco-innovation demonstrate the influence of political agenda setting in this context and show the ways forward with the main focus on regulation effects (Chaps. 3, 4, and 5). The contributions of current empirical research (Chaps. 6, 7, 8, 9, 10, 11, 12, and 13) reveal the new developments in eco-innovation research concerning research targets like sustainability, the focus on specific industries, the generation of new databases and the consideration of interaction effects. Altogether, it should provide practitioners, policy makers and researchers with the necessary information to further develop future strategies in eco-innovation in particular in the context of sustainable transition.

References

Ambec, S., & Lanoie, P. (2008). Does it pay to be green? A systematic overview. *Academy of Management Perspectives, 22*(4), 45–62.

Barbieri, N., Ghisetti, C., Gilli, M., & Nicolli, F. (2016). A survey of the literature on environmental innovation based on main path analysis. *Journal of Economic Surveys, 30*, 596–623.

Berkhout, F. (2011). Eco-innovation: Reflections on an evolving research agenda. *International Journal of Technology, Policy and Management, 11*(3/4), 191–197.

Borrás, S., & Edquist, C. (2013). The choice of innovation policy instruments. *Technological Forecasting and Social Change, 80*(8), 1513–1522.

Cleff, T., & Rennings, K. (1999). Determinants of environmental product and process innovation. *Environmental Policy and Governance, 9*(5), 191–201.

De Brito, M. P., Carbone, V., & Blanquart, C. M. (2008). Towards a sustainable fashion retail supply chain in Europe: Organisation and performance. *International Journal of Production Economics, 114*(2), 534–553.

Demirel, P., & Kesidou, E. (2011). Stimulating different types of eco-innovation in the UK: Government policies and firm motivations. *Ecological Economics, 70*(8), 1546–1557.

Driessen, P., & Hillebrand, B. (2002). Adoption and diffusion of green innovations. In G. C. Bartels & W. J. A. Nelissen (Eds.), *Marketing for sustainability: Towards transactional policy-making* (pp. 343–355). Amsterdam: IOS Press.

Flanagan, K., Uyarra, E., & Laranja, M. (2011). Reconceptualising the 'policy mix' for innovation. *Research Policy, 40*(5), 702–713.

Ghisetti, C., & Pontoni, F. (2015). Investigating policy and R&D effects on environmental innovation: A meta-analysis. *Ecological Economics, 118*, 57–66.

Horbach, J. (2005). Methodological aspects of an indicator system for sustainable innovation. In J. Horbach (Ed.), *Indicator systems for sustainable innovation. Sustainability and innovation* (pp. 1–19). Heidelberg: Physica.

Horbach, J. (2008). Determinants of environmental innovation – New evidence from German panel data sources. *Research Policy, 37*, 163–173.

Horbach, J. (2016). Empirical determinants of eco-innovation in European countries using the community innovation survey. *Environmental Innovation and Societal Transitions, 19*, 1–14.

Horbach, J. (2018). Determinants of eco-innovation at the firm level. In F. Boons & A. McMeekin (Eds.), *Handbook of sustainable innovation*. Cheltenham: Edward Elgar forthcoming.

Horbach, J., Rammer, C., & Rennings, K. (2012). Determinants of eco-innovations by type of environmental impact – The role of regulatory push/pull, technology push and market pull. *Ecological Economics, 78*, 112–122.

Hull, C. E., & Rothenberg, S. (2008). Firm performance: The interactions of corporate social performance with innovation and industry differentiation. *Strategic Management Journal, 29* (7), 781–789.

Kemp, R. (2010). Eco-innovation: Definition, measurement and open research issues. *Economia Politica, 27*, 397–420.

Kemp, R., & Pearson, P. (2008). *Final report MEI project about measuring eco-innovation*. Maastricht. Accessed March, 2018, from https://www.merit.unu.edu/meide10/

Khanna, M., Deltas, G., & Harrington, D. R. (2009). Adoption of pollution prevention techniques: The role of management systems and regulatory pressures. *Environmental and Resource Economics, 44*, 85–106.

Klewitz, J., & Hansen, E. G. (2013). Sustainability-oriented innovation of SMEs: A systematic review. *Journal of Cleaner Production, 65*, 57–75.

Klewitz, J., Zeyen, A., & Hansen, E. G. (2012). Intermediaries driving eco-innovation in SMEs: A qualitative investigation. *European Journal of Innovation Management, 15*(4), 442–467.

Lanoie, P., Laurent-Lucchetti, J., Johnstone, N., & Ambec, S. (2011). Environmental policy, innovation and performance: New insights on the Porter hypothesis. *Journal of Economics & Management Strategy, 20*(3), 803–842.

Magro, E., & Wilson, J. R. (2013). Complex innovation policy systems: Towards an evaluation mix. *Research Policy, 42*(9), 1647–1656.

Margolis, J. D., Elfenbein, H. A., & Walsh, J. (2007). Does it pay to be good? A meta-analysis and redirection of research on the relationship between corporate social and financial performance. *Ann Arbor, 1001*, 48109–41234. https://doi.org/10.2139/ssrn.1866371.

Markard, J., & Truffer, B. (2008). Technological innovation systems and the multi-level perspective: Towards an integrated framework. *Research Policy, 37*(4), 596–615.

McWilliams, A., & Siegel, D. (2000). Corporate social responsibility and financial performance: Correlation or misspecification? *Strategic Management Journal, 21*(5), 603–609.

Meadows, D. H., Meadows, D. L., Randers, J., & Behrens, W. W. (1972). The limits to growth. *New York, 102*, 27.

OECD. (2005). *Oslo manual: Guidelines for collecting and interpreting innovation data* (3rd ed.). Accessed March, 2018, from http://www.oecd-ilibrary.org/science-and-technology/oslo-manual_9789264013100-en

OECD. (2010). The innovation policy mix. In *OECD science, technology and industry outlook 2010* (pp. 251–279). Paris: OECD.

Orlitzky, M., Schmidt, F. L., & Rynes, S. L. (2003). Corporate social and financial performance: A meta-analysis. *Organization Studies, 24*(3), 403–441.

Orsato, R. (2009). *Sustainability strategies: When does it pay to be green?* (1st ed.). London: Palgrave Macmillan, INSEAD Business Press XXV–268.

Peng, X., & Liu, Y. (2016). Behind eco-innovation: Managerial environmental awareness and external resource acquisition. *Journal of Cleaner Production, 139*, 347–360.

Porter, M. E., & van der Linde, C. (1995). Towards a new conception of the environment-competitiveness relationship. *Journal of Economic Perspectives, 9*(4), 97–118.

Rehfeld, K. M., Rennings, K., & Ziegler, A. (2007). Integrated product policy and environmental product innovations: An empirical analysis. *Ecological Economics, 61*, 91–100.

Rennings, K. (2000). Redefining innovation – Eco-innovation research and the contribution from ecological economics. *Ecological Economics, 32*, 319–332.

Rennings, K., Ziegler, A., Ankele, K., & Hoffmann, E. (2006). The influence of different characteristics of the EU environmental management and auditing scheme on technical environmental innovations and economic performance. *Ecological Economics, 57*(1), 45–59.

Rexhäuser, S., & Rammer, C. (2014). Environmental innovations and firm profitability: Unmasking the porter hypothesis. *Environmental and Resource Economics, 57*(1), 145–167.

Rogge, K. S., Kern, F., & Howlett, M. (2017). Conceptual and empirical advances in analysing policy mixes for energy transitions. *Energy Research & Social Science, 33*, 1–10.

Rogge, K. S., & Reichardt, K. (2016). Policy mixes for sustainability transitions: An extended concept and framework for analysis. *Research Policy, 45*(8), 1620–1635.

Smith, M., & Crotty, J. (2008). Environmental regulation and innovation driving ecological design in the UK automotive industry. *Business Strategy and the Environment, 17*(6), 341–349.

Triguero, A., Moreno-Mondéjar, L., & Davia, M. A. (2013). Drivers of different types of eco-innovation in European SMEs. *Ecological Economics, 92*, 25–33.

United Nations, General Assembly. (2015). *Transforming our world: The 2030 agenda for sustainable development*. Resolution adopted by the General Assembly on 25 September 2015, (A/RES/70/1).

Wagner, M. (2007). On the relationship between environmental management, environmental innovation and patenting: Evidence from German manufacturing firms. *Research Policy, 36* (10), 1587–1602.

Wagner, M. (2008). Empirical influence of environmental management on innovation: Evidence from Europe. *Ecological Economics, 66*(2–3), 392–402.

WCED (World Commission on Environment and Development). (1987), *Report of the world commission on environment and development: Our common future*. London. Accessed March, 2018, from http://www.un-documents.net/our-common-future.pdf

Ziegler, A., & Nogareda, J. S. (2009). Environmental management systems and technological environmental innovations: Exploring the causal relationship. *Research Policy, 38*(5), 885–893.

Chapter 2
Changing Patterns in Eco-Innovation Research: A Bibliometric Analysis

Serdar Türkeli and René Kemp

2.1 Introduction

Eco-innovation has become a recognized concept, defined by various scholars (Fussler and James 1996; Rennings 2000; Kemp and Pearson 2007; Andersen 2008; Reid and Miedzinski 2008; Horbach et al. 2012) and institutions (EC 2006, 2011, 2012; EEA 2007; ASEIC 2011[1]; OECD 2010, 2012; UNEP 2014). This field of research has a 20-year history, although you might say that the field was established approximately 30 years ago with the Brundtland Report, published in 1987. Klewitz and Hansen (2014) indicate that a wide debate has emerged on eco-innovation (e.g. eco-design, cleaner production) and sustainability-oriented innovations (SOIs), that is, the integration of ecological and social aspects into products, processes, and organizational structures (Klewitz and Hansen 2014).

By November 2016, 47[2] countries around the world have produced scientific publications out of eco-innovation research. Considering 193 states in total, circa one-fourth of the countries around the world (~24%) show interest in eco-innovation scientific knowledge production. To some extent, this percentage might be considered low; however, these 47 countries represent approximately 70% of the total GHG emissions including the emissions stemming from land-use change and forestry according to World Resource Institute (WRI) 2012 data. In Fig. 2.1, other than Indonesia, we observe that all countries responsible for historical (1850-onwards) cumulative CO_2 emissions, also take part in publishable/published eco-innovation

[1]Citing European Commission.

[2]Based on Web of Science data, query: ("eco-inno*" OR "ecoinno*" OR "eco inno*") OR TITLE: ("eco-inno*" OR "ecoinno*" OR "eco inno*") Timespan: 1988–2016. Indexes: SCI-EXPANDED, SSCI, A&HCI, ESCI. November 18, 2016.

S. Türkeli (✉) · R. Kemp
United Nations University-MERIT, Maastricht University, Maastricht, The Netherlands
e-mail: turkeli@merit.unu.edu; r.kemp@maastrichtuniversity.nl

J. Horbach, C. Reif (eds.), *New Developments in Eco-Innovation Research*, Sustainability and Innovation, https://doi.org/10.1007/978-3-319-93019-0_2

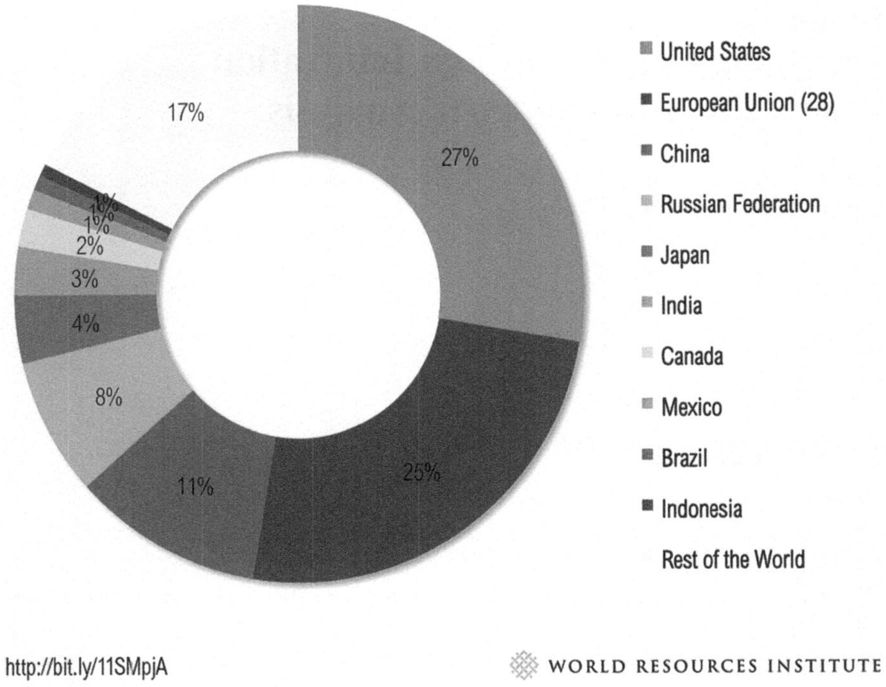

Fig. 2.1 Cumulative CO_2 emissions, 1850–2011 (% of world total)

scientific knowledge production (e.g. scientific publications appearing in peer-reviewed SSCI, SCI journals).

By November 2016, each continent or major world region has a scientific publication which directly relates to eco-innovation. In 2015 in North Africa, researchers from Algeria produced the country's first WoS-listed scientific publication in collaboration with co-authors from Canada on eco-design and eco-efficiency (Cherifi et al. 2015). In 2016, researchers in the Middle East published an article about a smart plug system for monitoring and controlling household energy consumption using a mobile application in the United Arab Emirates (Ghazal 2015). In the last 2–3 years, researchers from Latin America, Colombia [on eco-labels, (Prieto-Sandoval et al. 2016)], Chile [on eco-design/theory of inventive problem solving, (Vidal et al. 2015)] and Venezuela [on eco-efficiency in SMEs, (Fernandez-Vine 2013)] produced their first WoS-listed scientific publications in collaboration with co-authors from Spain. In Latin America, Brazil has also produced eco-innovation research outputs since 2010, and also in collaboration with Spain co-authors (e.g. Duran-Romero et al. 2015) and with French co-authors (e.g. Bossle et al. 2016) in the last two years. The role of eco-innovation research in Spain and its diffusion to Latin America could be an interesting point for further research and could serve to enhance the EU-Latin America scientific cooperation in knowledge production. Linguistic proximity might have a role in facilitating such cooperation.

In Europe, most of the production of scientific knowledge in eco-innovation takes place in Spain (n: 66). Italy (n: 47) and the UK (n: 43) form the second group based on number of publications, Germany (n: 35) and the Netherlands (n: 34) form the third group which is followed by France (n: 30). However, concerning the complexity of scientific knowledge production, these bulk numbers could mislead any interpretation or argumentation since a comprehensive multi-method analysis is needed to understand the field. Therefore, the research question of this chapter is: *What are the new developments and shifts in eco-innovation research? What are the findings of previous reviews? Why do these developments take place, and how are they governed?*

The chapter proceeds as follows. Section 2.2 provides a meta-review in eco-innovation research (n: 24 reviews). In Sect. 2.3 we list the data sources and methods of analysis used in this chapter. We discuss our findings in Sect. 2.4 in six sub-sections: 2.4.1 Variety and selection of different labels, 2.4.2 Comparative content analysis of eco-innovation and environmental innovation with respect to co-occurrences with certain keywords (n: 34 keywords), 2.4.3 A network analysis of temporal dynamics and influence of the authors' keywords in eco-innovation reviews, 2.4.4 Analysis of authors' keywords of overall eco-innovation literature (n: 350), 2.4.5 A multi-level analysis of eco-innovation literature (authors (co-citation analysis), organizations (bibliometric coupling), journals (citation analysis), countries (bibliometric coupling) and 2.4.6 the classical political economy of eco-innovation scientific knowledge (finance, production, distribution, and scientific use of eco-innovation literature), and we provide concluding knowledge production and policy remarks in Sect. 2.5.

2.2 A Review of the Reviews: Empirical Background

In this section, we provide a review of the reviews published for eco-innovation research. As a result of our initial analysis, these reviews can be classified into six main themes:

1. Methodological reviews
2. Firm-centered reviews
3. Technological—industrial/sectoral reviews
4. Science-centered reviews
5. Diffusion-centered reviews
6. Policy (tools and instrument)-centered reviews

2.2.1 Methodological Reviews

In this set, we have four key publications. Kemp and Pontoglio (2011) examine the innovation effects of environmental policy instruments in four kinds of literature:

(1) theoretical models on incentives for eco-innovation, (2) econometric studies based on observed data, (3) survey analysis based on stated information and (4) technology case studies. The aim of their review is to critically examine the methods and the results. The authors argue that even when case study results are specific and difficult to generalize, case studies are a necessary source of empirical evidence about the impact of policy and the factors responsible for this impact. Case studies point to issues that are neglected in the theoretical and econometric literature such as the specifics of the innovation context and policy interaction effects. The review article makes a plea for multi-method analysis.

On the other hand, the aim of the review of del Rio et al. (2016) is to provide a critical review of the literature on the econometric analyses of firm-level determinants of eco-innovation. The authors claim that their review reveals some important gaps in eco-innovation. The main gap detected is the lack of an integrated theoretical framework which merges the insights from different approaches (del Rio et al. 2016). In this respect, this critical review is in line with Kemp and Pontoglio (2011) in calling for multi-method research and analysis. Another issue for the econometric literature is that the influence of some variables (e.g. demand-pull and cost-savings) is still unsettled, whereas others (e.g. internal and international factors) have hardly been included in most of the analyses. Studies comparing the drivers of eco-innovation versus general innovation are relatively scarce with respect to those on the drivers of eco-innovation in general (del Rio et al. 2016, p. 4). Furthermore, analyses of the relevance of different determinants of eco-innovation for distinct eco-innovator and eco-innovation types have largely been missing (del Rio et al. 2016). Moreover, studies on middle-income and developing countries are still scarce (del Rio et al. 2016, p. 6). Besides that, the econometric analyses have relied on microeconometric methods based on cross-section data (mostly logit and probit models); the use of panel data is virtually absent. Detailed econometric analyses on the distinct drivers and barriers to eco-innovation in different sectors and regions have not been performed so far (del Rio et al. 2016, p. 8). Whether the position of the firm in the value chain and the market structure influence the propensity to eco-innovate are largely unexplored topics (del Rio et al. 2016).

By using path analysis, Barbieri et al. (2016) review the literature on environmental innovation (EI) and systematize it by identifying the main directions in which the literature on EI has developed over time (Barbieri et al. 2016). In order to do so, the authors use two algorithms to analyze a citation network of journal articles and books (Barbieri et al. 2016). This main path analysis reveals that EI literature revolves around (1) determinants of EI; (2) economic effects of EI; (3) environmental effects of EI, and (4) policy inducement of EI (Barbieri et al. 2016).

Diaz-Garcia et al. (2015) consider that although eco-innovation is still a young field of research, it has been a field of increasing concern for multiple actors: policy makers, academics and practitioners. The authors provide an overview of the existing body of literature on eco-innovations by identifying the most relevant publications (n: 384) in the field based on a Scopus search and selecting the discipline Social Sciences and Humanities (Diaz-Garcia et al. 2015). From their analysis, authors indicate that there is a clear increase in the relevance of eco-innovation within academia and several thematic trends arise in eco-innovation research, with *drivers of eco-innovation* being the most popular. The authors choose to develop a multilevel framework of eco-innovation drivers. Their literature review has a specific focus on systematizing the findings of the studies within this theme of eco-innovation.

Aforementioned methodological reviews reveal multi-domain, multi-method, multi-actor and multi-level characteristics of eco-innovation and eco-innovation research.

2.2.2 Firm-Centered Reviews

Bossle et al. (2016) provide a review article about the environmental behavior of firms. Their article examines how the business literature has researched eco-innovations related to the drivers that boost companies' adoption, which drivers and motivations for companies' adoption of eco-innovation exist and how results from the literature can help to define a conceptual framework of eco-innovation drivers and motivations. To address these questions, the authors conduct a systematic review. Differently from Diaz-Garcia et al. (2015) their data source is based on peer-reviewed articles from the ISI Web of Knowledge and consists of 96 full papers, of which 35 are matched with the specific target of analysis that focused on: (1) eco-innovation concepts and approaches; (2) methods and main findings; and (3) drivers and motivation for adoption of eco-innovation (Bossle et al. 2016). Their results support the multi-actor view that there is a growing interest in eco-innovation; not only from a managerial, but also from an academic perspective (54% of papers were published after 2010). While various methods are used in the selected articles supporting the 'multi-method characteristic' of eco-innovation research, evidence shows that leading firms are protagonists in developing new technologies (Bossle et al. 2016). The authors point out the importance of internal factors which companies can manage (going beyond mere compliance with external factors, over which companies have little or no control) in fully adopting eco-innovation. Although specific or one-off actions were enough to recognize eco-innovation in some cases, authors indicate that, to boost performance, companies need to improve their focus on eco-innovation as an explicit goal of their strategies.

Seeing sustainable innovation as a non-linear, recursive and self-organized process that can be studied as a complex adaptive system, the aim of Inigo and Albareda (2016) is to understand how firms engage in new processes, strategies and behaviors for sustainable innovation. The authors define five ontological sustainable innovation components: (1) operational, (2) collaborative, (3) organizational, (4) instrumental, and (5) holistic. Their analysis yields three complex adaptive system phenomena, which explain three main patterns of how each component interact and interconnect: (1) non-linearity (which explains the connection between the components, including positive and negative feedback loops and the rates of change, disorder, chaos and stability between them), (2) self-organization (which increases the order or regularity between the components interaction and even generates a new order and configuration with the different components behaving autonomously) and (3) emergence (which involves radically new processes and component interactions due to new and radical experimentation, rule-breaking and disruptive sustainable technologies). These three phenomena are deemed mutually dependent and specific to each company. They are explained in three extended examples, including three interaction mechanisms describing the dynamics of five components. The authors end the paper with research questions regarding the role of firms in sustainable innovation and sustainable system transitions and call for more interdisciplinary work, due to the systemic nature of sustainable innovation at the firm level. Their conclusions are in line with the earlier research reviewed in this chapter on the issue.

Ketata et al. (2015) extend the scope to future generations by defining sustainable innovation as an innovation that considers environmental and social issues as well as the needs of future generations. While over the past decade, sustainable innovation has occupied a top-ranking position on the agenda of many firms, and although sustainable innovation provides considerable new opportunities for many firms according to the authors, it goes along with an increased complexity. This increased complexity in turn requires certain (1) organizational routines and (2) organizational capabilities to deal with the upcoming challenges (Ketata et al. 2015). The authors explore which specific driving forces increase the degree of sustainable innovation within a firm's innovation activities by testing several driving forces empirically for more than 1100 firms in Germany. Ketata et al. (2015) find that firms need to invest in internal absorptive capacities to draw both broadly and deeply from external sources for innovation. Investments in employee training also turn out to be more important than technological R&D expenditures (Ketata et al. 2015). Investments in employee training as an internal factor supports the findings of Bossle et al. (2016).

On a broader scale, Bocken et al. (2014) offer guidance towards a new industrial sustainability agenda by stating that eco-innovations, eco-efficiency and corporate social responsibility practices define much of these current industrial sustainability agenda. While important, the authors deem these practices as insufficient in themselves to deliver the holistic changes necessary to achieve long-term social and environmental sustainability. Their main research question is how corporate innovation can be encouraged to significantly change the ways in which companies operate to ensure greater sustainability. The authors conclude that answer to this question is *sustainable business models (SBMs).* These models incorporate a triple bottom line approach and consider a wide range of stakeholder interests, including environment and society (Bocken et al. 2014). In this respect, SBMs are seen as important in driving and implementing corporate innovation for sustainability; helping to embed sustainability into business purpose and processes; and serving as a key driver of competitive advantage (Bocken et al. 2014). Therefore, the authors collate many innovative approaches contributing to delivering sustainability through business models under a unifying theme of *business model innovation.* Their literature and business practice review identifies a wide range of examples of mechanisms and solutions that can contribute to *business model innovation for sustainability.* Eight sustainable business model archetypes are introduced to describe the groups of mechanisms and solutions that may contribute to building up the business model for sustainability (Bocken et al. 2014). The aim of these eight archetypes is actually to develop a common language that can be used to accelerate the development of sustainable business models in research and practice. These archetypes indeed are similar to components of a circular economy. The original list is "(1) maximising material and energy efficiency; (2) creating value from 'waste'; (3) substituting with renewables and natural processes; (4) delivering functionality rather than ownership; (5) adopting a stewardship role; (6) encouraging sufficiency; (7) re-purposing the business for society/environment; and (8) develop scale-up solutions" (Bocken et al. 2014). SBM aims at integrating the systemic nature of sustainable innovation and business innovation at the firm level.

Klewitz and Hansen (2014) state that since the Brundtland report in 1987 a wide debate has emerged on eco-innovation (e.g. eco-design, cleaner production) and sustainability-oriented innovations (SOIs), that is, the integration of ecological and social aspects into products, processes, and organizational structures While prior research has often dealt with SOIs in large firms, the last decade has begun to generate broad knowledge on the specificities of SOIs in small and medium sized enterprises (SMEs) as they are increasingly recognized as central contributors to sustainable development (Klewitz and Hansen 2014). However, the authors indicate that this knowledge is scattered across different disciplines, research communities and journals and analyze the heterogeneous picture research has drawn within the past 20 years with a focus on the innovation practices including different types of SOIs and strategic sustainability behaviors of SMEs through an interdisciplinary, systematic review in a time frame between 1987 and 2010 (Klewitz and Hansen 2014). They bibliographically and thematically analyze 84 key journal articles and find that—first, SME strategic sustainability behavior ranges from resistant, reactive, anticipatory, and innovation-based to sustainability-rooted. Secondly, they identify innovation practices at the product, process, and organizational level. Third, their results show that research is still strong on eco-innovation rather than on innovation from a triple bottom line perspective (economic, social, and environmental dimensions), that is, SOIs of SMEs. Their main theoretical contribution is the development of an integrated framework on SOIs of SMEs where they delineate how distinct strategic sustainability behaviors can explain contingencies in types of innovation practices. Furthermore, the authors argue that SMEs with more proactive behaviors possess higher capabilities for more radical SOIs changing the innovation process itself. Thus, they propose that interaction with external actors (e.g. customers, authorities, research institutes) can ultimately increase the innovative capacity of SMEs for SOIs. Finally, they identify major research gaps with regard to radical SOIs, streamlined innovation methods and the role of SMEs in industry transformation and in sustainable supply chains. The authors also argue that there is a need for a stronger theoretical debate on SOIs of SMEs. In this sense, Klewitz and Hansen (2014) discuss multi-domain, multi-method, multi-actor and multi-level characteristics of eco-innovation at the SME/firm level.

De Medeiros et al. (2014) confirm that the growing awareness regarding environmental sustainability has fully reached business reality. The authors emphasize both demand and supply sides, consumers and companies alike, are looking for alternatives to mitigate pressing environmental demands resulting from continuous population and economic growth. However, they see companies as more disadvantaged because of the increasingly competitive scenarios where innovation is regarded as a survival need in many markets. Therefore, they aim to provide a systematic academic research guide for companies to succeed in environmentally sustainable product innovation. In this context, they conduct a systematic literature review on environmentally sustainable product innovation to (1) consolidate existing research and aggregate findings of different studies on environmentally sustainable product innovation through an interpretative framework of published literature on the topic, and (2) map critical success factors that drive the success of product innovation developed in this new logic of production and consumption. According to their findings, they identify four main critical success factors for environmentally sustainable product innovation: (1) market, law and regulation knowledge; (2) interfunctional collaboration; (3) innovation-oriented learning; and (4) R&D investments. The factors identified in their research and corresponding variables are discussed with professionals, as a further test. This resulted in preliminary approval of the framework they developed and identification of the most important variables within each factor.

2.2.3 *Technological: Industrial/Sectoral Reviews*

2.2.3.1 Iron and Steel Industry

The main goal of Burchart-Korol et al. (2016) is to present the most significant technological innovations aiming at the reduction of greenhouse gas emissions in steel production. The authors choose a sectoral field in which they could emphasize the importance of reduction of greenhouse gases and dust pollution. They selected the iron and steel industry and introduce a review of new solutions that are constantly sought-after. This review article actually presents the most recent innovative technologies which may be applied in the steel industry. The significance of CCS (CO_2 Capture and Storage) and CCU (CO_2 Capture and Utilization) in the steel industry forms the main part of their discussion.

2.2.3.2 Transport

From consumer intentions and adoption behavior perspective, Rezvani et al. (2015) problematize that, in spite of the purported positive environmental consequences of electrifying the light-duty vehicle fleet, the number of electric vehicles (EVs) in use is still insignificant. The authors propose one reason for these modest adoption figures which is the mass acceptance of EVs relies on consumers' perceptions of EVs. Their review presents an overview of the drivers for and barriers to consumer adoption of plug-in EVs and an overview of the theoretical perspectives that have been utilized for understanding consumer intentions and adoption behavior towards EVs, and a future research agenda. Their study reveals the importance of the demand side in eco-innovation trajectories.

2.2.3.3 IT

From perceived usefulness and satisfaction perspective Ghazal et al. (2016) examine the smart plug system for monitoring and controlling household energy consumption using a mobile application. The smart plug system is deemed an essential component in smart grids because of the fact that it provides real-time high-resolution information for distribution companies to aid them in decision-making (Ghazal et al. 2016). Their study is about the per capita energy consumption in the United Arab Emirates (UAE), which is one of the highest in the world (Ghazal et al. 2016). The authors state that the energy sector is the center of most ecological problems facing the world, and that eco-efficiency and eco-innovations are at the top of the sustainability agenda in most countries. For example, the UAE "Green Economy for Sustainable Development" (2012–2021) aims to position the country as a center for the export and re-export of green products and technologies. Their research aims to measure the perceived usefulness and satisfaction of the smart plug system and its mobile application in the UAE by including environmental concern as an additional variable to a well-established information systems success model. Their findings suggest that the smart plug system provides users with convenient access to information regarding their personal energy consumption. This allows consumers to control their per capita energy consumptions via the mobile application at very low costs. The practical implication is that per capita energy consumption is likely to significantly decrease with wide adoption of this smart plug system in the UAE (Ghazal et al. 2016). By using structural equation modeling, the authors conclude that environmental concern has an indirect impact on perceived satisfaction with the product; and both an indirect and a direct impact on the perceived usefulness of the smart plug system.

2.2.3.4 Food/Agriculture

According to Viaggi (2015), studies on the effects of research and innovation in agriculture have been largely characterized by efforts to make a connection between expenditure and productivity. According to the author, a number of issues have challenged the ability of productivity to measure the effects of research, namely, in recent years, increasing efforts towards improving the environmental performance of the farming sector. The author suggests that besides environmental concerns, however, a number of recent concepts have emerged that are shaping the current research and policy agenda which could result in a revision of the productivity concepts used to evaluate the impact of research. The objective of Viaggi (2015) is thus to discuss these issues and their implications for studies on the impact of research and innovation and address, in particular, (1) the development of the bioeconomy and related concepts such as the circular economy, resource efficiency and bio-refinery; (2) the connection with entrepreneurship and eco-innovation; (3) changing tools in research assessment, in particular, the widespread use of Life Cycle Assessment (LCA); and (4) the evolving concepts of sustainability and ecosystem services by arguing that while the traditional notion of productivity, intended as output/input ratio, maintains (and even strengthens) its role on the aggregate, a more analytical interpretation of the pathways towards research impact is needed, which requires a broadened view of productivity and its determinants.

In related industries, Mirabella et al. (2014) problematize the production of food waste that covers all the food life cycle: from agriculture, up to industrial manufacturing and processing, retail and household consumption by indicating that the large amount of waste produced by the food industry, in addition to being a great loss of valuable materials, also raises serious management problems, both from the economic and environmental point of view. The authors state that in developed countries, 42% of food waste is produced by households, while 39% losses occur in the food manufacturing industry, 14% in food service sector and remaining 5% in retail and distribution. Given these trends, the authors introduce concepts such as *cradle to cradle* which are increasingly used industrial ecology, and *circular economy* which are considered important principles for eco-innovation, ultimately aiming at a *zero waste economy* where waste is used as raw material for new products and applications. Many residues are argued to have the potential to be reused into other production systems, through e.g. biorefineries (Mirabella et al. 2014). Thus, their review focuses on the use of food waste coming from food manufacturing (FWm) through an extensive literature review. This review is used to present feasibility and constraints of applying industrial symbiosis in recovering waste from food processing. It focuses on recycling (excluding energy recovery) of the solid and liquid waste from food processing industry, as well as on the main uses of functional ingredients derived from processes, mainstream sectors of applications, e.g. in the nutraceutical and pharmaceutical industry.

2.2.3.5 Tourism

Lucchetti and Arcese (2014a) apply the Industrial Ecology (IE) concept to tourism. Their review consists of a theoretical review focused on IE for investigating the best way to implement industrial ecology in the tourism activities. This is a rare study and there are no further specific reviews on tourism.

Research reviewed above indicates that newly emerging, broader systemic frameworks, such as industrial ecology and circular economy are starting to provide context for industry-specific eco-innovation research.

2.2.4 Science-Based Reviews

Lang-Koetz et al. (2010) provide an overview of different sciences and technology fields, products and strategies with resource efficiency potential, such as nanotechnologies, material sciences, manufacturing technologies, process technologies and cross-cutting issues. The special focus is on applications from nanotechnology e.g. functional surfaces or new 'smart' materials with special functionalities. The authors show how companies can use the so-called *Resource Efficiency Technology Radar* method to identify and evaluate technologies with respect to resource efficiency to incorporate them into their development activities.

From natural sciences perspective, Artes et al. (2009) state that the minimal processing industry for fruit and vegetables needs an appropriate selection of raw materials and operationally improved sustainable strategies to reduce losses in order to provide high quality and safe commodities. According to the authors, the most important target for keeping overall quality of such commodities is a decrease in *microbial spoilage flora* as they cause both decay and safety issues. Thus, every step in the production chain influencing microbial load, the implementation of an accurate disinfection program, should be the main concern of commercial processing (Artes et al. 2009). Yet the only step that reduces microbial load throughout the production chain is indicated as washing disinfection by proper handling and optimizing existing techniques or a combination of them (Artes et al. 2009). However, the authors point out that while chlorine is a common efficient sanitation agent, there is the risk of undesirable by-products upon its reaction with organic matter and this may lead to new regulatory restrictions in the future. Besides, its efficacy is poor for some products. Consequently the minimal processing industry demands safer antimicrobial washing solutions, innovative sanitizers on fresh-cut plant commodities e.g. O(3), UV-C radiation, intense light pulses, super high O(2), N(2)O and noble gases, alone or in combination, which are considered promising treatments (Artes et al. 2009). This change, from use of conventional to the use of innovative sanitizers, however, requires further knowledge of the benefits and restrictions as well as a practical outlook (Artes et al. 2009).

2.2.5 Diffusion-Centered Reviews

Hojnik and Ruzzier (2016) provide an overview of the emerging literature on the drivers of eco-innovation and they suggest separating the drivers associated with the phases of development and diffusion and identifying particular drivers based on different eco-innovation types. The authors find that research in this area primarily adopts the resource-based and institutional theories as its theoretical foundations and that the prevailing effects identified are those of regulations and market pull factors. Product eco-innovation, process eco-innovation, organizational eco-innovation, and environmental R&D investments appear to be driven by common drivers, such as regulations, market pull factors, environmental management systems, and cost savings, and to be positively associated with company size (Hojnik and Ruzzier 2016). The majority of the studies in their literature review employs a quantitative research methodology and focus on the diffusion stage of eco-innovation.

Byrka et al. (2016), on the other hand, indicate that despite the very positive attitude towards eco-innovations and sustainability in general as measured by market surveys, the actual market penetration of green products and practices generally falls behind the expectations. In their review, the authors argue that difficulty of engagement is of critical importance when modeling diffusion of eco-innovations as used in the Campbell Paradigm. Such a notion of difficulty possesses three desired properties: (1) parsimony—which is represented by a single value, (2) interpretability—which can be regarded as an estimator of the otherwise complex notion of behavioral cost, and (3) applicability—which can be easily measured through market surveys (Byrka et al. 2016). By using simulation and an agent-based model spanned on different social network structures, the authors show that innovation adoption may exhibit abrupt changes in market penetration as a result of even small changes in difficulty. The latter may be of particular interest to policymakers who have to make strategic decisions when introducing socially—but not necessarily individually—desired products and practices, like dynamic or green electricity tariffs (Byrka et al. 2016).

According to Karakaya et al. (2014), understanding of the diffusion of eco-innovations recently has gained more importance given the fact that some eco-innovations are already at a mature stage in comparison to the literature in the field of eco-innovations which often focuses on policy, regulations, technology, market and firm specific factors rather than diffusion. Their review aims to clarify the concept of diffusion of eco-innovation and provide an overview of this emerging literature by identifying the most cited relevant publications and corresponding research streams, their strengths and limitations in the concept of diffusion of eco-innovations. Their results confirm the multi-disciplinary character of eco-innovation by insights from different research streams in different disciplines and the emergence of the literature on diffusion of eco-innovations.

Qi et al. (2013) raise the question of diffusion in a specific context and problematize the challenge of environmental degradation in China by indicating that a growing number of firms have begun to integrate environmental management systems into their business strategies and develop green innovation strategies. Based on the stakeholder theory, their review attempts to explore the influences of stakeholder types on the implementation of green process and green product innovation. Empirical results show that foreign customers play a significant role in driving companies to adopt a strategy of green process and green product innovation (Qi et al. 2013). While their review shows that for foreign-invested enterprises, the effect is limited to the adoption of green process innovation, community stakeholders and regulatory stakeholders have no significant effect on the corporate green process and green product innovation.

2.2.6 Policy (Tools and Instruments)-Centered Reviews

From policy perspective, Prieto-Sandoval et al. (2016) analyze the importance of ecolabels as an eco-innovation tool that can contribute to the sustainable design, production and consumption of products. Their research has a dual objective. The first objective is to build a theoretical framework that explains the relationship between ecolabels and eco-innovation, their determinants (demand, supply; and institutional and political influences) and the dimensions that arise from them. Secondly, according to their framework, they conduct a systematic literature review to identify the trends and opportunities in eco-labeling which they treat as a multidimensional topic that they analyze from empirical, geographical and sectoral perspectives. The authors claim that their main contributions are a proposal for cyclical eco-labeling innovation process, an understanding of the eco-labeling dimensions, and ecolabel performance in the market. Their systematic literature review reveals that ecolabels have been mainly explored in food sectors and, in developed countries, and researchers tend to assess their performance from the dimension of market dynamics.

A review of the innovation effects of environmental policy instruments is offered by Kemp and Pontoglio (2011). The conclusion of the review is that policy instruments cannot be usefully ranked with regard to their effects on eco-innovation, and that the often expressed view that market-based approaches such as pollution taxes and emission trading systems are better for promoting eco-innovation is not brought out by the case study literature or by survey analysis, and seems only warranted for non-innovative, or marginally-innovative changes. What the case study literature shows is that the specifics of policy and the situation in which they are applied are all-important for environmental innovation outcomes. Increasingly this finding is also acknowledged in the economic literature, but the common wisdom still is that market based instruments are superior in soliciting innovative responses. Regulation is generally viewed as stimulating merely the diffusion of environmental technology but the authors show that there is more evidence of regulations stimulating radical innovations than of market-based instruments doing so.

2.3 Data and Method

2.3.1 Data

2.3.1.1 Data Source

Our data source is the Thomson Reuters Web of Science (WoS) Core Collection. In bibliometric studies, WoS is one of the three main data sources that are commonly used. The other data and information sources are Scopus and Google Scholar. Amongst them, WoS is the strictest source with respect to its inclusion and eligibility criteria for scientific publications in its core collection (e.g. peer reviewed SCI, SSCI index journals). Scopus is relatively more flexible in the inclusion of conference proceedings/papers. Google Scholar is the most generous in listing even online self-publications. We chose the WoS Core Collection. In the dataset for eco-innovation literature we have 350 publications (time period: 1988–2016 November 18). A Scopus-based study on eco-innovation literature in 2015 by Diaz-Garcia et al. (2015) has an n: 384 publication set, thus we are confident about the representativeness of our data source and set. Several studies also found a notable match between the results from the WoS and Scopus (Gavel and Iselid 2008). Gavel and Iselid (2008) is also cited by Diaz-Garcia et al. (2015).

2.3.1.2 Retrieving Data

Quality of bibliometric analysis is highly dependent on the quality of the constructed search query (Türkeli et al. 2018). Search queries should be properly designed. For instance, using "eco-innovation" as a search term can miss the "eco-innovative" wording. For this purpose, a syntax rule is used, which is using the "*" suffix, such as "eco-inno*" to be able to capture different variants, e.g. eco-inno*vation*, eco-inno*vative*. A search term can have different syntaxes, such as "ecoinno*" or "eco inno*". These variants should all be included in the search query. The same applies to "environmental innovation". Using the search term "environmental* inno*" would capture "environmentally-friendly innovation". After gathering the data, manual control is also necessary. For instance, for the case of "environmental innovation" an example of a false finding can be exemplified with a 2004-pubished article of American Journal of Preventive Medicine: "Understanding how environmental attributes can influence particular physical activity behaviours is a public health research priority. Walking is the most common physical activity behaviour of adults; *environmental innovations* may be able to influence rates of participation" (Owen et al. 2004). Thus, we performed manual control to clean our data sets.

2.3.1.3 Limitations of the Retrieved Data

It is also possible to miss the scientific publications which are related to the term of interest but do not use the exact or slightly different variants of the term of interest appearing in the title, abstract and keywords of the article. Publications in the new Elsevier journal *Environmental Innovation and Societal Transitions* are not included in the analysis because the journal is not (yet) an official Web of Science journal. Another limitation is that studies of ecological modernization and sustainability transitions are largely outside the scope of the analysis, even though they form relevant bodies of research, especially the literature on sustainability transitions[3]

[3]In these fields researchers study processes of transformative change in the form of the alternative systems of energy, mobility, agriculture, water and waste management and the institutionalization

(Adams et al. 2016; Bergek and Berggren 2014; Chua 1999; Faber and Frenken 2009; van Berkel 2007). Our analysis is scoped to the publications included in the Web of Science which include the terms of interest of this chapter (eco-innovation, environmental innovation, sustainable innovation) and syntax-based variants in the title, abstract or keywords. Relevant studies which do not self-identify themselves with the search terms used are not included in our analyses.

2.3.2 Method

Bibliometric studies concentrate on five different types of analysis: Co-authorship, citation, co-citation, bibliometric coupling, and co-occurrences of terms:

- *Co-authorship analysis:* the relatedness of items is determined based on their number of co-authored publications (e.g. multi-level scientific knowledge production relations, unit of analysis: author, organization, country level) (Van Eck and Waltman 2010).
- *Co-occurrence analysis:* the relatedness of items is determined based on the number of publications in which they occur together (e.g. content analysis of keywords) (Van Eck and Waltman 2010). We used this co-occurrence analysis for both authors' keywords in the reviews (n: 24) and in the overall literature (n: 350).
- *Citation analysis:* the relatedness of items is determined based on the number of times they cite each other (e.g. connectivity analysis) (Van Eck and Waltman 2010). We used this method to analyze the impact of the main journals in the field.
- *Bibliographic coupling:* the relatedness of items is determined based on the number of references they share (e.g. community detection) (Van Eck and Waltman 2010). Bibliographic coupling of organizations (network visualization use legends with an average year of publications). Bibliographic coupling of countries (term map visualization).
- *Co-Citation analysis:* the relatedness of terms is determined based on the number of times they are cited together (e.g. *recognized connectivity analysis*) (Van Eck and Waltman 2010).

Co-authorship analysis can use three different units of analysis, author, organization, or country level. This can construct a multi-level analysis (see sub-section 2.4.4). We also utilized two types of network graph visualizations: with respect to the average year of publication, and with respect to received citations. Co-occurrence (term) maps or networks can be used to conduct content analysis

---

of environmental protection at the levels of government, business, consumers and science and technology. Reviews of the literature on sustainability transitions and the related literature on technological innovation systems that offer green benefits are provided by Markard et al. (2012) and Markard and Truffer (2008) and by Mol et al. (2009) and Mol and Sonnenfeld (2000) for the literature on ecological modernization.

(see sub-section 2.4.2). We used network analysis to calculate the centrality of e.g. bibliographic coupling of cited authors. We used eigenvector centrality to assign this centrality, which is a commonly used network statistics/measure to measure the influence of a node, here the author. The data for analyzing the classical political economy aspects of eco-innovation scientific knowledge: funding, production, dissemination and the use are gathered as follows: Funding agency information is retrieved from WoS data (field tags: FX, FU), HistCite12.3.17 is used for detecting internal and external references to the eco-innovation literature (tags: LCS, GCS), main source journals/dissemination information is retrieved from WoS data (field tag: SO), the use information (citing articles) is retrieved from WoS citation report analysis. Reviews are detected based on the classification of WoS. TCpY (Times cited per year) and TC/NR (Times cited divided by number of references) criteria are calculated by the authors. VosViewer 1.6.5 (Van Eck and Waltman 2010), HistCite 12.3.17 (Garfield 2009), and Gephi 0.9 (Bastian et al. 2009) are the main tools we use to conduct our analyses.

2.4 Findings and Discussions

2.4.1 Variety/Diversity and Selection

Analyzing the details of knowledge production is quite informative for both policy research and development because knowing about the evolving characteristics of a research field is valuable information for policy intelligence. First of all, the characteristics of a research field do not depend on the choice of a single author or an institution, although they might be quite influential, the characteristics depend on the aggregation of the choices of the authors/institutions accumulated over time. In this respect, knowledge production is a multi-agent game under conceptual and constructed dynamic structures. Analyzing variety and selection dynamics is important to be able to study the political economy of environmentally-positive innovations. Especially, from the perspective of ideas (of scholars) that generate policy-related conceptual frameworks, applied policy programs based on these frameworks, or from the perspective of interests (of authors, organizations, firms, associations) and institutions (e.g. UN, EC, OECD) that take up the label within this variety, and altogether ideas, interests and institutions constitute the realm of the research field.

Table 2.1 presents the number of scientific publications produced under different conceptual structures, in short, variety and selection of labels. The biggest overlap is between eco-innovation and environmental innovation (n: 71). From Table 2.1, we observe the evolutionary dynamics of generated variety in labeling the knowledge production in environmentally-positive innovations; such as eco-innovation (n: 350/leading), followers are sustainable innovation (n: 347), environmental innovation (n: 328), green innovation (n: 172), innovation for sustainable development (n: 39), clean tech innovation (n: 17).

Table 2.1 Varieties and selections of labels (n: number of publications in WoS core collection)

Labels	Eco-innovation	Sustainable innovation	Environmental innovation	Green innovation	Innovation for sustainable development	Low carbon innovation	Clean tech innovation	Innovation for green economy
Eco-innovation	350	15	71	22	0	2	1	1
Sustainable innovation		347						
Environmental innovation	71	15	328	27	2	1	1	0
Green innovation				172				
Innovation for sustainable development					39			
Low carbon innovation						23		
Clean tech innovation							17	
Innovation for green economy								2

Web of Science Query: TI OR TS = ((“eco-inno*” OR “ecoinno*” OR “eco inno*”); TI OR TS = ((“sustainab* inno*” OR “inno* for sustainab*” OR “sustainability-driven inno*” OR “inno* for environment* sustainab*”); TI OR TS = ((“inno* for green economy” OR “green economy inno*”)); TI OR TS = (“inno* for sustainable development” OR “sustainable development inno*”; TI OR TS = (“clean* tech* inno*” OR “clean* inno”); TI OR TS = ((“low-carbon inno*” OR “lowcarbon inno*” OR “low carbon inno”); TI OR TS = ((“green inno*” OR “green tech* inno*”); TI OR TS = ((“environmental inno*” OR “environmentally* inno*”); Indexes = SCI-EXPANDED, SSCI, A&HCI, ESCI Timespan = 1988–2016

However, we cannot really observe how a label is selected over the other labels swept out in Kuhnian terms, as a singled out paradigm. Multiple epistemes do exist, authors continue to produce at times under multiple labels, yet eco-innovation becomes the dominant label for studying the field. Knowledge production under eco-innovation is approximately two times larger than the knowledge production under environmental innovation in the last 3 years (see Tables 2.2 and 2.3).

The most cited paper identified in our analysis is the article of Klaus Rennings titled "Redefining eco-innovation" published in Ecological Economics in 2000. This article is foundational to the field of eco-innovation and environmental innovation. The second most cited paper is the article "Determinants of Environmental innovation" in US manufacturing industries by Brunnermeier and Cohen published in the Journal of Environmental Economics and Management in 2003 (see Table 2.4).

The third most cited paper is the article "Strategic niche management and sustainable innovation journeys" by Schot and Geels published in Technology Analysis and Strategic Management in 2008. The last article belongs to the field of sustainability transitions, which is large and expanding literature (which is not scrutinized in this chapter).[4]

2.4.2 *Comparative Content Analysis: Eco-Innovation and Environmental Innovation*

We analyzed 34 keyword co-occurrences. Between eco-innovation and environmental innovation research, the similar patterns for the share of co-occurrences with certain keywords (theory, supply, demand etc.) do not apply for the keywords *consumption, product, and regulation* (see Table 2.5 and 2.6)

Eco-innovation research is more product (47 vs. 31%) and consumption (12 vs. 4%) based. It relates less to regulation (18 vs. 31%) and slightly relates more to market (29 vs. 25%) than environmental innovation does. Other co-occurrences are quite similar.

Highly-cited policy themes in the eco-innovation literature are co-innovation policy, environmental policy and transition management or ecological structural policy. Recently published policy themes in this research field are SME policy, biofuel policy, policy mixes.

[4]A bibliometric analysis of the sustainability transition literature is provided in Chappin and Ligtvoet (2014). The study identified René Kemp, Frank Geels, Jan Rotmans, Johan Schot, Arie Rip and Adrian Smith as the most important researchers in that field in terms of citations.

Table 2.2 Eco-innovation

Field: Publication years	Record count	% of 350	Bar chart
1996	1	0.286	
2000	1	0.286	
2001	1	0.286	
2002	1	0.286	
2003	1	0.286	
2004	1	0.286	
2005	1	0.286	
2006	2	0.571	
2007	4	1.143	
2008	4	1.143	
2009	7	2.000	
2010	14	4.000	
2011	30	8.571	
2012	29	8.286	
2013	27	7.714	
2014	54	15.429	
2015	76	21.714	
2016	96	27.429	

2.4.3 A Network Analysis of Temporal Dynamics and Influence of the Authors' Keywords in Eco-innovation Reviews (n: 24 Reviews)

Eco-innovation research has more reviews than environmental innovation research (24 vs. 14). In 2016, nine reviews are published. In 2015, five reviews are published and in 2014 six reviews are published (see Table 2.7).

Table 2.7 above is sorted by the "number of times cited divided by number of references" criterion, a criterion which penalizes the times cited per year if a large number of references is used in the article. Using citations per year as a metric, *sustainable business model archetypes, innovation effects of environmental policy instruments, valorization of food manufacturing waste, sustainable sanitation techniques, SMEs, electric vehicle adoption research, diffusion of eco-innovations, product innovation*s are the leading reviews.

We observe that product/process innovation, environmental policy instruments and research methodology related reviews have an average publication year of 2013, they are relatively old. *Circular economy, bio-economy, smart cities* related reviews have 2015 as average publication year, more recent. Publications concentrate on the review of *specific technologies, complex systems, systems-thinking, innovation drivers, and determinants* have 2016 as average publication year which means that these topics are either relatively new (for the case of specific technologies e.g. *wireless sensor networks, mobile computing, CCS, CCU*) or substantially

Table 2.3 Environmental innovation

Field: Publication years	Record count	% of 328	Bar chart
1991	1	0.305	
1992	2	0.610	
1994	1	0.305	
1995	1	0.305	
1996	4	1.220	
1997	3	0.915	
1998	3	0.915	
1999	1	0.305	
2000	4	1.220	
2001	5	1.524	
2002	9	2.744	
2003	10	3.049	
2004	7	2.134	
2005	6	1.829	
2006	6	1.829	
2007	11	3.354	
2008	18	5.488	
2009	20	6.098	
2010	24	7.317	
2011	19	5.793	
2012	16	4.878	
2013	32	9.756	
2014	34	10.366	
2015	43	13.110	
2016	48	14.634	

Table 2.4 Common set, Top 3 articles w.r.t. received citations

1. **Redefining innovation - eco-innovation research and the contribution from ecological economics**
By: Rennings, K
ECOLOGICAL ECONOMICS Volume: 32 Issue: 2 Pages: 319-332 Published: FEB 2000
Full Text? | Full Text from Publisher | View Abstract
Times Cited: 307 *(from Web of Science Core Collection)*
Usage Count

2. **Determinants of environmental innovation in US manufacturing industries**
By: Brunnermeier, SB; Cohen, MA
JOURNAL OF ENVIRONMENTAL ECONOMICS AND MANAGEMENT Volume: 45 Issue: 2 Pages: 278-293 Published: MAR 2003
Full Text? | Full Text from Publisher | View Abstract
Times Cited: 215 *(from Web of Science Core Collection)*
Usage Count

3. **Strategic niche management and sustainable innovation journeys: theory, findings, research agenda, and policy**
By: Schot, Johan; Geels, Frank W.
TECHNOLOGY ANALYSIS & STRATEGIC MANAGEMENT Volume: 20 Issue: 5 Pages: 537-554 Published: 2008
Full Text? | View Abstract
Times Cited: 211 *(from Web of Science Core Collection)*
Usage Count

Table 2.5 Comparative content analysis through co-occurrences of certain keywords

Co-occurrences	**Theory**	**Empirical**	**Data**	**History**	**Future**			
Eco-innovation n: 350	60	84	74	6	50			
	17.10%	24.00%	21.10%	1.7%	14.3%			
Environmental innovation n: 323	56	100	102	8	35			
	17.07%	30.49%	31.10%	2.44%	10.67%			
Co-occurrences	**Production**	**Supply**	**Industry**	**Science**	**Technology**	**Energy**	**Radical**	**Product**
Eco-innovation n:350	63	46	154	28	185	88	20	**165**
	18.00%	13.10%	44.00%	8.00%	53.10%	25.1%	5.7%	**47.1%**
Environmental innovation n:328	47	45	151	17	191	72	21	**103**
	14.33%	13.72%	46.04%	5.18%	58.23%	21.95%	6.40%	**31.40%**
Co-occurrences	**Consumption**	**Demand**	**Household**	**Adoption**	**Diffusion**			
Eco-innovation n:350	**43**	53	9	52	41			
	12.30%	15.10%	2.60%	14.90%	11.70%			
Environmental innovation n:328	**14**	54	4	66	62			
	43.27%	16.46%	1.22%	20.12%	18.90%			

Table 2.6 Comparative content analysis through co-occurrences of certain keywords

Co-occurrences	Local	Regional	National	International
Eco-innovation n: 350	11	19	20	26
	3.1%	5.4%	5.7%	7.4%
Environmental innovation n: 328	20	15	20	36
	6.10%	4.57%	6.10%	10.98%

Co-occurrences	Policy	Instrument	Tai	Regulation	Price	Standards	Institutions	Politics
Eco-innovation n:350	139	32	10	**63**	18	19	35	12
	39.70%	9.1%	2.90%	**18.00 %**	5.10%	5.40%	10.00%	3.40%
Environmental innovation n:328	167	35	7	**104**	20	21	32	14
	50.91%	10.67%	2.13%	**31.71 %**	6.10%	6.40%	9.76%	4.27%

Co-occurrences	Market	State	Social	Voluntary
Eco-innovation n: 350	103	33	63	7
	29.40%	9.40%	18.00%	2.00%
Environmental innovation n: 328	83	31	56	17
	25.30%	9.45%	17.07%	5.18%

Table 2.7 A review of reviews

Authors	Title	Source	NR	TC	TcpY	TC/ NR	PY	VL	IS
Bocken, NMP; Short, SW; Rana, P; Evans,S	A literature and practice review to develop sustainable business model archetypes	Journal of Cleaner Production	95	84	28.00	0.88	2014	65	
Kemp, R; Pontoglio, S	The innovation effects of environmental policy instruments-Atypical case of the blind men and the elephant	Ecological Economics	70	60	10.00	0.86	2011	72	
Artes, F;Gomez, P; Aguayo, E;Escalona, V; Artes-Heri	Sustainable sanitation techniques for keeping quality and safety of fresh-cut plant commodities	Postharvest Biology and Technology	148	116	14.50	0.78	2009	51	3
Mirabella, N;Castellani, V;Sala,S	Current options for the valorization of food manufacturing waste: a review	Journal of Cleaner Production	126	56	18.67	0.44	2014	65	
Karakaya, E; Hidalgo, A; Nuur, C	Diffusion of eco-innovations: A review	Renewable & Sustainable Energy Reviews	64	18	6.00	028	2014	33	
Rezvani, Z;Jansson, J; Bodin, J	Advances in consumer electric vehicle adoption research: A review and research agenda	Transportation Research Part D	68	19	9.50	0.28	2015	34	
Klewitz, J; Hansen, EG	Sustainability-oriented innovation of SMEs: a systematic review	Journal of Cleaner Production	181	42	14.00	0.23	2014	65	
de Medeiros, JF; Ribeiro, JLD; Cortimiglia, MN	Success factors for environmentally sustainable product innovation: a systematic literature review	Journal of Cleaner Production	103	18	6.00	0.17	2014	65	
Lucchetti, MC;Arcese, G	Tourism Management and Industrial Ecology: A Theoretical Review	Sustainability	40	4	1.33	0.10	2014	6	8
Burchart-Korol, D; Pichlak, M; Kruczek, M	Innovative technologies for greenhouse gas emission reduction in steel production	Metalurgija	22	2	2.00	0.09	2016	55	1
Qi, GY; Zeng, SX; Tam, CM; Yin, HT; Zou, HL	Stakeholders' Influences on Corporate Green Innovation Strategy: A Case Study of Manufacturing Firms	Corporate Social Responsibility and Environmental Management	113	9	2.25	0.08	2013	20	1
Bossle, MB; de Barcellos, MD; Vieira, LM;Sauvee, L	The drivers for adoption of eco-innovation	Journal of Cleaner Production	67	4	4.00	0.06	2016	113	
Lang-Koetz, C; Pasterski, N; Rohn, H	Identifying New Technologies, Products and Strategies for Resource Efficiency	Chemical Engineering & Technology	34	2	029	0.06	2010	33	4

(continued)

Table 2.7 (continued)

Authors	Title	Source	NR	TC	TcpY	TC/ NR	PY	VL	IS
Diaz-Garcia, C; Gonzalez-Moreno, A; Saez-Martinez,	Eco-innovation: insights from a literature review	Innovation-Management Policy & Practice	77	4	2.00	0.05	2015	17	1
del Rio, P; Penasco, C; Romero-Jordan, D	What drives eco-innovators? A critical review of the empirical literature based on econometric methods	Journal of Cleaner Production	75	2	2.00	0.03	2016	112	
Hojnik, J; Ruzzier, M	What drives eco-innovation? A review of an emerging literature	Environmental Innovation and Societal Transitions	83	2	2.00	0.02	2016	19	
Kammerlander, M;5chanes, K;Hartwig, F;Jager, J; O	A resource-efficient and sufficient future mobility system for improved well-being in Europe	European Journal of Futures Research	67	1	0.50	0.01	2015	3	1
Byrka, K;Jędrzejewski, A;Sznajd-Weron, K; Weron, R	Difficulty is critical: The importance of social factors in modeling diffusion of green products and practi	Renewable & Sustainable Energy Reviews	110	1	1.00	0.01	2016	62	
Prieto-Sandoval, V; Alfaro, JA; Mejia-Villa, A;Ormaza	ECO-labels as a multidimensional research topic: Trends and opportunities	Journal of Cleaner Production	115	0	0.00	0.00	2016	135	
Inigo, EA;Albareda,L	Understanding sustainable innovation as a complex adaptive system: a systemic approach to the firm	Journal of Cleaner Production	256	0	0.00	0.00	2016	126	
Barbieri, N; Ghisetti, C; Gilli, M; Marin, G; Nicolli, F	A survey of the literature on environmental innovation based on main path analysis	Journal of Economic Surveys	110	0	0.00	0.00	2016	30	3
Ghazal, M;Akmal, M; lyanna,S;Ghoudi, K	Smart plugs: Perceived usefulness and satisfaction: Evidence from United Arab Emirates	Renewable & Sustainable Energy Reviews	50	0	0.00	0.00	2016	55	
Viaggi, D	Research and innovation in agriculture: beyond productivity?	Bio-based and Applied Economics	52	0	0.00	0.00	2015	4	3
Ketata, l;Sofka,W; Grimpe,C	The role of internal capabilities and firms' environment for sustainable innovation: evidence for Germany	R & D Management	113	0	0.00	0.00	2015	45	1

Fig. 2.2 Co-occurrence analysis: temporal network of the authors' keywords used in eco-innovation reviews (n: 24 reviews)

re-reviewed in the latest years (for the case of drivers and determinants) (see Fig. 2.2).

In terms of received citations, *business model innovation, sustainable production, value creation and industrial sustainability* are the leading review keywords and followed by *environmental policy instruments* and *research methodology*. *Stakeholders, sustainable consumption, sustainable entrepreneurship, sustainability-oriented innovation, sustainability strategy* and *SMEs* form a third group of highly-cited review keywords (see Fig. 2.3).

2.4.4 Overall Eco-Innovation Research

In overall eco-innovation literature (n: 350 publications), we observe that other than "business model innovation", "sustainable production", and "review" keywords, *open innovation, patent analysis, regulation, networks, stakeholders, rebound effect, creativity, diffusion, consumer behavior, sustainable consumption, environmental management system* keywords received the highest number of normalized citations (see Fig. 2.4).

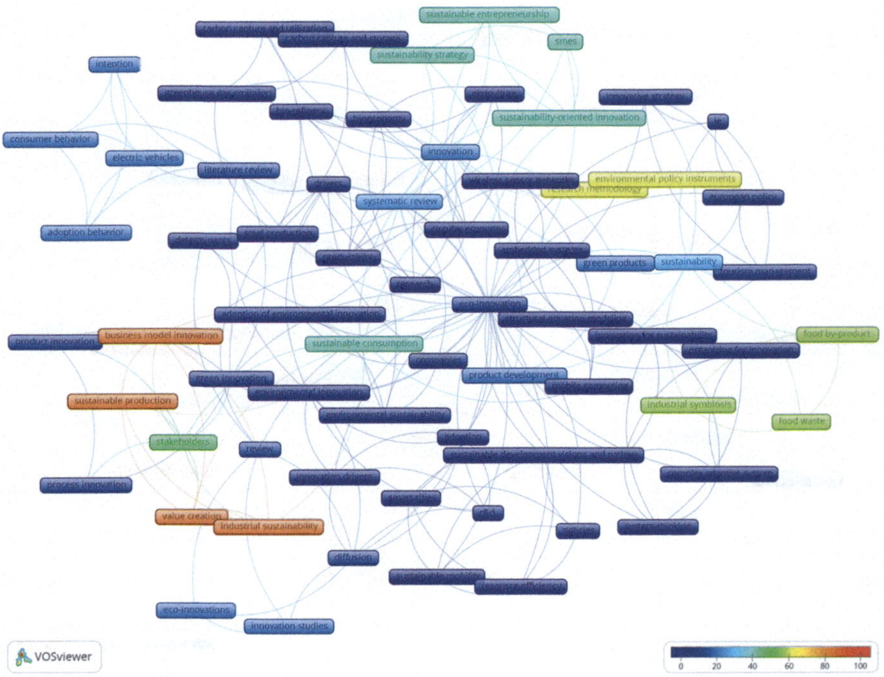

Fig. 2.3 Co-occurrences analysis: received citation-scored network of the authors' keywords in eco-innovation reviews (n: 24 reviews)

For the same network graph, overlay visualization brings forward the temporal breakdown with respect to the average publication year of the keywords, so emerging themes and new directions in overall eco-innovation research become observable (n: 350) (see Fig. 2.5).

Table 2.8 lists the keywords from the most frequently cited publications published in the specific year. In the last three years, the most frequently cited publications each year have *themes related to the circular economy.*

2.4.5 A Multi-level Centrality Analysis: Authors, Organizations, Journals, and Countries

In Fig. 2.6, we see that Kemp R., Rennings K., Porter M. and Horbach J. are the most central authors in bibliographic coupling. OECD and European Commission also emerge as central institutions. Kemp R's, location is at a science-policy interface, between policy institutions and academic research. Authors, who cite Kemp R., are more likely to cite OECD and Rennings K. Porter M. and Rennings K. are the two authors most likely to be cited together. Authors who cite the OECD are also more

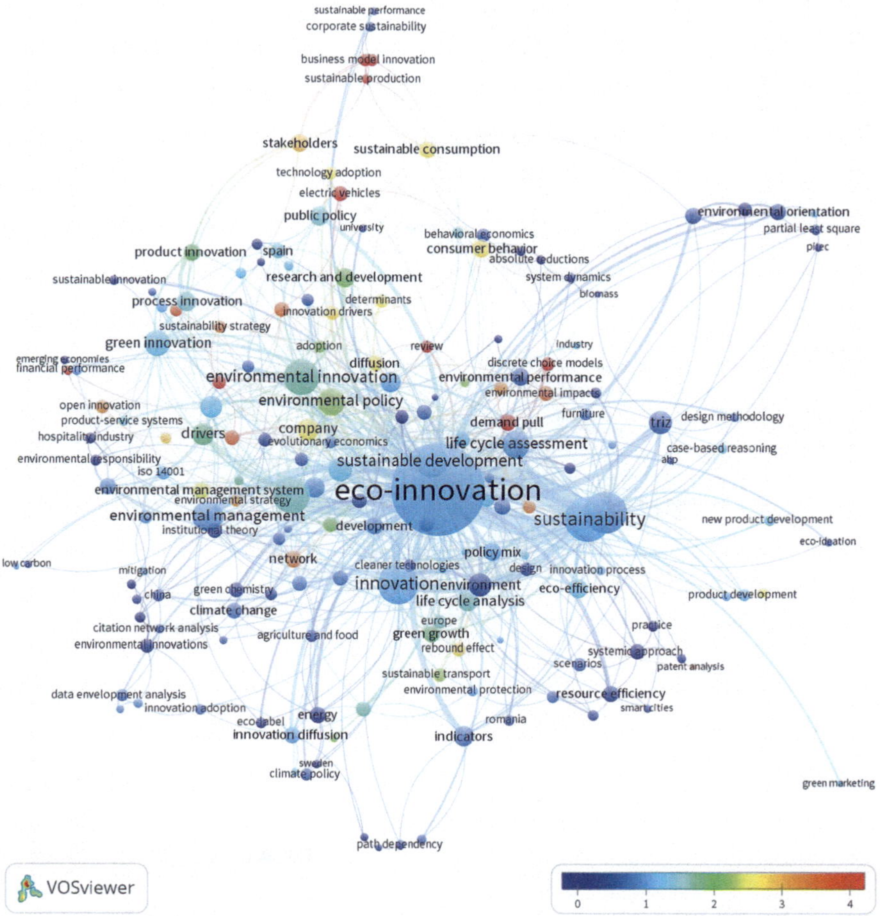

Fig. 2.4 Co-occurrence analysis/no of keywords: 178 (two or more occurrences) authors' keywords, legend: average normalized citations (n: 350)

likely to cite the European Commission. Eigencentrality is the influence of the author.

Emerging organizations are in red (2016). The location of the organization in Fig. 2.7 indicates that the publications produced by these organizations are more likely to share common references. The thickness of links represents the references the organizations share (the thicker links signify more shared references). Location information is useful for community detection (see Fig. 2.7).

Although the number of eco-innovation publications from the Journal of Cleaner Production is larger than the ecological economics and research policy published outputs, in terms of (normalized) received citations, the latter two journals are more influential. We discuss these findings in sub-section 2.4.5 classical political economy of eco-innovation scientific knowledge/dissemination (links: publications in the

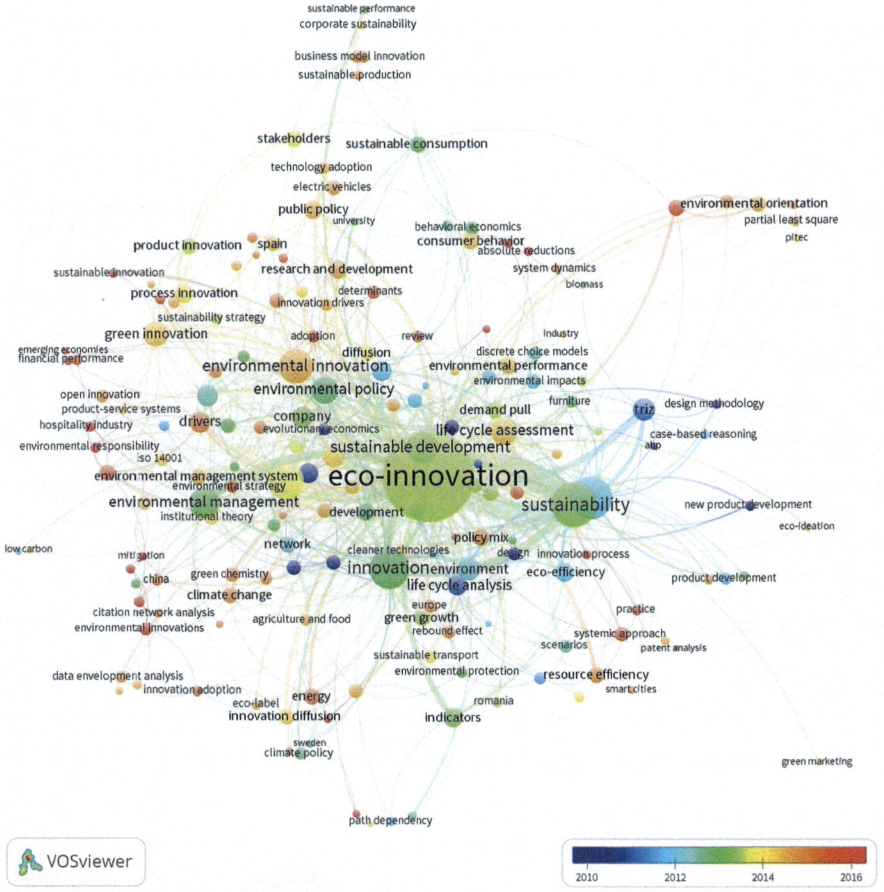

Fig. 2.5 Co-occurrence analysis/no of keywords: 178 (two or more occurrences) authors' keywords, legend: average publication year (n: 350)

journals citing each other: more publications citing each other thicker the link) (see Fig. 2.8).

The eco-innovation publications from England and Spain are most likely to be cited together. Shading changes with respect to normalized citations received by the country (see Fig. 2.9).

Table 2.8 Highly-cited keywords published in the last 5 years (Circular economy-related themes in bold)

Year	Most-cited keywords (year stratified)
2016	• Sustainable supply chain management; Relationships; Governance; Modelling/ REVERSE LOGISTICS; CONCEPTUAL-FRAMEWORK; GREEN; FUTURE; IMPLEMENTATION; INDUSTRIES **• Patent analysis; Waste heat recovery; Internal combustion engine; Automotive; Technology monitoring/ORGANIC RANKINE CYCLES; WORKING FLUIDS; POWER; CITATIONS; THERMOELECTRICS; TRAJECTORIES; SELECTION; VEHICLES; INDUSTRY; ENERGY** • Environmental innovation; environmental strategy; environmental management systems; network;/SMEs MANAGEMENT SYSTEMS; PERFORMANCE; CORPORATE; DETERMINANTS; TECHNOLOGY; STRATEGIES; IMPACT; FIRMS
2015	• Consumer behaviour; Electric vehicles; Adoption behavior; Intention; Literature review/ PRO-ENVIRONMENTAL BEHAVIOR; ECO-INNOVATION ADOPTION; ATTITUDES; EMOTION; DETERMINANTS; METAANALYSIS; TECHNOLOGY; EXPERIENCE; INTENTION; APPRAISAL • Environmental innovation; Open innovation; Absorptive capacity; Open eco-innovation mode/RESEARCH-AND-DEVELOPMENT; ENVIRONMENTAL INNOVATIONS; ABSORPTIVE-CAPACITY; MANUFACTURING FIRMS; TECHNOLOGY-PUSH; INTERNATIONAL DIFFUSION; TECHNICAL CHANGE; DEMAND-PULL; PERFORMANCE; DETERMINANTS • Eco-innovation; Biofuels; Demand-pull; Technology-push; Environmental policy; Patents/RESEARCH-AND-DEVELOPMENT; ENVIRONMENTAL-POLICY; RENEWABLE ENERGY; PANEL-DATA; INTERNATIONAL DIFFUSION; PATENT CITATIONS; TECHNICAL CHANGE; CLIMATE-CHANGE; ECONOMICS; MODELS **• Electronic waste; Consumer behaviour; Design characteristics; Machine learning;/ ECO-INNOVATION ADOPTION; E-WASTE; PERSONAL COMPUTERS; MANAGEMENT; HOUSEHOLDS; FACILITY; OPTIONS; SYSTEMS; CHINA; COST**
2014	• Business model innovation; Industrial sustainability; Value creation; Stakeholders; Sustainable consumption; Sustainable production;/PRODUCT-SERVICE SYSTEMS; OF-THE-ART; RESEARCH AGENDA; CONSUMER-GOODS; START-UPS; INNOVATION; INDUSTRIES; TECH **• Food waste; Food by-product; Industrial symbiosis; Sustainability;/LACTIC-ACID PRODUCTION; SUPERCRITICAL CARBON-DIOXIDE; PROCESSING BY-PRODUCTS; POTATO PEEL EXTRACT; NATURAL ANTIOXIDANT; DIETARY FIBER; OLIVE OIL; FUNCTIONAL-PROPERTIES; CHEESE MANUFACTURE; PHENOLIC-COMPOUNDS** • Public policy; Technology adoption; Electric vehicles; Eco-innovation;/ALTERNATIVE-FUEL VEHICLES; MARKET FAILURES; TECHNOLOGICAL DISCONTINUITIES; POLICY; DEMAND; INNOVATION; EMERGENCE; ECONOMICS; BARRIERS; SUPPORT • Sustainability-oriented innovation; Eco-innovation; SMEs; Systematic review; Sustainability strategy; Sustainable entrepreneurship;/MEDIUM-SIZED ENTERPRISES; SUPPLY CHAIN MANAGEMENT; PROACTIVE ENVIRONMENTAL STRATEGY; RESOURCE-BASED VIEW; OF-THE-ART; CLEANER-PRODUCTION; SMALL BUSINESSES; RESEARCH AGENDA; ECO-EFFICIENCY; SOCIAL-RESPONSIBILITY

(continued)

Table 2.8 (continued)

Year	Most-cited keywords (year stratified)
2013	• Environmental innovation (eco-innovation); Drivers of innovation; SMEs; Product innovation; Process innovation; Organizational innovation;/ENVIRONMENTAL-MANAGEMENT SYSTEMS; PRODUCT DEVELOPMENT; PERFORMANCE; FIRMS; GREEN • Eco-Innovation; industry; discrete choice models;/ENVIRONMENTAL INNOVATION; EMPIRICAL-ANALYSIS; FIRMS; PATTERNS; BEHAVIOR • Sustainability strategies; global value chains; upgrading; home furnishing; eco-innovation; environmental management;/SUPPLY-CHAIN; MANAGEMENT; INNOVATION; PERSPECTIVE; CLUSTERS; TRADE • Manufacturing firms; Innovation management; Eco-innovation; Open-innovation; Product-Service Systems • Eco-innovation; Energy efficiency; Metrics; Sustainable transport; Indicators; S/I curves; DECISION-MAKING; VEHICLES; SYSTEM • Eco-efficiency; Environmental public policy;/SME CLEANER PRODUCTION; WESTERN-AUSTRALIA; DELPHI METHOD; BARRIERS; SMES; PERFORMANCE; STRATEGIES; POLICIES
2012	• Eco-innovation; Environmental impacts; Discrete choice models; Regulation; Cost savings; Demand pull; Environmental policy;/MANAGEMENT-SYSTEMS; EMPIRICAL-ANALYSIS; GERMANY; PERFORMANCE; POLICY; FIRM; US • Eco-innovations; Environmental regulations; Organizational capabilities;/RESEARCH-AND-DEVELOPMENT; ENVIRONMENTAL INNOVATION; POLLUTION-CONTROL; PANEL-DATA; ISO-14001; GERMANY; MANAGEMENT; ADOPTION; US; DETERMINANTS • Geographic proximity; green growth; innovation; industrial ecology; network; sustainability;/QUANTITATIVE ASSESSMENT; ECOLOGY; SUSTAINABILITY; NETWORKS; FIRMS; PARADIGM; URBAN • Clean innovations; Private; Development; Diffusion; Policy mix; Demand-inducing instruments;/DIFFUSION; US

2.4.6 *Classical Political Economy of Eco-Innovation Scientific Knowledge*

In this subsection, we analyze the classical political economy aspects of eco-innovation scientific knowledge. (1) Financing (*who funds the eco-innovation scientific research? What is the structure of funding?*), (2) Production (*who produces the eco-innovation scientific knowledge?*), (3) Distribution (*Who disseminates the literature?*), and the use/consumption (*who cites the eco-innovation literature?*).

2.4.6.1 Finance

Who Funds the Eco-Innovation Research?

A multi-level and multi-domain co-funding structure is emerging. Lately, in the Netherlands, eco-innovation research is funded by the following organizations listed

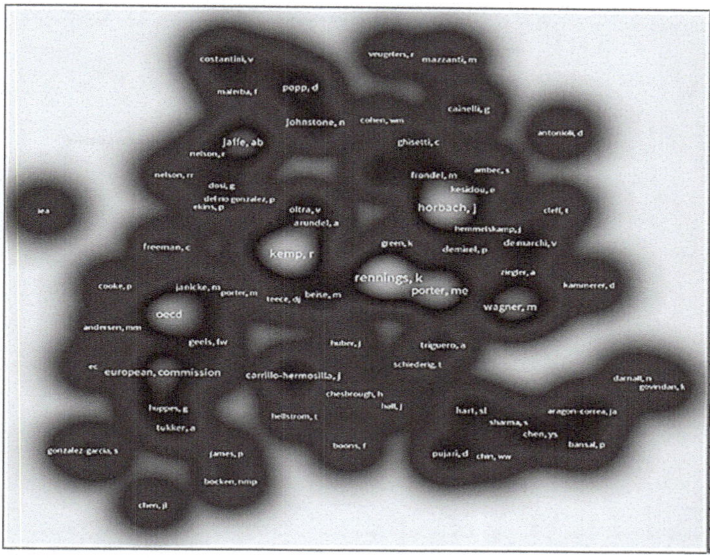

label	eigencentrality
oecd	1
kemp, r	0.996229
rennings, k	0.996229
wagner, m	0.996229
horbach, j	0.993585
porter, me	0.993585
beise, m	0.991174
brunnermeier, sb	0.991174
demirel, p	0.991174
jaffe, ab	0.991174
carrillo-hermosilla	0.990754
johnstone, n	0.98952
de marchi, v	0.988079
frondel, m	0.984464
oltra, v	0.98341
cainelli, g	0.98137
popp, d	0.980039
ghisetti, c	0.978824
european, comm	0.978406
triguero, a	0.977265
rehfeld, km	0.976009
del rio, p	0.972289
kesidou, e	0.970635
arundel, a	0.963489
dosi, g	0.963254

Fig. 2.6 Authors (Co-citation analysis, unit of analysis: cited first author, weight: total citations received per author)

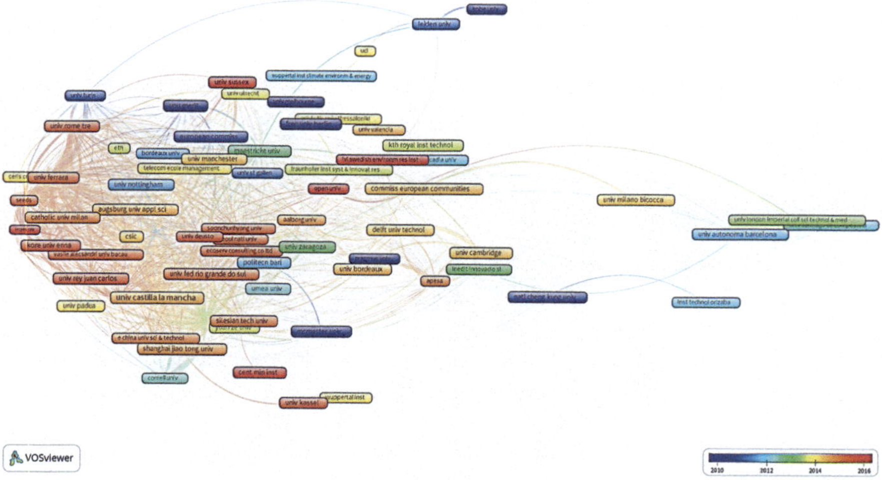

Fig. 2.7 Organizations: bibliographic coupling, legend: avg. publication year

in Table 2.9. Presence of international funding other than EU is the main differentiation point of the Netherlands (n: 34) from Spain (n: 66) (see Table 2.10).

The most cited article published in 2016 is "*The drivers for adoption of eco-innovation*" by Bossle, MB.; de Barcellos, MD.; Vieira, LM.; Sauvee, L. The funding agencies for this article are the National Council for Scientific and Technological Development (CNPq) of Brazil (its parent organization: Ministry of Science

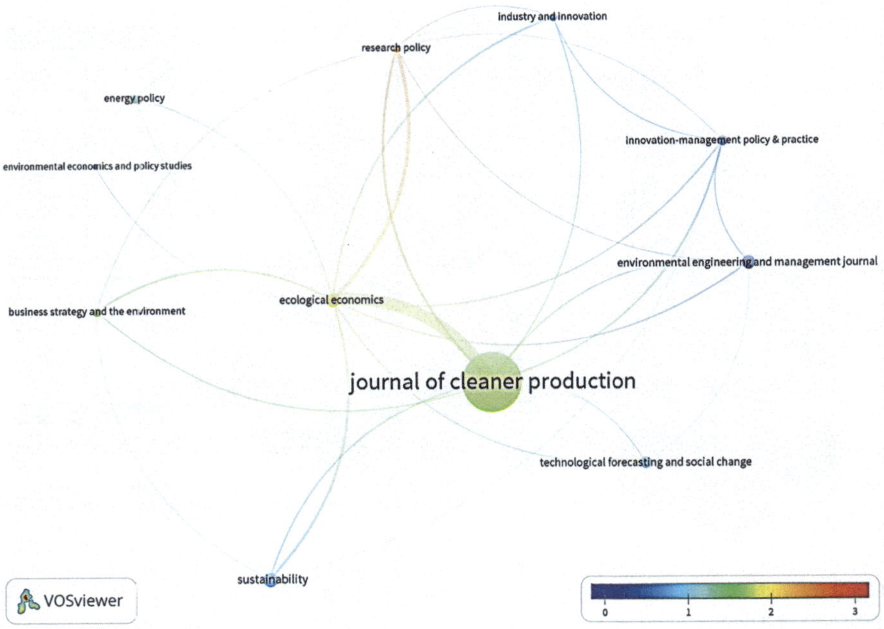

Fig. 2.8 Journals: citation analysis: legend: avg. normalized citations, size of nodes: number of publications

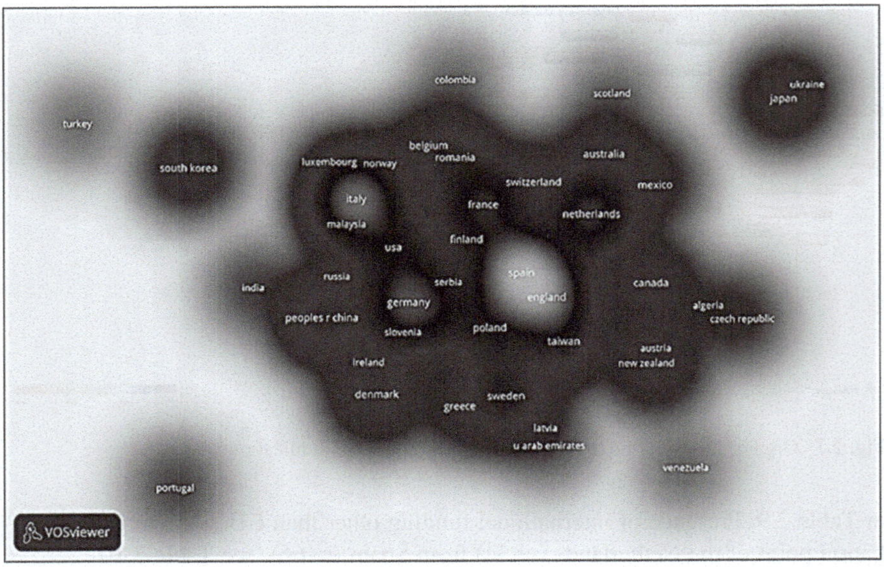

Fig. 2.9 Countries: bibliographic coupling of countries, normalized citations received by the country

Table 2.9 Multi-level funding structure in the Netherlands

Supranational	EU (2), PROJECT TIGI EUROPEAN COMMISSION (1), EC (1), REWARD A PROJECT IN THE ECO-INNOVATION PROGRAM FROM EACI (1)
National	TNO (2), NETHERLANDS ORGANIZATION FOR SCIENTIFIC RESEARCH (2), NWO ACTS PROGRAMME (1)
Companies	MYSHINEBOX (1)
	RESCOM (1)
	SAWTOOTH SOFTWARE SEQUIM WA (1)
Association (Voluntary)	BUNDESVERBAND DER DUNGERMISCHER E V (1)
International	UNIVERSITY OF WISCONSIN MADISON CENTER FOR SUSTAIN-ABILITY AND THE GLOBAL ENVIRONMENT (1), NATIONAL SCI-ENCE FOUNDATION (1)
	GERMAN MINISTRY OF EDUCATION AND RESEARCH BMBF (1), FEDERAL MINISTRY OF EDUCATION AND RESEARCH BMBF GERMANY AS PART OF THE RESEARCH PROJECT CLIMATE CHANGE FINANCIAL MARKETS AND INNOVATION CFI (1)
	UK EPSRC THROUGH THE CENTRE FOR INNOVATION AND ENERGY DEMAND CIED (1)
	SWEDISH RESEARCH COUNCIL (1)

and Technology) and CAPES, Foundation of the Brazilian Ministry of Education (a public *foundation* within the *Ministry of Education*). It is an article written in cooperation with France, Institute Polytechnic LaSalle Beauvais. In this respect, the article is a successful example of effective funding of the international research in the field. The Finnish Funding Agency for Technology and Innovation (TEKES) funded the article titled "Technology competition in the internal combustion engine waste heat recovery: a patent landscape analysis" by Karvonen, M.; Kapoor, R.; Uusitalo, A.; Ojanen, V., which is also one of the most cited articles published in 2016. The most cited articles published in 2015 are funded by the Swedish Energy Agency and the Swedish Foundation for Humanities and Social Sciences; this is an example of multi-domain co-funding (see Table 2.11). The article was titled "*Advances in consumer electric vehicle adoption research: A review and research agenda*". Similar to the year 2016, adoption related studies attract citations. Adoption related studies have a tendency to receive multi-domain co-funding.

2.4.6.2 Who Produces the Eco-Innovation Scientific Knowledge?

Authors in the Europe Union produced 79% of the eco-innovation publications (n: 277) (see Fig. 2.10). In the EU, most of the production of scientific knowledge in Eco-Innovation (in total n: 350) takes place in Spain (n: 66). Italy (n: 47) and the UK (n: 43) form the second group, and Germany (n: 35) and the Netherlands (n: 34) the third group, which is followed by France (n: 30) (Fig. 2.10). However, in terms of recognition and amplification, local citations received (here local citations means the

Table 2.10 Multi-level funding structure in Spain

Supranational	EUROPEAN REGIONAL DEVELOPMENT FUND SPANISH NATIONAL PLAN FOR SCIENTIFIC RESEARCH DEVELOPMENT AND TECHNOLOGICAL INNOVATION (1), PROJECT TIGI FP7 EUROPEAN COMMISSION (1), EUROPEAN COMMISSION (1), PROJECT TIGI EUROPEAN COMMISSION (1), EU (1), EDUCATION AUDIOVISUAL AND CULTURE EXECUTIVE AGENCY EACEA OF EUROPEAN COMMISSION UNDER ERASMUS MUNDUS ACTION 1 PROGRAM (1)
National	MINISTRY OF ECONOMY AND COMPETITIVENESS OF SPAIN (2), SPANISH GOVERNMENT (1), SPANISH SCIENCE AND INNOVATION MINISTRY (2), SPANISH MINISTRY OF SCIENCE AND TECHNOLOGY (1), MINISTRY OF ECONOMY AND COMPETITIVENESS (1), SPAIN S MINISTRY OF SCIENCE AN INNOVATION (1), SPANISH ECONOMY AND COMPETITIVENESS MINISTRY (6), SPANISH MINISTRY OF ECONOMY AND COMPETITIVENESS (4), SPANISH MINISTRY OF SCIENCE AND INNOVATION (2), SPANISH MINISTRY OF EDUCATION CULTURE AND SPORT (1), MINISTERIO DE CIENCIA E INNOVACION (1), SPANISH MINISTRY OF EDUCATION (2), SPANISH MINISTRY FOR EDUCATION AND SCIENCE (1)
Regional	REGIONAL GOVERNMENT OF ARAGON SPAIN (1), XUNTA DE GALICIA (2), TORRELAVEGA LOCAL GOVERNMENT (1), VALENCIAN REGIONAL GOVERNMENT (1), REMEDINAL3 CM FROM GOVERNMENT OF MADRID REGION (1), UNIVERSIDAD DE NAVARRA BANCO SANTANDER (1), DEPARTMENT OF EDUCATION UNIVERSITIES AND RESEARCH OF THE BASQUE GOVERNMENT (1), CSIC THROUGH JAE JUNTA PARA LA AMPLIACION DE ESTUDIOS PREDOC PROGRAM (1)
Universities	UNIVERSITAT POLITECNICA DE VALENCIA (4), UNIVERSIDAD DE NAVARRA (1), UNIVERSIDAD DE LA SABANA (1), PROJECT PRIN MIUR ITALIAN MINISTRY OF EDUCATION UNIVERSITY AND RESEARCH (1)
Companies	ENISA EMPRESA NACIONAL DE INNOVACION (1), INFONOMIA (1), IBERDROLA (1)
Councils and Foundations	CONSEJO SUPERIOR DE INVESTIGACIONES CIENTIFICAS CSIC (1), CONSEJO ECONOMIC Y SOCIAL OF CASTILLA LA MANCHA (1), UPV PAID 06 2011 FIRST PROJECTS (1), FUNDACION SENECA DE LA REGION DE MURCIA IN SPAIN (1), CENTRE DE DIFUSIO TECNOLOGICA DE LA FUSTA I MOBLE DE CATALUNYA CENFIM (1)

Funding agencies of highly cited articles published in 2015 and 2016: co-funding, supra/international, multi-level, multi-domain

citations received from the eco-innovation publications), we observe Germany is the leading country with its publications receiving 260 citations, followed by the UK (106 citations). Publications in Spain (92 citations) and Italy (81 citations) have fewer citations. We need to correct for this when assessing their impact on the production of eco-innovation knowledge. Netherlands is in the third group with

Table 2.11 Funding agencies of the most cited articles published in 2015

Funding agencies	Types	Received citations
Swedish Energy Agency; Swedish Foundation for Humanities and Social Sciences	**Multi-domain co-funding**	19
Project PRIN-MIUR—Italian Ministry of Education, University and Research [2010S2LHSE_002]	**Legislation based co-funding:** Progetti di Ricerca di Interesse Nazionale—PRIN) "*With the changes to legislation on the funding of "projects of national interest", the Italian Ministry of Education, Universities and Research (MIUR) has introduced a new mechanism for the allocation of funds based on co-funding, group research work and peer evaluation.*" (Research Italy 2010)	17
Regione Lazio; Roma Tre University; INEA in the project BIOESEGEN—MIPAF [17532/7303/10]; European Union [266959]; Italian Ministry of Education, University and Research	**Multi-level funding**	10
National Science Foundation—USA [CMMI-1435908]	**Multi-domain funding**: *Department of Industrial & System Engineering*, Buffalo, NY 14260 USA; Northeastern University, *Healthcare Systems Engineering Institute*, Boston, MA 02115 USA; SUNY Buffalo, Buffalo, NY 14260 USA; *PC Rebuilder & Recyclers* (company), Chicago, IL 60651 USA	8
EU [283002]	**Supranational funding:** [Vivanco, David Font; van der Voet, Ester] Leiden Univ, Inst Environm Sci CML, NL-2300 RA Leiden, Netherlands; [Kemp, Rene] Maastricht Univ, ICIS, NL-6200 MD Maastricht, Netherlands	7

respect to the number of publications and also in the third group with respect to the citations their publications have received (69 citations).

In terms of global citations (here global citations mean citations received from the publications external to the eco-innovation literature), we observe that Germany is still with 785 citations the leader, followed by Spain (665 citations) and the Netherlands (502 citations). These findings indicate that the bulk numbers in terms of the number of publications can be misleading in assessing the contribution of a country to the eco-innovation research field. Outside Europe, we observe that the eco-innovation research in Canada (55 citations) and Taiwan (34 citations) are mostly frequently cited by eco-innovation scientific research community, followed by publications in the US, China, and Japan.

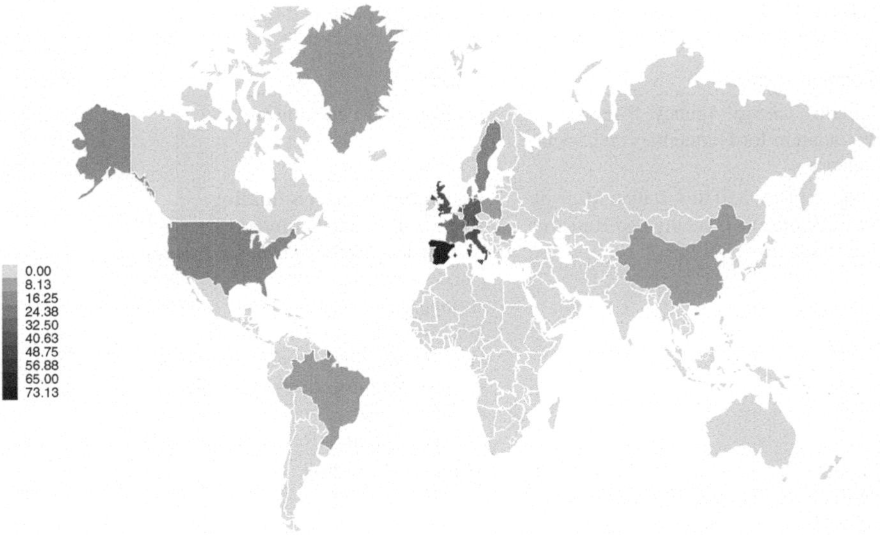

Fig. 2.10 Eco-innovation research and publications

In this respect, we detect the structure of eco-innovation research worldwide, in core, semi-periphery, and periphery countries in the making of eco-innovation scientific knowledge. The policymakers of the latter two categories might want to establish co-funding and co-authorship cooperation with the core countries which produce, reproduce and amplify the eco-innovation scientific knowledge.

Core countries for the eco-innovation scientific knowledge production are the ones which receive both local and global citations for their eco-innovation publications (29 countries).

- In the EU: *Germany, UK, Spain, Italy, France, Netherlands, Belgium, Sweden, Austria, Romania, Finland, Poland, Slovenia, Denmark*
- In Europe: +*Norway, Switzerland*
- In Northern America: *Canada, USA, Mexico*
- In East Asia: *Taiwan, China, Japan, South Korea,*
- In Africa: *Algeria*
- In Latin America: *Brazil*
- In Oceania: *Australia, New Zealand*

Semi-periphery countries are the ones which receive no local citations (no particular role in producing eco-innovation scientific knowledge) but receive global citations (amplification role) (nine countries: *Czech Republic, Greece, Ireland, Malaysia, Portugal, Serbia, Slovakia, Turkey, and Venezuela*).

Periphery countries are the ones which receive neither local nor global citations for their eco-innovation research (seven countries: *Chile, Colombia, Latvia, Luxembourg, Russia, South Africa, and UAE*).

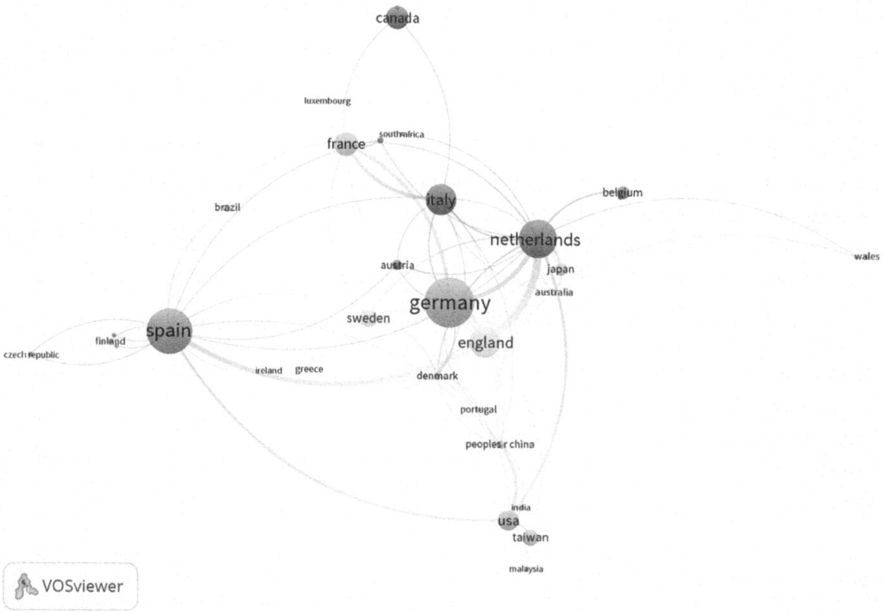

Fig. 2.11 Co-authorship network: unit of analysis: country level, weights: citations received

Location of nodes: the countries that are closer are more likely to engage in co-authorship (see Fig. 2.11).

2.4.6.3 Who Disseminates the Eco-Innovation Research?

Twenty percent of the eco-innovation research is disseminated by the Journal of Cleaner Production. The publisher is ELSEVIER Sci. Ltd. Elsevier is a global business, headquartered in Amsterdam, The Netherlands and has offices worldwide, part of RELX Group. According to RELX website, RELX Group is owned by two parent companies: RELX PLC (which owns 52.9% of RELX Group.) is the London Stock Exchange listed vehicle for holding shares in RELX Group. RELX NV (47.1% of RELX Group) is the Amsterdam Stock Exchange listed vehicle for holding shares in RELX Group, publishes over 2000 journals. For its 71 eco-innovation publications (see Table 2.12), the total number of citing articles without self-citations is 562 publications. Average citations per item is 12.1. Ecological economics, on the other hand, for its 14 eco-innovation publications, has the total number of citing articles without self-citations as 427 publications, with an average citation per item approaching to 52.00.

The second in line (see Table 2.12) is *Sustainability,* an international, cross-disciplinary, scholarly yet open access journal of environmental, cultural, economic, and social sustainability, published monthly online by the Multidisciplinary Digital

Table 2.12 Main journals for eco-innovation scientific knowledge dissemination (>1%)

Field: Source titles	Record count	% of 350	Bar chart
Journal of Cleaner Production	71	20.286	
Sustainability	15	4.286	
Ecological Economics	14	4.000	
Environmental Engineering and Management Journal	14	4.000	
Technological Forecasting and Social Change	12	3.429	
Innovation Management Policy Practice	10	2.857	
Business Strategy and the Environment	9	2.571	
Energy Policy	9	2.571	
Industry and Innovation	8	2.286	
Research Policy	8	2.286	
Environmental Economics and Policy Studies	5	1.429	
International Journal of Life Cycle Assessment	4	1.143	
Journal of Industrial Ecology	4	1.143	
Metalurgija	4	1.143	

Publishing Institute (MDPI). MDPI is an academic open-access publisher with headquarters in Basel, Switzerland, additional offices are located in Beijing and Wuhan (China), Barcelona (Spain) and in Belgrade (Serbia), and publishes 160 diverse peer-reviewed, scientific, open access, electronic journals. Although open, there are only 24 publications citing 15 articles published in Sustainability, with an average citation per item, 1.60.

Research policy, on the other hand, for its eight eco-innovation publications, receives an average citation per item 16.62; while the total number of citing articles without self-citations is 93.

These findings indicate that the dissemination effect of scientific publications is not only related to the bulk number of publications but also the received citations. Such information could guide funding agencies or universities in their publication dissemination or performance criteria.

2.4.6.4 Who Uses/Cites Eco-Innovation Research?

Total number of citing articles without self-citations (from Web of Science Core Collection) for the body of the eco-innovation scientific research is 1829. Eco-innovation research is mostly cited by the researchers in the USA (n: 262) and England (n: 221). From East Asia, we can also observe that the researchers in China (n: 166) and Taiwan (n: 72) are the most active in citing the eco-innovation literature. Engagement of South Korea (n: 44) and Japan (n: 29) is rather low (see Table 2.13). There is an increasing interest in China: 2007, n: 1, 0.602%; 2008, n: 1; 0.602%; 2009, n: 2, 1.205%; 2010, n: 6, 3.614%; 2011, n: 6, 3.614%; 2012, n: 11, 6.627%; 2013, n: 18, 10.843%; 2014, n: 23, 13.855%; 2015, n: 51, 30.723%; 2016 (until November 18), n: 47, 28.313%. Forty-three of these 166 records (25.904%) do not

Table 2.13 Citing countries (Top 25)

Field: Countries/territories	Record count	% of 1829	Bar chart
USA	262	14.325	
England	221	12.033	
Spain	184	10.060	
Peoples R China	166	9.076	
Germany	161	8.803	
Italy	154	8.4–20	
Netherlands	129	7.053	
France	101	5.522	
Sweden	98	5.358	
Brazil	87	4.757	
Australia	78	4.265	
Taiwan	72	3.937	
Canada	71	3.882	
South Korea	44	2.406	
Denmark	43	2.351	
Portugal	42	2.296	
Finland	41	2.242	
Switzerland	40	2.187	
Belgium	33	1.804	
Poland	33	1.804	
India	31	1.695	
Norway	31	1.695	
Japan	29	1.586	
Austria	26	1.422	
Greece	25	1.367	

contain data in the funding organization field. Yet approximately half of these citing articles are funded by five funding sources: National Natural Science Foundation of China, n: 62, 37.349%; Fundamental Research Funds for the Central Universities, n: 8, 4.819%; Ministry of Education of China, n: 8, 4.819%; National Social Science Foundation of China, n: 5, 3.012%; Natural Science Foundation of China, n: 4, 2.410%.

The most cited article which cites eco-innovation literature from USA is "*An organizational theoretic review of green supply chain management literature*" by Sarkis, Joseph (Clark University, Graduate School of Management); Zhu, Qinghua and Lai, Kee-hung, which is published in International Journal of Production Economics, Volume: 130 Issue: 1 March 2011. This article is also the most cited eco-innovation literature citing article written by a researcher in China (Zhu, Qinghua, Dalian University of Technology, School of Management). The publications which cite the eco-innovation literature and produced in the USA are mostly written in cooperation with China.

2.5 Conclusions

In this chapter, we analyzed patterns in the nature of eco-innovation research. For this, we conducted different types of bibliometric analyses on the Web of Science Core Collection data. The analyses revealed that eco-innovation is examined from different perspectives. These are (1) supply-side perspectives focusing on firms, industries (e.g. drivers for and barriers to eco-innovation); (2) technology-centered research (e.g. carbon capture and storage, electric vehicles, smart plugs); (3) science-based research (e.g. new materials); (4) sectoral studies (e.g. steel and iron industry, transport, information technology, food, agriculture, tourism); (5) the knowledge support element in eco-innovation (e.g. skills and training); (6) demand-side analyses (e.g. diffusion and adoption dynamics of individuals, households, firms), and (7) a policy influence perspective (the impact of policy instruments e.g. eco-labels, policy mixes). Concept-wise, we observe that the concepts of industrial ecology, industrial symbiosis, and circular economy are gaining importance as analytical lenses. Our analyses reveal differences between "eco-innovation" and "environmental innovation" research in that the latter pays more attention to policy influences and is less consumption-oriented. We also identified a shift from analyzing the impact towards supply and demand side research, a shift from environmental innovations towards the generative processes and dilemmas for sustainability-oriented innovations, and a rise in publications from less developed parts of the world.

Acknowledgement We want to thank Christiane Reif and Jens Horbach for offering comments and suggestions to an earlier version of the chapter. We are grateful to Alison Cathles for chapter proofreading.

References

Adams, R., Jeanrenaud, S., Bessant, J., Denyer, D., & Overy, P. (2016). Sustainability-oriented innovation: A systematic review. *International Journal of Management Reviews, 18*(2), 180–205.

Andersen, M. M. (2008, June). Eco-innovation–towards a taxonomy and a theory. In *25th celebration DRUID conference*.

Artes, F., Gomez, P., Aguayo, E., Escalona, V., & Artes-Hernandez, F. (2009). Sustainable sanitation techniques for keeping quality and safety of fresh-cut plant commodities. *Postharvest Biology and Technology, 51*(3), 287–296.

ASEIC. (2011). Opgeroepen op September 2, 2016, van About ASEIC: http://www.aseic.org/IntroR.do

Barbieri, N., Ghisetti, C., Gilli, M., Marin, G., & Nicolli, F. (2016). A survey of the literature on environmental innovation based on main path analysis. *Journal of Economic Surveys, 30*(3), 596–623.

Bastian, M., Heymann, S., & Jacomy, M. (2009). Gephi: An open source software for exploring and manipulating networks. *Icwsm, 8*, 361–362.

Bergek, A., & Berggren, C. (2014). The impact of environmental policy instruments on innovation: A review of energy and automotive industry studies. *Ecological Economics, 106*, 112–123.

Bocken, N. M. P., Short, S. W., Rana, P., & Evans, S. (2014). A literature and practice review to develop sustainable business model archetypes. *Journal of Cleaner Production, 65*, 42–56.

Bossle, M. B., de Barcellos, M. D., Vieira, L. M., & Sauvee, L. (2016). The drivers for adoption of eco-innovation. *Journal of Cleaner Production, 113*, 861–872.

Brunnermeier, S. B., & Cohen, M. A. (2003). Determinants of environmental innovation in US manufacturing industries. *Journal of Environmental Economics and Management, 45*(2), 278–293.

Burchart-Korol, D., Pichlak, M., & Kruczek, M. (2016). Innovative technologies for greenhouse gas emission reduction in steel production. *Metalurgija, 55*(1), 119–122.

Byrka, K., Jedrzejewski, A., Sznajd-Weron, K., & Weron, R. (2016). Difficulty is critical: The importance of social factors in modeling diffusion of green products and practices. *Renewable & Sustainable Energy Reviews, 62*, 723–735.

Chappin, E. J., & Ligtvoet, A. (2014). Transition and transformation: A bibliometric analysis of two scientific networks researching socio-technical change. *Renewable and Sustainable Energy Reviews, 30*, 715–723.

Chua, S. (1999). Economic growth, liberalization, and the environment: A review of the economic evidence. *Annual Review of Energy and the Environment, 24*, 391–430.

de Medeiros, J. F., Ribeiro, J. L. D., & Cortimiglia, M. N. (2014). Success factors for environmentally sustainable product innovation: A systematic literature review. *Journal of Cleaner Production, 65*, 76–86.

del Rio, P., Penasco, C., & Romero-Jordan, D. (2016). What drives eco-innovators? A critical review of the empirical literature based on econometric methods. *Journal of Cleaner Production, 112*, 2158–2170.

Diaz-Garcia, C., Gonzalez-Moreno, A., & Saez-Martinez, F. J. (2015). Eco-innovation: Insights from a literature review. *Innovation-Management Policy & Practice, 17*(1), 6–23.

Environment Agency. (2007). http://ew.eea.europa.eu. Opgeroepen op July 2016, van what is eco-innovation? http://web.archive.org/web/20131107101137/http:/ew.eea.europa.eu/Industry/innovation/faq296659

European Commission. (2006). Decision no 1639/2006/EC of the European Parliament and of the Council of 24 October 2006 establishing a competitiveness and innovation framework programme (2007 to 2013). Official Journal of the European Union.

European Commission. (2011). *Attitudes of European entrepreneurs towards eco-innovation: Analytical report. Flash Eurobarometer 315, Directorate-General Environment.* Brussels: Eurobarometer.

European Commission. (2012). Eco-innovation the key to Europe's future competitiveness. European Commission: Luxembourg.

Faber, A., & Frenken, K. (2009). Models in evolutionary economics and environmental policy: Towards an evolutionary environmental economics. *Technological Forecasting and Social Change, 76*(4), 462–470.

Fussler, C., & James, P. (1996). *A breakthrough discipline for innovation and sustainability.* London: Pitman Publishing.

Garfield, E. (2009). From the science of science to Scientometrics visualizing the history of science with HistCite software. *Journal of Informetrics, 3*(3), 173–179.

Gavel, Y., & Iselid, L. (2008). Web of science and scopus: A journal title overlap study. *Online Information Review, 32*(1), 8e21.

Ghazal, M., Akmal, M., Iyanna, S., & Ghoudi, K. (2016). Smart plugs: Perceived usefulness and satisfaction: Evidence from United Arab Emirates. *Renewable & Sustainable Energy Reviews, 55*, 1248–1259.

Hojnik, J., & Ruzzier, M. (2016). What drives eco-innovation? A review of an emerging literature. *Environmental Innovation and Societal Transitions, 19*, 31–41.

Horbach, J., Rammer, C., & Rennings, K. (2012). Determinants of eco-innovations by type of environmental impact—The role of regulatory push/pull, technology push and market pull. *Ecological Economics, 78*, 112–122.

Inigo, E. A., & Albareda, L. (2016). Understanding sustainable innovation as a complex adaptive system: A systemic approach to the firm. *Journal of Cleaner Production, 126*, 1–20.

Karakaya, E., Hidalgo, A., & Nuur, C. (2014). Diffusion of eco-innovations: A review. *Renewable & Sustainable Energy Reviews, 33*, 392–399.

Kemp, R., & Pearson, P. (2007). Final report of measuring eco-innovation (MEI project), Bruxelles.

Kemp, R., & Pontoglio, S. (2011). The innovation effects of environmental policy instruments—A typical case of the blind men and the elephant? *Ecological Economics, 72*, 28–36.

Ketata, I., Sofka, W., & Grimpe, C. (2015). The role of internal capabilities and firms' environment for sustainable innovation: Evidence for Germany. *R & D Management, 45*(1), 60–75.

Klewitz, J., & Hansen, E. G. (2014). Sustainability-oriented innovation of SMEs: A systematic review. *Journal of Cleaner Production, 65*, 57–75.

Lang-Koetz, C., Pastewski, N., & Rohn, H. (2010). Identifiying new technologies, products and strategies for resource efficiency. *Chemical Engineering & Technology, 33*(4), 559–566.

Lucchetti, M. C., & Arcese, G. (2014). Tourism management and industrial ecology: A theoretical review. *Sustainability, 6*(8), 4900–4909.

Markard, J., Raven, R., & Truffer, B. (2012). Sustainability transitions: An emerging field of research and its prospects. *Research Policy, 41*(6), 955–967.

Markard, J., & Truffer, B. (2008). Technological innovation systems and the multi-level perspective: Towards an integrated framework. *Research Policy, 37*, 596–615.

Mirabella, N., Castellani, V., & Sala, S. (2014). Current options for the valorization of food manufacturing waste: A review. *Journal of Cleaner Production, 65*, 28–41.

Mol, A., & Sonnenfeld, D. (2000). *Ecological modernisation around the world: Perspectives and critical debates*. London: Frank Cass.

Mol, A., Sonnenfeld, D., & Spaargaren, G. (Eds.). (2009). *The ecological modernisation reader. Environmental reform in theory and practice*. London: Routledge.

OECD. (2010). *Eco-innovation in industry – Enabling green growth*. La Vergne, TN: OECD Publishing.

OECD. (2012). *The future of eco-innovation: The role of business models in green transformation*. OECD Publishing.

Prieto-Sandoval, V., Alfaro, J. A., Mejía-Villa, A., & Ormazabal, M. (2016). ECO-labels as a multidimensional research topic: Trends and opportunities. *Journal of Cleaner Production, 135*, 806–818.

Qi, G. Y., Zeng, S. X., Tam, C. M., Yin, H. T., & Zou, H. L. (2013). Stakeholders' influences on corporate green innovation strategy: A case study of manufacturing firms in China. *Corporate Social Responsibility and Environmental Management, 20*(1), 1–14.

Reid, A., & Miedzinski, M. (2008). *Eco-innovation: Final report for sectoral innovation watch* (Systematic Eco-Innovation Report).

Rennings, K. (2000). Redefining innovation—eco-innovation research and the contribution from ecological economics. *Ecological Economics, 32*(2), 319–332.

Rezvani, Z., Jansson, J., & Bodin, J. (2015). Advances in consumer electric vehicle adoption research: A review and research agenda. *Transportation Research Part D: Transport and Environment, 34*, 122–136.

Schot, J., & Geels, F. W. (2008). Strategic niche management and sustainable innovation journeys: Theory, findings, research agenda, and policy. *Technology Analysis & Strategic Management, 20*(5), 537–554.

Türkeli, S., Kemp, R., Huang, B., Bleischwitz, R., & McDowall, W. (2018). Circular economy scientific knowledge in the European Union and China: A bibliometric, network and survey analysis (2006–2016). *Journal of Cleaner Production, 197*, 1244–1261.

UNEP. (2014). Eco-innovation manual.

van Berkel, R. (2007). Eco-efficiency in primary metals production: Context, perspectives and methods. *Resources Conservation and Recycling, 51*(3), 511–540.

Van Eck, N. J., & Waltman, L. (2010). Software survey: VOSviewer, a computer program for bibliometric mapping. *Scientometrics, 84*(2), 523–538.

Viaggi, D. (2015). Research and innovation in agriculture: Beyond productivity. *Bio-based and Applied Economics, 4*(3), 279–300.

Chapter 3
On the Economic Returns of Eco-Innovation: Where Do We Stand?

Claudia Ghisetti

3.1 Introduction

Of paramount importance for policymakers is to properly assess the economic consequences of sustainable production choices aimed at reducing environmental negative externalities. Such an assessment is relevant at different levels of aggregation, starting from the firm level analysis, to understand whether "going green" brings about certain economic gains or instead it is counterproductive, and moving to the meso (sectors) or macro (country) levels of analysis, to understand whether any aggregate effects are at stake and to which direction they point.

Overall, this understanding would call (or not call) for the need of designing proper industrial or innovation policies that include specific environment-oriented components. This understanding has to be combined with clear evidence on the underlying determinants of greener production choices (e.g. Horbach et al. 2012), which are stated to differ from standard technological innovations as they are subject to a knowledge and an environmental externality, a so-called *double externality* (Jaffe et al. 2005; Rennings 2000; Popp et al. 2010). More importantly, this has to be combined with evidence on how to properly stimulate environmental innovations (see for a review Ghisetti and Pontoni 2015), which policy instruments are effective for this regulatory push-pull (e.g. Ghisetti 2017) and how to remove barriers that are detrimental to their uptake (Marin et al. 2015).

The current chapter draws on existing literature on the topic which is certainly crucial and at the bases of the current work (e.g. definitions, determinants, barriers, policy-stimulus, relation standard innovation and "crowding-out" of eco-innovations)

C. Ghisetti (✉)
European Commission, DG Joint Research Centre, Unit I.1, Modelling, Indicators & Impact Evaluation, Ispra, VA, Italy
e-mail: claudia.ghisetti@ec.europa.eu

J. Horbach, C. Reif (eds.), *New Developments in Eco-Innovation Research*, Sustainability and Innovation, https://doi.org/10.1007/978-3-319-93019-0_3

and it focuses mainly (and only) on the expected economic effects of the adoption of environmental innovations, or eco-innovations (henceforth EI).

Section 3.2 revises the existing literature on the economic returns of EI in terms of productivity and profitability (Sect. 3.2.1), in terms of the typology of EI under scrutiny and their heterogeneous effects (Sect. 3.2.2) and in terms of the employment effects (Sect. 3.2.3).

New empirical evidence on the economic effects of EI at the aggregate level is provided in Sect. 3.3 on a dataset created by the pooling of two consecutive waves of the Community Innovation Survey. The Section also describes the data, the empirical model and discusses the main results.

Section 3.4 concludes and outlines future research directions.

3.2 Eco-Innovation and Economic Returns

One of the first contribution arguing in favor of the potential positive effects of EI comes from the seminal paper by Porter and Van der Linde (1995), postulating that environmental regulation is not necessarily detrimental for firms' performance, rather if regulations are well designed, regulation-induced EI may induce positive effects in the long-run, leading to "win-win" solutions. In a dynamic context, the loss of competitiveness potentially associated with compliance costs can be partially offset by the "induced innovation" improvements.

This hypothesis is further examined in the paper by Jaffe and Palmer (1997) who articulate the hypothesis in its narrow, weak and strong characterization. The first posits that only certain types of environmental policies—namely the well-designed ones—can improve innovation and competitiveness. The second strengthens the regulatory effects on innovation and relaxes the direct effect on competitiveness. The third postulates that efficiency gains achieved by the "induced innovation" can completely offset the loss of competitiveness caused by compliance costs.

A broad strand of empirical literature has been focused on assessing the competitiveness effects of environmental regulation, which indirectly or directly passes through EI.

Among those contributions, to mention a few, Rexhäuser and Rammer (2013) outline a negative although weak effect on firm's profitability displayed by EI introduced in response to environmental regulation on a sample of German manufacturing firms.

Peneder et al. (2017) focus on German, Austrian and Swiss firms and assess the competitiveness effects of selected policy instruments (e.g. energy-related taxes, subsidies, standards and negotiated agreements) through a system of equations finding an overall null effect of EI on competitiveness.

Manello (2017) analyses the link between environmental and economic performance measured as Total Factor Productivity (TFP) on a sample of Italian and German chemical firms using data from the European Pollution Release and Transfer Register (E-PRTR) finding a confirmation to the Porter Hypothesis: firms facing higher compliance costs have reacted with productivity enhancing investments.

Costantini and Mazzanti (2012) assess the strong and the narrow version of the Porter Hypothesis on trade competitiveness in the manufacturing sector, finding support that EI are able to stimulate "green" exports.

Rubashkina et al. (2015) empirically test for the weak and the strong versions of the Porter Hypothesis to disentangle the effect of environmental regulation on 17 European countries at the sectoral level on the years 1997–2009. Their findings support the weak hypothesis that regulation stimulates innovation. However, they fail to support the strong version, as no productivity gains or losses are associated with the regulation, even when accounting for the endogeneity of the regulation.

Using as a proxy for environmental regulation a measure of environmental taxation, Franco and Marin (2017) study the effect of regulation on innovation and productivity on manufacturing sectors in eight European countries. Not only the study provides additional evidence on the strong and weak version of the Porter Hypothesis, but it also includes an analysis on the effects occurring indirectly in other sectors, induced by the stringency of the regulation occurring in their strongly interrelated sectors. In particular, Franco and Marin (2017) distinguish between the stringency of environmental taxation of the sector and the one of taxation in upstream and downstream sectors. Overall they find support for both weak and strong version of the Porter Hypothesis, even after controlling for endogeneity. Moreover, it is the first study to highlight that environmental regulation stringency in other sectors may impact on certain sectors in the presence of strong inter-sectoral linkages. More precisely, their findings suggest that environmental regulation in downstream sectors induce the corresponding upstream sector to innovate and this improves as well their productivity.

More recently, Dechezleprêtre and Sato (2017) largely discuss the endogeneity issue arising from environmental policies being correlated with the unobserved determinants of the outcome variable of interest, including supply chain linkages, firm-specific factors, political institutions, and the stringency of other regulations. Furthermore, they stress the possible reverse causality issue when using aggregated data that occurs if/when policies are set strategically by the policymakers, e.g. by exempting certain sectors from environmental regulations to stimulate their growth, exports, production, employment etc.

Overall, although mixed evidence is found, relatively high support is found to the stimulus environmental regulation exerts on innovation, and relatively less support is found with respect to the competitiveness returns of such regulation.

In the same direction of the Porter Hypothesis, also the natural-resource-based view of the firm, hypothesizes that firms' profitability can be positively affected by EI. The channel through which EI can be beneficial is quite different, as in this approach the positive economic gains are given by the competitive advantages that are created once accounting for the natural environment surrounding the firm. For a firm the inclusion of environmental aspects is a pro-active reaction to resources depletion which may be threatening firm's resources (Hart and Dowell 2011). This reaction is, in turn, able to foster the development of strategic resources and dynamic capabilities (Aragon-Correa and Sharma 2003; Hart and Milstein 2003) that are later associated to positive economic returns (Hart 1995). Notwithstanding the possible competitive advantages of greener production choices, their search costs may be high (King and Lenox 2002) and their economic benefits are discussed to be underestimated by managers, eventually limiting the scope of the associated economic gains (Berchicci and King 2007).

Expectations on positive economic returns associated with environmentally responsible choices are also discussed by the literature on the so-called Corporate Social Responsibility (CSR). This literature associates the positive returns of greener production choices to improvements in market's evaluation of the firm, access to new markets or cost reduction driven by increased resource efficiency (Ambec et al. 2008; Hart 1997; Margolis and Walsh 2003; Orlitzky et al. 2011; Porter and Kramer 2002, 2006). This positive economic return passes through innovation, as innovation is the moderator between CSR and economic returns (Martinez-Conesa et al. 2017). Furthermore, the concept of strategic CSR has been put forth, defined as "any 'responsible' activity that allows a firm to achieve a sustainable competitive advantage, regardless of motive" (McWilliams and Siegel 2011: 1480) such that it allows for the private provision of (environmental) public goods.

All in all, in a recent review paper Barbieri et al. (2016) summarize literature on the economic effects of EI as it follows: "EI may influence in an asymmetric way short-term measures of profitability (e.g. stock market returns, profits) and long-term performance (e.g. productivity, international competitiveness, survival, and firm growth)" (Barbieri et al. 2016: 609).

3.2.1 Profitability and Productivity Returns

A quite vast amount of contributions have assessed the short-term economic returns of environmental choices at the firm level.

Using different profitability measures (namely Returns on Equity (ROE), Returns on Assets (ROA), Returns on Sales (ROS) or Tobins'q index) and with a focus on firms operating in different countries and sectors, a very mixed picture emerges when trying to answer the question whether it pays to be green.

Certain studies find support that it pays to be green (Cheng et al. 2014; Dowell et al. 2000; Hart and Ahuja 1996; Russo and Fouts 1997; King and Lenox 2001, 2002; Salama 2005) even once accounting for the simultaneity arising from the joint determination of environmental and economic performances through a three stages least squares approach (Al-Tuwaijri et al. 2004). Similarly, Lanoie et al. (2011) analyze the relationship between environmental regulatory stringency, environmental performance and financial performance on seven OECD countries and find support that environmental R&D positively affects business performance, even though the costs to comply with environmental regulation are not fully offset.

Certain studies find instead empirical support for a short-term detrimental effect (Cordeiro and Sarkis 1997; Sarkis and Cordeiro 2001; Wagner et al. 2002). Moreover, other studies failed to depict any significant correlation (Elsayed and Paton 2005; Freedman and Jaggi 1992; Amores-Salvadó et al. 2014) and severe misspecification problems in the previous findings were discussed by Elsayed and Paton (2005).

As a matter of fact, the meta-analysis of the literature by Horváthová (2010) summarizes that 15% of the studies found a negative return of going green, 55% a positive return, and 30% found no significant effect.

A more recent meta-analysis by Endrikat et al. (2014) re-stresses the inconsistency of previous findings and the lack of consensus on the relationship between financial and environmental performance and focuses on how previous studies have dealt with the direction of the causality between the two. The conclusion is that overall the relationship is positive and that the natural-resource-based-view of the firm finds a confirmation: firms' proactive environmental activities increase their efficiency, reduce costs and increase revenues thus leading to improved financial performance (Hart 1995; Hart and Dowell 2011).

By focusing on a longer-term economic output, more specifically productivity, Marin and Lotti (2017) assess the effects of environmental patents on productivity on a panel of Italian manufacturing firms, by applying a Crepon-Duguet-Mairesse CDM model (Crépon et al. 1998). Results go in the direction that it does not necessarily pay to be green, as the productivity returns of EI are smaller than the ones related to non-environmental technologies. Furthermore, EI tend to crowd out non-environmental innovations which may be even more profitable (Marin 2014). In the same vein, Greenstone et al. (2012) evaluate the impact of the Clean Air Act on total factor productivity (TFP) on US, finding a negative effect of its increased stringency on TFP.

Soltmann et al. (2014) perform a sectoral analysis on 12 OECD country-sectors (Austria, Denmark, Finland, France, Germany, Italy, Japan, the Netherlands, Sweden, Switzerland, the United Kingdom and the United States) in the period 1980–2009 and approximating EI through patent applications in environmental technologies (through the OECD ENV-TECH classification). They model total value added as the outcome variable in an augmented Cobb-Douglas production function. Their results suggest for the presence of an U-shape relation between environmental patents and value added at the industry level: for most industries, increases in EI negatively affect performance. Similarly, Riillo (2017) in a study on Italian SMEs finds a U-shaped relationship between green management and firm performance (measured in terms of labor productivity) leading the author to conclude that greener firms in energy-intensive sectors show no significant difference in performance than other firms.

To conclude, the effects of EI on a different outcome, i.e. firm growth, have been assessed on a panel of Italian firms (Leoncini et al. 2017). Firm growth is found to be more affected by green technologies than non-green ones, with the exception of struggling and rapidly growing firms. Furthermore, firm experience, approximated by firm's age, positively moderates the effects of green technologies on growth.

3.2.2 For Whom and When It Pays to Be Green

A synthesis on these mixed findings in the evidence on economic returns of greener production choices is found in Telle (2006), who argues that the real challenge would be to unveil when or for whom it can pay to go green, rather than posing the too general question whether it pays or not to be green.

Coherently to this view, Russo and Fouts (1997) suggest that firms' profitability depends on whether firms choose to introduce end-of-pipe technologies or to redesign their production processes and services: the first are not associated to any changes in firms' resources nor capabilities and are thus not expected to produce any positive economic return in the short run. A confirmation to that is found in Cleff and Rennings (1999), who analyze the categories of end-of-pipe technologies versus cleaner production technologies, finding support that the first are mostly introduced to comply with environmental regulation while the latter are more often introduced to improve economic returns via increased market share or cost reduction.

Li (2014) analyses the linkages between institutions, EI and firms performance on a sample of Chinese manufacturing firms finding that the effects of EI exerted on financial performance are positively moderated by the level of resources the firm commits: increases in resources committed to achieve environmental practices are associated to improved financial performance.

Still related to the Telle (2006) conclusion, the contribution by Ghisetti and Rennings (2014) argues that the question whether it pays to be green has to be better qualified in terms of the typologies of environmental innovations to be considered, as different typologies of EI may lead to heterogeneous profitability effects. Through an empirical analysis on German manufacturing firms performed on two waves of the German Community Innovation Survey, the authors find support on the different degrees of returns on sales (operating revenues from sales) depending on the typology of EI. On the one side EI aimed at improving resource and energy efficiency have a positive influence on financial performance, but, on the other side, those aimed at reducing externalities, such as harmful materials and air, water, noise and soil pollution, are associated to a worsening of the financial performance. However, these effects are only depicted for innovation having a high environmental impact, suggesting for the presence of an environmental threshold.

Miroshnychenko et al. (2017) provide a novel and global empirical overview on the financial returns of green practices by analysing a panel of publicly-traded companies in 58 countries over 13 years. Their findings do confirm that the question "does it pay to be green" has to be better defined in terms of green practice considered. More precisely, what they define as internal green practices (pollution prevention and green supply chain management) are found to be the major drivers of financial performance, whereas product development is secondary and the adoption of environmental management schemes (namely ISO 14001) negatively impacts financial performance, measured as Tobin's q and Returns on Equity (ROE). Results are confirmed when controlling for the potential endogeneity of green practices and corporate financial performance measures.

Similar differential effect depending on the typology of strategy pursued but on a longer term economic measure is found by Antonietti and Marzucchi (2014) who outline that firm's productive efficiency (TFP) is positively affected by "green" tangible investment strategies and that this, in turn, impacts positively exports towards countries facing stringent environmental regulations.

Overall, it is confirmed that the question "does it pay to be green" needs to be better qualified.

3.2.3 Employment Effects and "Green Jobs"

A recent strand of literature has focused on unveiling the job creation/destruction potential of sustainability transition, following the recent discussion on "green jobs" (ILO 2013; OECD 2014).

Among those, Horbach (2010) outlines a positive job creation effect of EI which is greater than the one for standard innovations in a sample of German firms operating in the environmental sector.

Coherently, Horbach and Rennings (2013) stress that employment growth in German innovative firms is stronger for firms adopting material and energy-saving innovations, while the effect changes for other typologies of EI.

On the contrary, Cainelli et al. (2011) on a sample of Italian firms find a negative effect of EI on employment growth in the short term.

Mixed evidence is also depicted in Licht and Peters (2013) who focus on 16 European countries CIS data finding a positive and significant effect of product EI on employment growth of product innovations but with no significant difference with respect to standard innovations.

More recently, Gagliardi et al. (2016) study the employment effects of EI on Italian manufacturing firms and find that EI positively and significantly affect job creation, and also that it happens with a greater effect than standard innovations. Results hold true when overcoming for endogeneity of technological change by IV approach.

Also Kunapatarawong and Martínez-Ros (2016) find support of a positive employment effect of EI in an empirical analysis conducted on Spanish firms in the period 2007–2011, and this effect is stronger for firms operating in "dirtier" sectors.

A closer look into the effects of environmental regulation on the labour market is discussed by Vona et al. (2018), whose analysis on "green jobs" digs into the skills associated to certain occupations and it assesses the demand for green skills that is associated to changes in environmental regulation. The study unveils, for the US, whether general skills are different from those of the workers that are displaced to a different occupation in response to an environmental regulation.

Overall, their findings suggest that the general skill composition of "dirty" and "green jobs" is closer than to other occupations and the latter are rarely more complex than the first. Consequently, environmental regulation only has a limited effect on employment and on the labor market.

Complementary evidence is discussed in Consoli et al. (2016), who compare the work content of green and "non-green jobs" finding relevant differences that may guide implications on how to properly redirect the labor market toward "green (er) jobs". The work content of "green jobs" is indeed of a less routinized nature than "non-green jobs" and "green jobs" are found be grounded on a higher intensity of human capital, longer work experience and stronger on-the-job training.

3.3 New Evidence on Community Innovation Survey 2008 and 2014 Data

This section aims at providing original evidence on the evolution of EI adoption across time, sectors and countries, and lately, on the effects this adoption has on economic outcomes.

3.3.1 Data and Descriptive Evidence

Different datasets have been combined to serve this purpose. The Community Innovation Survey, a widely explored firm level innovation survey conducted by Eurostat in EU 28 Member States, is the core dataset of the analysis. The two waves 2008 and 2014 have included ad hoc sections on "Innovations with environmental benefits". Sectoral data for EU member states on these two waves have been combined with Eurostat national accounts data (Eurostat: nama_10_a64), with Eurostat structural business statistics (Eurostat: sbs_na_ind_r2) and with air emission intensity data by NACE sectors (Eurostat: env_ac_aeint_r2).

Due to the voluntary nature of the ad hoc section on innovations with environmental benefits, some countries have no available information on EI adoption in 2008 (Greece and Slovenia), some countries have no available information on EI adoption in 2014 (Belgium, France, Ireland and the Netherlands) and some countries are missing in both waves (Denmark, Norway, Spain and the United Kingdom). This led to the exclusion of those 10 countries from the analysis, which is based on the remaining 18 EU countries.

Some additional cleanings were required in order to combine the different sectoral aggregations, which were not perfectly overlapping across the four different data-sources. At the end of the cleaning, only manufacturing sectors have been considered and the following sectoral aggregation (Nace Revision 2) are exploited (Table 3.1).

Table 3.1 Sectors included in the analysis

Nace	Description
C10–C12	Manufacture of food, beverages and tobacco products
C13–C15	Manufacture of textile, wearing apparel and leather and related products
C16–C18	Manufacture of wood and of products of wood and cork, except furniture, of paper and paper products, printing and reproduction of recorded media
C19	Manufacture of coke and refined petroleum products
C20	Manufacture of chemicals and chemical products
C21	Manufacture of basic pharmaceutical products and pharmaceutical preparations
C22–C23	Manufacture of rubber and plastic products, and of other non-metallic mineral products
C24–C25	Manufacture of basic metals, of fabricated metal products, except machinery and equipment
C26	Manufacture of computer, electronic and optical products
C27	Manufacture of electrical equipment
C28	Manufacture of machinery and equipment n.e.c.
C29-C30	Manufacture of motor vehicles and other transport equipment
C31-C33	Manufacture of furniture, repair and installation of machinery and equipment and other manufacture

Table 3.2 Main EI variables

For the firm	
ECOMAT	Measures the share of enterprises with reduced material or water use per unit of output;
ECOPOL	Measures the share of enterprises with reduced air, water, noise or soil pollution;
ECOREC	Measures the share of enterprises with recycled waste, water or materials;
ECOSUB	Measures the share of enterprises that replaced materials with less polluting or hazardous ones.
For the end-users	
ECOENU	Measures the share of enterprises in the sector adopting innovation with reduced energy use or CO2 footprint by end user;
ECOPOS	Measures the share of enterprises with reduced air, water, noise or soil pollution by end user;
ECOREA	Measures the share of enterprises that facilitated recycling of product after use by the end user.

The core variables of interest relate to the share of firms in the sector having adopted innovations leading to specific environmental benefits either to the firm adopting it or to the end users who bought the new eco-innovative product. In particular, the following typologies of eco-innovations are under scrutiny (Table 3.2).

The final potential sample for the empirical analysis would consist of 13 (sectors) * 18 (countries) * 2 (years) = 468.

However, the core variables of interest have an average share of missing values across years, sectors and countries of around 20%, thus reducing the sample to

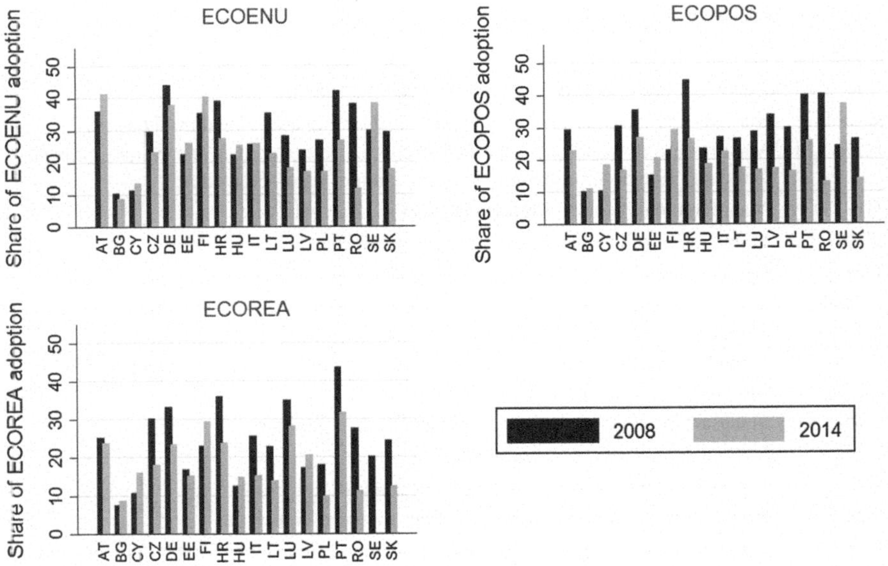

Fig. 3.1 Eco-innovation adoption for end users, by sector

377 observations. In addition, some missing variables among the controls were depicted, leading the final empirical sample to 264.

A closer look into the evolution on the sectoral adoption of those innovations between 2008 and 2014 already offers very interesting evidence, either when it is performed by comparing sectors, or when it is performed at the country level.

Figure 3.1 summarizes the rate of adoption of EI for the end users and it is quite unexpectedly evident that it has substantially decreased for the three typologies of EI (ECOENU, ECOPOS and ECOREA) and for all the sectors, with the only exception of "Manufacture of coke and refined petroleum products" (C19), which experienced an increase in such a share in the case of ECOENU adoption. Quite similar evidence emerges in Fig. 3.2 as the only sector experiencing an improvement in the share of EI adoption by firms is again "Manufacture of coke and refined petroleum products" (C19), for innovations typologies ECOMAT and ECOPOL. Interestingly enough, this sector is also the one with the highest share of adoption of such innovations (ECOMAT and ECOPOL) in the 2 years considered (2008 and 2014).

Figure 3.3 shows the distribution of the rate of adoption of EI with environmental benefits experienced by the end users and its evolution over time by country. It allows shedding light on countries experiencing the highest share of adoption of each typology of EI by year, as well as its evolution between 2008 and 2014.

It emerges a less clear-cut picture than the previously described one by sectors. As for the innovations with environmental benefits to the end users, Finland is the only country of the sample having augmented the share of EI adoption in all the three innovation typologies: ECOENU, ECOPOS and ECOREA. In all the three cases, however, the number of countries with a reduced share is greater (throughout the

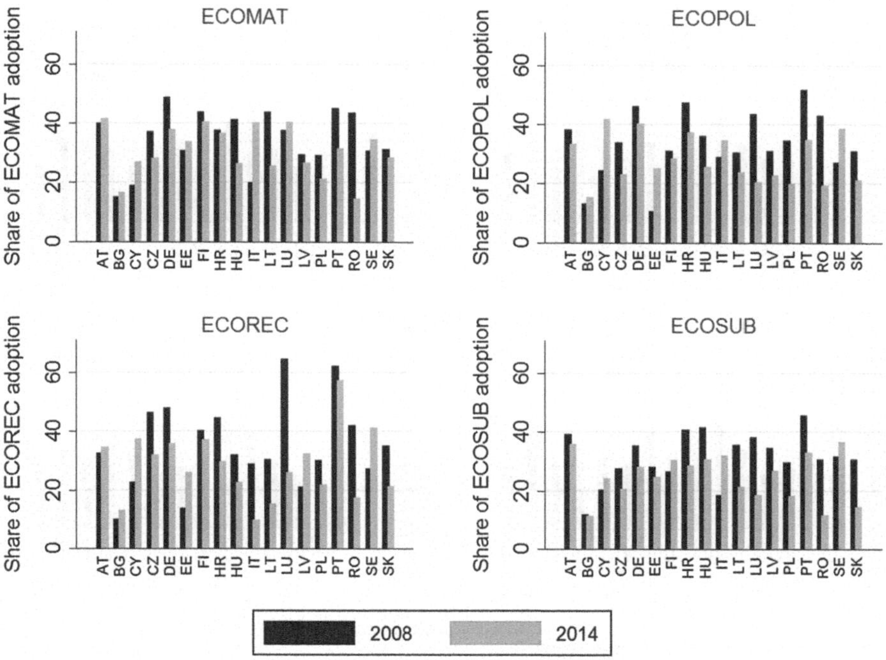

Fig. 3.2 Eco-innovation adoption by firms, by sector

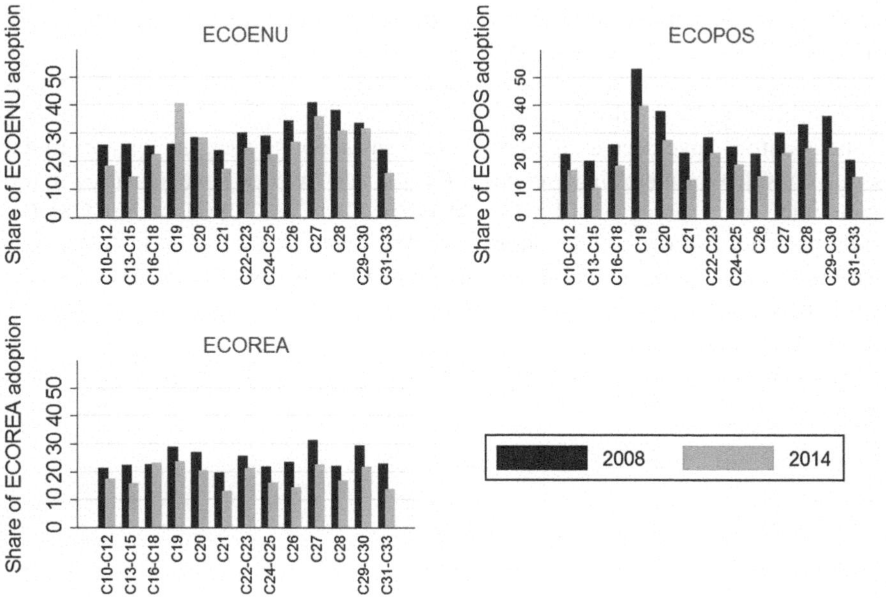

Fig. 3.3 Eco-innovation adoption for end users, by country

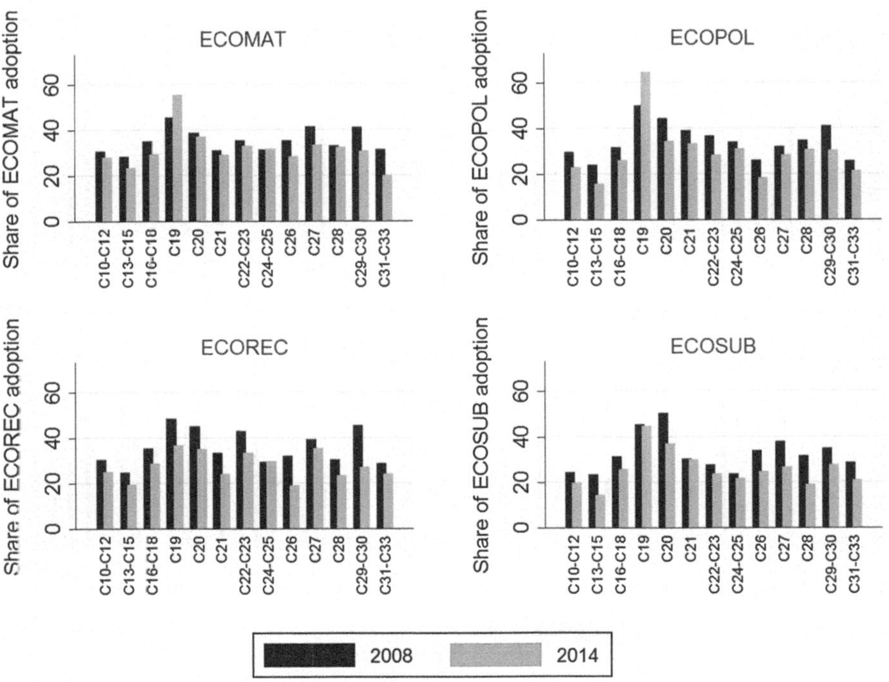

Fig. 3.4 Eco-innovation adoption by firms, by country

three categories) than the number of countries with an increased share, respectively 61% for ECOENU, 72% for ECOPOS and 78% for ECOREA.

Moving to the descriptive evidence on innovations with environmental benefits for the firm, Fig. 3.4 confirms some of the previous evidence, as again, the majority of the countries have decreased the share of firms adopting such innovations in the period considered. Interestingly enough, Cyprus and Sweden have increased in the share of the four typologies of EI (ECOMAT, ECOREC, ECOPOL and ECOSUB), while Italy and Estonia in three out of four. The Figure also gives some interesting evidence worthy further exploration: Italy doubled in the share of ECOMAT adoption between 2008 and 2014 and, contrarily, Luxembourg halved in the share of ECOREC adoption.

All in all, it is clear that the rate of adoption of the different typologies of EI have not increased over time.

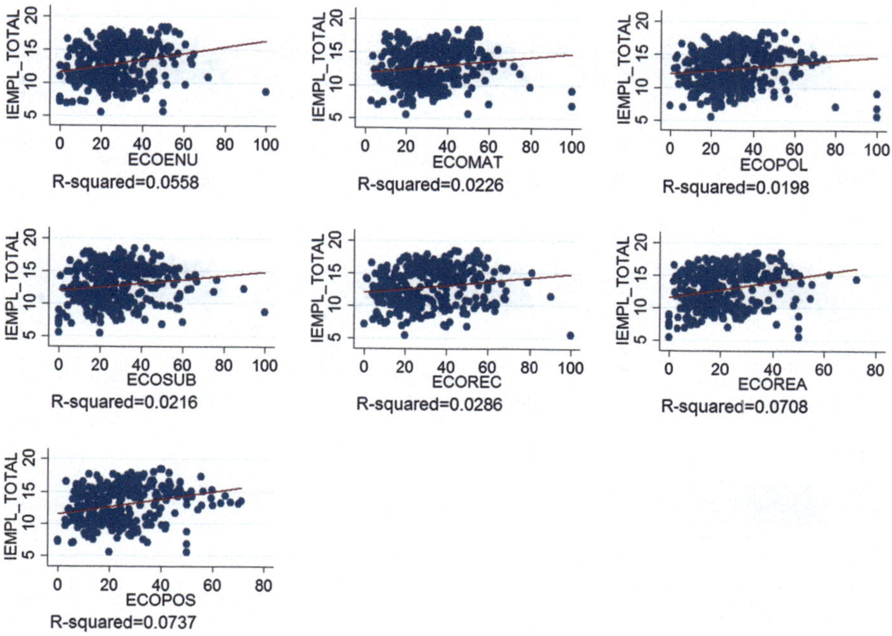

Fig. 3.5 Eco-innovation and employment, scatter plot and fit

3.3.2 Empirical Analysis

A subsequent step of the analysis of the Chapter is to relate such evidence to the one regarding the economic performance of the sectors and countries.

At first, visual correlations will be explored to better visualize the data. At second, an econometric analysis will be performed.

To start with the visual exploration, a scatter plot of the natural logarithm of sector-country employment in the 2 years considered and of each of the seven EI variables is constructed in Fig. 3.5 fit line and R-squared of a linear regression on the two variables are also reported in each scatter plot. Fit lines would lead to a positive correlation between each of the EI variables and employment. At the same time, all the reported R-squared are very weak and far below 0.10. A similar picture of inconclusive evidence is emerging even when the scatter plot is constructed for the 2 years separately.[1]

Similar inconclusive evidence is visualized in Fig. 3.6, representing the natural logarithm of Value added.

Figure 3.6 outlines a similar cloud of country-sector observation which is far from allowing depicting a linear and positive correlation. Indeed, although the fit line is

[1]For the sake of parsimony only scatter plots referring to 2008 and 2014 jointly are reported in this Chapter.

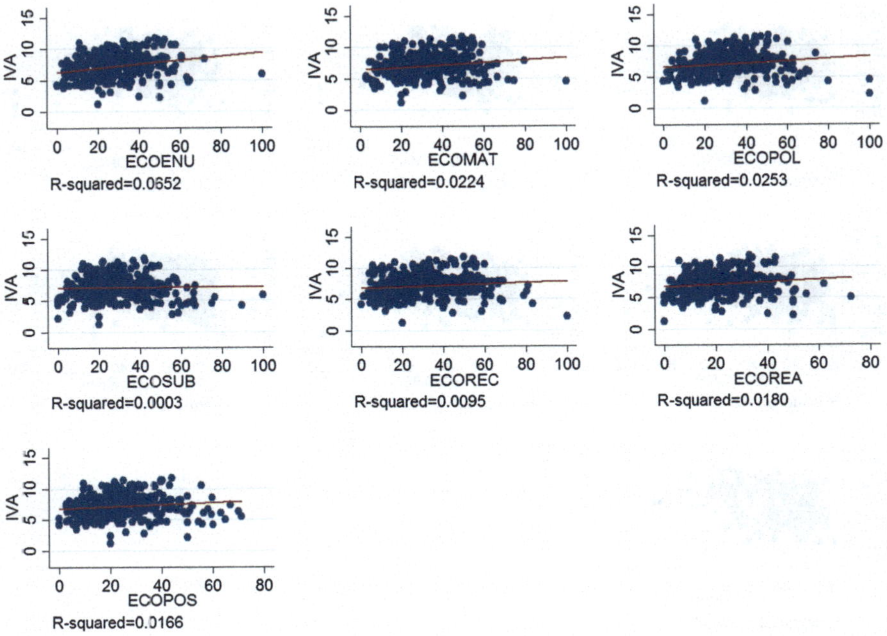

Fig. 3.6 Eco-innovation and value added, scatter plot and fit

slightly positive, R-squared are again dramatically low. Comparable evidence emerges when constructing separate scatter plots for the two different years.

The consequent step of the analysis is to investigate whether this apparently weak, although positive correlations between EI and economic outcomes are confirmed in an econometric multivariate setting.

Following previous literature conclusion on the need to differentiate between the heterogeneous economic effects that different typologies of EI may imply (e.g. Ghisetti and Rennings 2014; Telle 2006), the current analysis unpacks EI and it separately assesses for their different economic returns.

The following econometric log-linear augmented Cobb-Douglas model (Cobb and Douglas 1928) is estimated through a pooled OLS with clustered standard errors by country:

$$\log(\mathrm{Y}_{\mathrm{i,t}}) = \alpha + \beta_1 \log(\mathrm{EI}_{\mathrm{i,t}}) + \beta_2 \log(\mathrm{L}_{\mathrm{i,t}}) + \beta_3 \log(\mathrm{K}_{\mathrm{i,t}}) + \beta_4 \log(\mathrm{T}_{\mathrm{i,t}}) \\ + \beta_5 \log(\mathrm{MNC}_{\mathrm{i,t}}) + \varepsilon_{i,t}$$

with $\mathrm{i} = 1 \rightarrow 264$; $\mathrm{t} = 2008$ or 2014

The economic output Y is the dependent variable and it is approximated by the natural logarithm of country-sector value added.

Capital input (K) is approximated by the natural logarithm of net investments in tangible capital (Eurostat: structural business statistics). A better alternative would have been to measure K through a capital formation variable. However, it would

Table 3.3 Variables descriptive statistics

	N	Mean	SD	Min	Max
logVA	264	7.15	1.47	1.22	10.94
T	264	49.09	21.16	1.98	100
MNC	264	13.44	15.56	0.17	100
K	264	484.60	744.60	1.20	7150
L	264	12.95	2.79	5.55	17.92
GHGint	264	1.01	2.35	0.00	19.27
ECOENU	261	26.23	12.81	2.70	71.6
ECOMAT	264	31.05	12.88	2.95	79.8
ECOPOL	262	28.71	13.43	1.89	70.75
ECOPOS	262	23.03	12.09	2.90	71.05
ECOREA	259	20.11	11.31	0.00	55.26
ECOREC	262	29.96	15.65	3.58	78.95
ECOSUB	264	26.93	12.72	4.20	76.32

have resulted in too many missing sector-country given the lack of data. Therefore, K is approximated by a second-best alternative, which is however well correlated to capital formation and hence able to control for its effect in the model.

Labor input (L) is approximated by the natural logarithm of the number of employees in the country-sector.

The Technological input is approximated by the share R&D active firms in the sector.

The model is then augmented by environmental innovations, EI, which are added separately depending on the typology of innovation considered out of the seven typologies surveyed in the Community Innovation Survey. Multicollinearity among them would not allow their joint inclusion in the model.

Specific sectoral conditions are also accounted for by controlling for the share of enterprises in the sector that belong to a foreign group (MNC).

The main variables summary statistics are reported in Table 3.3.

Also Soltmann et al. (2014) tested for the economic returns on value added of EI by conducting a similar analysis on cross sectoral data for three countries. The current analysis has the novelty of capturing EI through an innovation adoption count variable rather than through a patent approximation. This variable better fits the current analysis, as the main interest of the current analysis is to focus on the adoption of the innovation and its consequent market returns, rather than on new inventions. As the second element of novelty, the current analysis allows to differentiate between the typologies of EI and to disentangle their differential effects (in any) on the outcome variable.

Table 3.4 Log-linear model estimations

	(1)	(2)	(3)	(4)	(5)	(6)	(7)
T	0.489***	0.534***	0.498***	0.537***	0.536***	0.548***	0.534***
	(0.12)	(0.10)	(0.12)	(0.12)	(0.11)	(0.13)	(0.12)
MNC	−0.077	−0.090	−0.074	−0.093	−0.083	−0.091	−0.088
	(0.07)	(0.07)	(0.07)	(0.06)	(0.07)	(0.07)	(0.07)
K	0.439***	0.444***	0.417***	0.443***	0.437***	0.443***	0.443***
	(0.04)	(0.04)	(0.05)	(0.05)	(0.04)	(0.05)	(0.04)
L	0.595***	0.591***	0.623***	0.590***	0.579***	0.595***	0.595***
	(0.07)	(0.07)	(0.07)	(0.07)	(0.07)	(0.07)	(0.07)
ECOENU	0.108						
	(0.11)						
ECOMAT		0.014					
		(0.08)					
ECOPOL			0.127				
			(0.08)				
ECOPOS				0.010			
				(0.08)			
ECOREA					0.044		
					(0.07)		
ECOREC						−0.012	
						(0.10)	
ECOSUB							0.015
							(0.11)
Constant	−3.18***	−3.03***	−3.49***	−2.99***	−2.94***	−3.04***	−3.07***
	(0.83)	(0.83)	(0.82)	(0.81)	(0.83)	(0.74)	(0.78)
N	263	264	264	264	257	264	264
R^2	0.885	0.881	0.878	0.878	0.874	0.876	0.882
Adj. R^2	0.883	0.874	0.875	0.875	0.871	0.872	0.879
Year dummy	Y	Y	Y	Y	Y	Y	Y

Country clustered standard errors in parentheses * $p < 0.10$, ** $p < 0.05$, *** $p < 0.01$

3.3.3 Main Results and Discussion

Results of the econometric analysis are reported in Table 3.4, where each column (1) to (7) corresponds to a different EI variable included.

Whereas results regarding technology, capital and labor are consistent with previous literature, those on EI fail to outline the presence of any statistically significant effect.

The natural logarithm of R&D (T), of employees (L) and of investments in tangible capital (K) all have the expected positive and significant effect on value added. Instead, the share or multinational firm seems not to affect the sectoral value added.

Most importantly, none of the seven typologies of EI play a significant effect in explaining the value added. This will lead to conclude that at this level of analysis, i.e. the sectoral one, results from previous literature on the economic returns of EI are not confirmed.

As most of the previously discussed findings where conducted at the firm level of analysis, this result in a way is not contradictory as it is based on a different level of analysis, more aggregated. For this reason, the eventually positive firm net effect may have been counterbalanced by the eventually negative effects found in firms in the same sectors, thus leading to inconclusive results. At the same time, this is new and important evidence deserving further exploration. However, this finding is in contrast to the ones by Soltmann et al. (2014) who depicted a U-shaped relationship suggesting that for most sectors the effect was negative.

To rule out the possibility that previous results were omitting a relevant piece of information, Table 3.5 extends the model by adding an environmental policy variable as well as its interaction with EI variables. Such variable (GHGint) is constructed as the natural logarithm of the ratio between Greenhouse Gases emission in the country-sector and its value added. It is discussed to be a valuable approximation of environmental policy stringency at the sectoral level in the relevant literature (e.g. Costantini and Crespi 2008).

The reason for such an inclusion, lies in the reasonable expectations that EI may not play a direct effect on value added, rather an effect that is mediated by the presence of an environmental policy which lead firms envisaging opportunities which would have not been able to explore in the absence of a proper policy stimulus, as postulated in the so-called Porter Hypothesis.

Results, reported in Table 3.5 strongly confirm previous findings and exclude for the presence of a moderation effect played by environmental policies: none of the seven EI variables interacted with the environmental policy stringency has a statistically significant effect.

A final robustness check, given the 2 year panel nature of the data set an additional analysis provided in this Chapter is to apply a first difference data transformation estimation, to eliminate the individual effects and the influence of any time-invariant variable omitted (Wooldridge 2001).

After first-differencing all the included variables, and thus removing the constant, the estimated model becomes:

$$\begin{aligned}\Delta\left(\mathrm{Y}_{i,t-(t-1)}\right) = \beta_1\Delta\left(\mathrm{EI}_{i,t-(t-1)}\right) + \beta_2\Delta\left(\mathrm{L}_{i,t-(t-1)}\right) + \beta_3\Delta\left(\mathrm{K}_{i,t-(t-1)}\right) \\ + \beta_4\Delta\left(\mathrm{T}_{i,t-(t-1)}\right) + \beta_5\Delta\left(\mathrm{MNC}_{i,t-(t-1)}\right) + \Delta\, \varepsilon_{i,t-(t-1)}\end{aligned}$$

Clearly, the final sample would be half of the previous sample size (=132) which is reduced to 120 due to some missing values in control variables in $t-1$. Table 3.6 reports estimation results of the first difference estimation. None of the seven EI variables is found to display any significant effect on the value added, while it is confirmed the significant and positive effect of capital, as of the net investments in tangible assets.

Table 3.5 Log-linear model estimations augmented by environmental policy

	(1)	(2)	(3)	(4)	(5)	(6)	(7)
T	0.437***	0.440***	0.402***	0.446***	0.453***	0.431***	0.457***
	(0.11)	(0.10)	(0.10)	(0.10)	(0.11)	(0.13)	(0.12)
MNC	−0.097	−0.104	−0.088	−0.108	−0.096	−0.102	−0.105
	(0.07)	(0.07)	(0.07)	(0.07)	(0.07)	(0.07)	(0.07)
K	0.525***	0.546***	0.517***	0.540***	0.523***	0.535***	0.536***
	(0.05)	(0.05)	(0.05)	(0.06)	(0.05)	(0.05)	(0.05)
L	0.502***	0.479***	0.511***	0.481***	0.486***	0.491***	0.491***
	(0.08)	(0.08)	(0.08)	(0.08)	(0.08)	(0.08)	(0.07)
GHGint	0.048	−0.159	−0.116	−0.057	−0.047	−0.168*	0.058
	(0.12)	(0.12)	(0.11)	(0.09)	(0.12)	(0.09)	(0.10)
ECOENU	0.003						
	(0.13)						
ECOENU*GHGint	−0.041						
	(0.04)						
ECOMAT		0.062					
		(0.10)					
ECOMAT*GHGint		0.019					
		(0.03)					
ECOPOL			0.156				
			(0.11)				
ECOPOL*GHGint			0.006				
			(0.03)				
ECOPOS				0.014			
				(0.09)			
ECOPOS*GHGint				−0.012			
				(0.03)			

ECOREA					0.041		
					(0.10)		
ECOREA*GHGint					−0.013		
					(0.04)		
ECOREC						0.083	
						(0.12)	
ECOREC*GHGint						0.021	
						(0.03)	
ECOSUB							−0.054
							(0.10)
ECOSUB*GHGint							−0.046
							(0.03)
Constant	−2.241**	−2.343***	−2.710***	−2.160**	−2.211**	−2.453***	−2.07**
	(0.84)	(0.76)	(0.83)	(0.83)	(0.81)	(0.70)	(0.75)
N	263	264	264	263	257	264	264
R^2	0.894	0.891	0.889	0.888	0.883	0.886	0.892
Adj. R^2	0.8902	0.8878	0.8858	0.8845	0.8794	0.8829	0.8890
Year dummy	Y	Y	Y	Y	Y	Y	Y

Country clustered standard errors in parentheses $^{*}p < 0.10$, $^{**}p < 0.05$, $^{***}p < 0.01$

Table 3.6 First difference estimation results

	(1)	(2)	(3)	(4)	(5)	(6)	(7)
FD_RD	15.336	12.542	14.778	12.626	11.395	3.806	13.048
	(27.76)	(25.96)	(26.98)	(25.96)	(28.57)	(26.72)	(26.03)
FD_MNC	−3.121	0.637	−0.562	0.470	6.917	9.678	−7.021
	(21.73)	(20.94)	(21.10)	(20.89)	(23.55)	(20.78)	(21.64)
FD_INVESTMENT	0.562**	0.571**	0.559**	0.575**	0.582**	0.567**	0.517**
	(0.24)	(0.24)	(0.24)	(0.24)	(0.24)	(0.23)	(0.24)
FD_EMPL	−0.000**	−0.000**	−0.000**	−0.000**	−0.000**	−0.000***	−0.000**
	(0.00)	(0.00)	(0.00)	(0.00)	(0.00)	(0.00)	(0.00)
FD_ECOENU	−20.412						
	(31.14)						
FD_ECOMAT		−4.172					
		(26.04)					
FD_ECOPOL			−14.011				
			(27.39)				
FD_ECOPOS				−5.804			
				(30.53)			
FD_ECOREA					17.776		
					(38.81)		
FD_ECOREC						38.142	
						(30.38)	
FD_ECOSUB							−31.815
							(30.91)
N	117	120	118	120	117	120	119
R^2	0.080	0.077	0.079	0.077	0.079	0.089	0.085
Adj. R^2	0.0393	0.0365	0.0382	0.0366	0.0375	0.0495	0.0448

3.4 Conclusion and New Research Lines

The Chapter has provided a discussion on the linkages between greener production choices, namely accounted for by the adoption of environmental innovations and the economic performance associated with that choice.

From the literature review, it emerged inconclusive evidence. There seems to be predominant a literature postulating that EI are associated with positive economic returns, either with respect to profitability or to productivity. However, a relevant set of studies still finds a negative or a neutral effect and these are not at all negligible.

As most of the previous studies were conducted at the firm-level, with very few exceptions (e.g. Soltmann et al. 2014) the Chapter has proposed new empirical evidence at the aggregate level, i.e. the sectoral one.

Such new empirical evidence has been proposed to unveil whether the adoption of eco-innovations has any aggregate effects.

The empirical strategy is based on both a visual qualitative analysis on EI distribution across sectors and countries, on some scatter plots and later on a multivariate econometric analysis conducted on a 2-year based panel data for 18 EU countries and 13 manufacturing sectors.

Overall it emerges that no statistically significant effect is found by EI on value added. In other words, it may pay to be green, at the firm level, when firms may get first-mover advantages, by entering new markets and meeting a new green demand or when firms may be able to reduce production costs and increase energy or material efficiency. However, at the aggregate level, such a micro effect seems to be counterbalanced by certain negative sectoral effects so that the overall net effect is not found to be significant in this study.

This finding calls for future research to investigate the micro-foundations of such a non-significant *meso* effect.

An additional research line would be to investigate what drove the drop in the sectoral adoption of eco-innovations in the year 2014, as the external policy context seems to have pushed in the opposite direction. Related to this, an understanding of the persistence of (eco)-innovators, their determinants and effects would be a valuable contribution. Additionally, the scatter plots outlined interesting evidence that would constitute a research line to explore the extreme cases spotted of e.g. Italy doubling EI adoption in the 2 years and Luxembourg halving it. Also in this context, understanding the micro-foundations of such evidence would provide new interesting findings.

Although no causal claim has been advanced in this Chapter, given the absence of significance in the variables of interest, not to be neglected is the possible endogeneity of the model estimated and presented. This has not been solved and consequently, results presented have to be read as indicative but cannot point to any causal connections, although indicative of a general absence of correlations.

Acknowledgment The author wishes to thank Klaus Rennings for the great contribution he gave to this scientific community and, personally, for the precious knowledge share and interactions he held with the author. The work has been done while the author worked at the Joint Research Centre of the European Commission. The scientific output expressed does not imply a policy position of the European Commission. Neither the European Commission nor any person acting on behalf of the Commission is responsible for the use which might be made of this publication.

References

Al-Tuwaijri, S., Christensen, T., & Ii, K. H. (2004). The relations among environmental disclosure, environmental performance, and economic performance: A simultaneous equations approach. *Accounting Organizations and Society, 29*, 447–471.

Ambec, S., Lanoie, P., Stefan, A., & Paul, L. (2008). Does it pay to be green? A systematic overview. *The Academy of Management Perspectives, 22*(4), 45–63.

Amores-Salvadó, J., Castro, G. M., & Navas-López, J. E. (2014). Green corporate image: Moderating the connection between environmental product innovation and firm performance. *Journal of Cleaner Production, 83*, 356–365.

Antonietti, R., & Marzucchi, A. (2014). Green tangible investment strategies and export performance: A firm-level investigation. *Ecological Economics, 108*, 150–161.

Aragon-Correa, J. A., & Sharma, S. (2003). A contingent resource-based view of proactive corporate environmental strategy. *Academy of Management Review, 28*(1), 71–88.

Barbieri, N., Ghisetti, C., Gilli, M., Marin, G., & Nicolli, F. (2016). A survey of the literature on environmental innovation based on main path analysis. *Journal of Economic Surveys, 30*(3), 596–623.

Berchicci, L., & King, A. (2007). 11 Postcards from the edge. *The Academy of Management Annals, 1*(1), 513–547.

Cainelli, G., Mazzanti, M., & Zoboli, R. (2011). Environmentally-oriented innovative strategies and firm performances in services: Micro evidence from Italy. *International Review of Applied Economics, 25*(1), 61–85.

Cheng, C. C., Yang, C. L., & Sheu, C. (2014). The link between eco-innovation and business performance: A Taiwanese industry context. *Journal of Cleaner Production, 64*, 81–90.

Cleff, T., & Rennings, K. (1999). Determinants of environmental product and process innovation. *European Environment, 9*, 191–201.

Cobb, C. W., & Douglas, P. H. (1928). A theory of production. *American Economic Review, 18* (sup), 139–165.

Consoli, D., Marin, G., Marzucchi, A., & Vona, F. (2016). Do green jobs differ from non-green jobs in terms of skills and human capital? *Research Policy, 45*(5), 1046–1060.

Cordeiro, J. J., & Sarkis, J. (1997). Environmental proactivism and firm performance: Evidence from security analyst earnings forecasts. *Business Strategy and the Environment, 6*(2), 104–114.

Costantini, V., & Crespi, F. (2008). Environmental regulation and the export dynamics of energy technologies. *Ecological Economics, 66*(2–3), 447–460.

Costantini, V., & Mazzanti, M. (2012). On the green and innovative side of trade competitiveness ? The impact of environmental policies and innovation on EU exports. *Research Policy, 41*(1), 132–153.

Crépon, B., Duguet, E., & Mairesse, J. (1998). Research, innovation and productivity: An econometric analysis at the firm level. *Economics of Innovation and New Technology, 7*(2), 115–158.

Dechezleprêtre, A., & Sato, M. (2017). The impacts of environmental regulations on competitiveness. *Review of Environmental Economics and Policy, 11*(2), 183–206.

Dowell, G., Hart, S., & Yeung, B. (2000). Do corporate global environmental standards create or destroy market value? *Management Science, 46*(8), 1059–1074.

Elsayed, K., & Paton, D. (2005). The impact of environmental performance on firm performance: Static and dynamic panel data evidence. *Structural Change and Economic Dynamics, 16*, 395–412.

Endrikat, J., Guenther, E., & Hoppe, H. (2014). Making sense of conflicting empirical findings: A meta-analytic review of the relationship between corporate environmental and financial performance. *European Management Journal, 32*(5), 735–751.

Franco, C., & Marin, G. (2017). The effect of within-sector, upstream and downstream environmental taxes on innovation and productivity. *Environmental and Resource Economics, 66*(2), 261–291.

Freedman, M., & Jaggi, B. (1992). An investigation of the long-run relationship between pollution performance and economic performance: The case of pulp and paper firms. *Critical Perspectives Account, 3*(4), 315–336.

Gagliardi, L., Marin, G., & Miriello, C. (2016). The greener the better? Job creation effects of environmentally-friendly technological change. *Industrial and Corporate Change, 25*(5), 779–807.

Ghisetti, C. (2017). Demand-pull and environmental innovations: Estimating the effects of innovative public procurement. *Technological Forecasting and Social Change, 125*, 178–187.

Ghisetti, C., & Pontoni, F. (2015). Investigating policy and R&D effects on environmental innovation: A meta-analysis. *Ecological Economics, 118*, 57–66.

Ghisetti, C., & Rennings, K. (2014). Environmental innovations and profitability: How does it pay to be green? An empirical analysis on the German innovation survey. *Journal of Cleaner Production, 75*, 106–117.

Greenstone, M., List, J., & Syverson, C. (2012). *The effects of environmental regulation on the competitiveness of U.S. manufacturing* (NBERWorking Paper n. 18392). National Bureau of Economic Research.

Hart, S. L. (1995). A natural-resource-based view of the firm. *Academy of Management Review, 20* (4), 986–1014.

Hart, S. L. (1997). Beyond greening: Strategies for a sustainable world. *Harvard Business Review, 75*, 66–76.

Hart, S., & Ahuja, G. (1996). Does it pay to be green? An empirical examination of the relationship between emission reduction and firm performance. *Business Strategy and the Environment, 5* (1996), 30–37.

Hart, S. L., & Dowell, G. (2011). A natural-resource-based view of the firm: Fifteen years after. *Journal of Management, 37*, 1464–1479.

Hart, S., & Milstein, M. B. (2003). Creating sustainable value. *Academy of Management Perspectives, 17*(2), 56–67.

Horbach, J. (2010). The impact of innovation activities on employment in the environmental sector – Empirical results for Germany at the firm level. *Jahrbucher Fur Nationalokonomie Und Statistik, 230*(4), 403–419.

Horbach, J., & Rennings, K. (2013). Environmental innovation and employment dynamics in different technology fi elds e an analysis based on the German Community Innovation Survey 2009. *Journal of Cleaner Production, 57*, 158–165.

Horbach, J., Rammer, C., & Rennings, K. (2012). Determinants of eco-innovations by type of environmental impact—The role of regulatory push/pull, technology push and market pull. *Ecological Economics, 78*, 112–122.

Horváthová, E. (2010). Does environmental performance affect financial performance? A meta-analysis. *Ecological Economics, 70*(1), 52–59.

ILO. (2013) *Methodologies for assessing green jobs* (Policy Brief). International Labour Organization.

Jaffe, A. B., & Palmer, K. (1997). Environmental regulation and innovation: A panel data study. *Review of Economics and Statistics, 79*(4), 610–619.

Jaffe, A. B., Newell, R. G., & Stavins, R. N. (2005). A tale of two market failures: Technology and environmental policy. *Ecological Economics, 54*(2–3), 164–174.

King, A. A., & Lenox, M. J. (2001). Does it really pay to be green? An empirical study of firm environmental and financial performance. *Journal of Industrial Ecology, 5*(1), 105–116.
King, A., & Lenox, M. (2002). Exploring the locus of profitable pollution reduction. *Management Science, 48*(2), 289–299.
Kunapatarawong, R., & Martínez-Ros, E. (2016). Towards green growth: How does green innovation affect employment? *Research Policy, 45*(6), 1218–1232.
Lanoie, P., Laurent-Lucchetti, J., Johnstone, N., & Ambec, S. (2011). Environmental policy, innovation and performance: New insights on the Porter hypothesis. *Journal of Economics & Management Strategy, 20*(3), 803–842.
Leoncini, R., Marzucchi, A., Montresor, S., Rentocchini, F., & Rizzo, U. (2017). 'Better late than never': The interplay between green technology and age for firm growth. *Small Business Economics*. https://doi.org/10.1007/s11187-017-9939-6 (Online first).
Li, Y. (2014). Environmental innovation practices and performance: Moderating effect of resource commitment. *Journal of Cleaner Production, 66*, 450–458.
Licht, G., & Peters, B. (2013). *The impact of green innovation on employment growth in Europe, WWWforEurope* (Working Paper No 50).
Manello, A. (2017). Productivity growth, environmental regulation and win–win opportunities: The case of chemical industry in Italy and Germany. *European Journal of Operational Research, 262*(2), 733–743.
Margolis, J. D., & Walsh, J. P. (2003). Misery loves companies: Rethinking social initiatives by business. *Administrative Science Quarterly, 48*(2), 268–305.
Marin, G. (2014). Do eco-innovations harm productivity growth through crowding out? Results of an extended CDM model for Italy. *Research Policy, 43*(2), 301–317.
Marin, G., & Lotti, F. (2017). Productivity effects of eco-innovations using data on eco-patents. *Industrial and Corporate Change, 26*(1), 125–148.
Marin, G., Marzucchi, A., & Zoboli, R. (2015). SMEs and barriers to Eco-innovation in the EU: Exploring different firm profiles. *Journal of Evolutionary Economics, 25*(3), 671–705.
Martinez-Conesa, I., Soto-Acosta, P., & Palacios-Manzano, M. (2017). Corporate social responsibility and its effect on innovation and firm performance: An empirical research in SMEs. *Journal of Cleaner Production, 142*, 2374–2383.
McWilliams, A., & Siegel, D. S. (2011). Creating and capturing value: Strategic corporate social responsibility, resource-based theory, and sustainable competitive advantage. *Journal of Management, 37*(5), 1480–1495.
Miroshnychenko, I., Barontini, R., & Testa, F. (2017). Green practices and financial performance: A global outlook. *Journal of Cleaner Production, 147*, 340–351.
OECD. (2014). *Issue note addressing social implications of green growth: Inclusive labour markets for green growth.* (Green Growth and Sustainable Development Forum). 13–14 November 2014.
Orlitzky, M., Siegel, D., & Waldman, D. (2011). Strategic corporate social responsibility and environmental sustainability. *Business & Society, 50*(1), 1–28.
Peneder, M., Arvanitis, S., Rammer, C., Stucki, T., & Wörter, M. (2017). *Competitiveness and ecological impacts of green energy technologies firm-level evidence for the DACH region Michael* (WIFO Working Paper Series 544/2017).
Popp, D., Newell, R. G., & Jaffe, A. B. (2010). Energy, the environment, and technological change. In B. H. Hall & N. Rosenberg (Eds.), *Handbook of the economics of innovation* (Vol. 2, pp. 873–937). Burlington, NJ: Academic Press.
Porter, M. E., & Kramer, M. R. (2002). The competitive advantage of corporate philanthropy. *Harvard Business Review, 80*, 56–68.
Porter, M. E., & Kramer, M. R. (2006). Strategy & society: The link between competitive advantage and corporate social responsibility. *Harvard Business Review, 84*(12), 78–92.
Porter, M. E., & Van der Linde, C. (1995). Toward a new conception of the environment-competitiveness relationship. *The Journal of Economic Perspectives, 9*(4), 97–118.

Rennings, K. (2000). Redefining innovation—eco-innovation research and the contribution from ecological economics. *Ecological Economics, 32*(2), 319–332.

Rexhäuser, S., & Rammer, C. (2013). Environmental innovations and firm profitability: Unmasking the porter hypothesis. *Environmental and Resource Economics, 57*(1), 145–167.

Riillo, C. A. F. (2017). Beyond the question "Does it pay to be green?": How much green? and when? *Journal of Cleaner Production, 141*, 626–640.

Rubashkina, Y., Galeotti, M., & Verdolini, E. (2015). Environmental regulation and competitiveness: Empirical evidence on the Porter Hypothesis from European manufacturing sectors. *Energy Policy, 83*, 288–300.

Russo, M., & Fouts, P. (1997). A resource-based perspective on corporate environmental performance and profitability. *Academy of management Journal, 40*(3), 534–559.

Salama, A. (2005). A note on the impact of environmental performance on financial performance. *Structural Change and Economic Dynamics, 16*, 413–421.

Sarkis, J., & Cordeiro, J. J. (2001). An empirical evaluation of environmental efficiencies and firm performance: Pollution prevention versus end-of-pipe practice. *European Journal of Operational Research, 135*, 102–113.

Soltmann, C., Stucki, T., & Woerter, M. (2014). The impact of environmentally friendly innovations on value added. *Environmental and Resource Economics, 62*(3), 457–479.

Telle, K. (2006). "It pays to be green" – A premature conclusion? *Environmental and Resource Economics, 35*(3), 195–220.

Vona, F., Marin, G., Consoli, D., & Popp, D. (2018). Environmental regulation and green skills: An empirical exploration. *Journal of the Association of Environmental and Resource Economists.* https://doi.org/10.1086/698859

Wagner, M., Van Phu, N., Azomahou, T., & Wehrmeyer, W. (2002). The relationship between the environmental and economic performance of firms: An empirical analysis of the European paper industry. *Corporate Social Responsibility and Environmental Management, 146*, 133–146.

Wooldridge, J. M. (2001). *Econometric analysis of cross section and panel data*. Cambridge, MA: MIT Press.

Chapter 4
Shaping System Innovation: Transformative Environmental Policies

Klaus Jacob

4.1 Introduction

Innovation is key for a sustainable development. Technologies are the interface of human activities with natural systems: The extraction of resources, production and use of products, recycling and disposal of waste is based on technologies. Innovative technologies that are more efficient in terms of resource use or emissions than standard are important contributions to improve the sustainability of economies and societies. Many examples demonstrate the potential of technological improvements. Suggestions for improving resource efficiency by factor 4 (Weizsäcker et al. 1995) or factor 10 (Hinterberger and Schmidt-Bleek 1999) are based on these potentials.

Environmental innovations that exploit these potentials are not emerging from market forces only. Various barriers and market failures to environmental innovation have been identified: The double externality of environmental innovation (Rennings 2000), lack of information about potentials for resource efficiency and split incentives across value chains, lock-in effects in resource intensive technologies point to the need of market correcting policies to enable environmental innovation (Bleischwitz and Jacob 2011). An innovation oriented environmental policy that encourages innovation and their broad diffusion is an ambitious program in itself—it requires a policy coherence across innovation, environmental and industrial policies (Haum et al. 2010) and can build on a wide range of environmental policy instruments that create a supply push and demand pull (Rennings et al. 2008).

The contribution is based on an extended concept for transformative environmental policies developed in Jacob et al. (2018) (forthcoming). The author is grateful to Louise Fitzgerald for commenting an earlier version.

K. Jacob (✉)
Freie Universität Berlin, Forschungszentrum für Umweltpolitik (ffu), Berlin, Germany
e-mail: klaus.jacob@fu-berlin.de

J. Horbach, C. Reif (eds.), *New Developments in Eco-Innovation Research*, Sustainability and Innovation, https://doi.org/10.1007/978-3-319-93019-0_4

Policies and innovation that increase efficiency of technologies are, however, not sufficient to achieve a sustainable development. Rebound and growth effects compensate reductions in emissions and resource depletion (e.g. Rennings 2014; Dimitropoulos et al. 2016). Rebound effects in a narrow sense would suggest an increase in demand for a given service or product resulting from cost reductions because of efficiency improvements. Such effects are observed for energy efficient technologies. Growth effects suggest an increase in overall demand for goods and services (and related emissions and resource use) resulting from increases in incomes (Semmling et al. 2015). Regardless of if they are rebound or growth effects, a strategy based on efficiency improvements by environmental innovation is not sufficient to achieve a sustainable use of resources.

Therefore, an innovation oriented environmental policy aiming for improvements in efficiency appears as insufficient to address the root causes of over depletion of resources and excess of planetary boundaries. Instead, it is called for comprehensive transformations that not only address technologies, but socio-technical and socio-economic systems as a whole (e.g. Geels 2002; WBGU 2011). Socio-technical and—economic systems are specific configurations of technologies, products, infrastructures, markets, social practices, related institutions and cultural values to serve needs of society (Geels 2002; Grießhammer and Brohmann 2016). The systems are characterized by the mutual interdependency and reinforcing stabilization of the various elements of the system and the relative independence of their environment. Examples for socio-technical or socio-economic systems are the systems for energy, nutrition, health, education, communication or mobility. Socio-technical systems are characterized by technologies (e.g. mobility, energy), while socio-economic systems are characterized by markets and functions (e.g. health, education, nutrition).

4.2 Transformation to Sustainability

A transformation of such systems would provide services to society in a fundamentally different way. History is rich in examples of transformations of different scales. Famous examples include the transformation of shipping from sail ships to steam powered ships, mobility from horse carts to automobiles, wood based energy to coal and later to gas based systems, preservation of food with smoke or salt to refrigeration (and the related changes in nutrition), etc. What is common to these examples is the fact that they are not limited to technologies (although technological innovation can be considered as a starting point in some of the cases), but also entail changes in social practices and cultural values. In particular, the conceptions of normality in institutions, infrastructures, markets are transformed. Transformations often (not always) enable new actors to replace the incumbents in providing services to society. In a nutshell: transformation can be considered as a passage from one state of equilibrium in a given system to another. As a result, society changes its perceptions of normality.

A *sustainability* transformation would change not only technologies, but also social practices, cultural values and institutions in a way that the concerns of a sustainable development would be prioritized (WBGU 2011). In such societies, the root causes of environmental degradation would be effectively addressed. The concerns of sustainable development would be embedded in cultural values, institutions and social practices. There is no example of such transformation so far, and it is open how such systemic change would take place. Past transformations rather lead to an increase in resource use and emissions. Change of this kind cannot be foreseen, let alone be planned for.

The study of past transformation suggests that transformative change can be distinguished from other forms of change not only in regards to the objects of transformation (i.e. socio-technical or socio-economic systems), but also in regards to the process of change. This process is typically described in a multilevel perspective (Geels 2002; Geels and Kemp 2012). Technologies, social practices, institutions and the other elements of socio-technical and -economic systems are subject of continuous innovation and improvements. Innovation in the sense of continuous improvement is an in-built mechanism of modern societies and it is largely driven by competition. Firms compete on reducing costs and developing new product features. However, innovation and competition is not limited to the economic sphere only: Policy makers compete on policy innovation, and similarly social practices are subject to fashions and social differentiation. Usually, such processes of innovation take place within the existing framework of technologies, infrastructures, institutions, perceptions of normality, etc. However, the existing actors pursue innovation within the established structures and paradigms for innovation. This includes incremental innovation for improving efficiency of given technology.

Transformative change is different from the regular process of innovation. Past examples of transformation suggest that transformative change starts from (technical, social and institutional) innovations that are established in niches (Schot and Geel 2008). These innovations and their niches entail a vision of an alternative system configuration to provide services to society, rather than promising a mere improvement of the existing status quo. They are often developed and promoted by actors from outside the established structures.

The initial phases of transformation are characterized by a high density of innovation that entails alternative visions. The existing systems are challenged from different, and often contradicting perspectives. The increasing intensity of transformative innovation is a result of the decreasing innovation capacity of the existing system: Kondratieff and later Schumpeter demonstrated that the innovation potential of a technological paradigm came to an end when the various options for new products and cost reduction had been exploited (Kondratieff 1926; Schumpeter 1942; Jänicke and Jacob 2013). When this takes place, the time is ripe for replacing the existing technological paradigm and starting a new long-term innovation cycle.

In the beginning of a transformation there is, however, no dominant alternative vision, but rather competing visions for an alternative system configuration. The current competing views on sustainability in agriculture may serve as an example. Some actors favor a regionalized, small-scale ecological agriculture, others a high tech agriculture in glasshouses powered by solar power and production of in-vitro meat, largely detached from natural systems, and yet others envision urban farming as a pathway towards a sustainable agriculture. All of these visions are meant to replace the current agricultural system, however, they are following contradicting and competing pathways. All of these visions do not only imply technological improvements of the existing system, but a replacement of the current system including their institutions, infrastructures, and social practices. Similar competing visions can be demonstrated for the energy sector (e.g. decentralized vs. centralized renewable energy), or mobility (e.g. autonomous, battery powered mobility versus mobility based largely on public transport, etc.).

These visions are realized, implemented and tested in niches. By means of demonstrating the feasibility of alternative system configuration, the legitimacy of the existing systems and their dominance is put into question and challenged. Actors that represent the incumbent system typically challenge claims that their services could be provided by alternative means. However, niches that successfully attract imitators or users and that are able to scale up their innovation, counter such arguments.

Once an innovation passes critical moments, e.g. a critical mass of applications has been reached, network effects are created, or once institutions are established that stabilize the niches, the pace of transformative change is accelerated. This is particularly the case once a co-evolution with other elements of a system begins. If an innovation and its diffusion changes the framework conditions for another element of the given system, and thereby positive feedback loops are established, innovation co-evolves in e.g. infrastructures, markets, institution or social practices. Such positive feedback mechanisms have been demonstrated for institutions, technology and markets of renewable energy technologies (Jänicke and Rennings 2011): Policies that promote renewable energy create markets for such technologies, which pull for technological innovation, and market growth and demonstrated feasibility of innovation demands for new policies. As a result, the level of policy ambition co-evolves with markets and technological advances.

Once such critical moments have been passed, a vision for an alternative system configuration is shared by an increasing number of actors, and gains acceptance. Such a shared vision delegitimizes current practices and systems. A widely accepted alternative vision provides guidance and directs the innovation processes, it coordinates actors and their activities and creates network effects and it legitimizes institutions that support the scaling up of an alternative configuration. The hitherto production and consumption becomes culturally stigmatized and social practices are adapted to the new system. Accordingly, the rate of change is accelerating.

As a result of this acceleration, existing practices, technologies, structures and the related actors are questioned on their legitimacy and suitability to provide their services to society. The incumbents become subject of exnovation—the opposite of innovation processes (Heyen 2016). This can be an emergent and sudden phenomenon, driven by rapid changes in social practices or by market forces. An example is the divestment movement from fossil fuels or (sudden) shifts in consumer preferences. However, such emergent phenomena may not be desirable as investments in capital or qualifications are suddenly devalued. This is why organized phase out processes are often advocated for, as was witnessed in hard coal or for nuclear power in Germany, and envisaged for lignite extraction.

A transformation comes to an end, once the alternative system configuration is established as a new equilibrium and accepted as normality. The intensity of innovation and diffusion flattens back to a normal level and new institutions, technologies, social practices, infrastructures, etc. are in place. Innovation takes place, but again as incremental improvements rather than disruptive innovation.

To summarize, transformations are characterized by

- Sudden and disruptive system innovation: Socio-technical or socio-economic systems are reconfigured in a way that societal needs are provided in a fundamentally different way.
- Transformation starts from innovation (including social and institutional innovation) that suggests an alternative system configuration. Innovations take place from outside the current system.
- In early phases, the process of innovation is undirected. Only once a vision for an alternative system configuration is widely accepted and co-evolutionary processes take place, transformation gains momentum and direction.
- Transformation typically implies that new actors (and their practices) replace the existing.
- Ultimately, it is characterized by phasing out of an incumbent system configuration.

4.3 The Role of Public Policies in Transformation

What is the role of public policies in the context of a transformation to sustainability? On first sight, public policy-making to govern transformation appears as an excessive demand. Policy-making is incremental and fragmented. Policy-making, at least in western liberal democracies, is restricted from being prescriptive on social practices and cultural norms (although it could be questioned if this is actually the case). The different domains of public policy-making are integral parts of socio-technical and economic systems rather than in the position to actually exert control over such systems. Increasingly, socio-technical and socio-economic systems are organized beyond different levels of policy-making spanning from the local to the national, European and international level. Against this backdrop, a holistic and strategic approach in governing such systems appears as unrealistic, if not undesirable.

Theories of political process underpin such assessments of the role of public policies in regards to steering socio-technical and—economic systems. An often quoted and empirically well underpinned conception of policies and policy-making processes describes the process as muddling through and incrementalism (Lindblom 1959; Knaggård 2014). Since there is no overall planning possible (and any attempt to provide such overall planning necessarily fails as there is no complete knowledge on side effects of such planning), policy-making is a stepwise trial and error process. There are good reasons for incrementalism, including flexibility, avoidance of conflict and flexibility. Muddling through implies that there is no central planning agent, but policies are negotiated between independent actors. Other theories that focus on institutions as constraints for policy makers underpin expectations on limited capacities to actually govern societal systems (e.g. Mayntz and Scharpf 1995). Similarly, theories that focus on power and interest (e.g. Maeße 2013) or policy learning (e.g. Lindberg 2013) would not expect disruptive system innovation and transformation to result from public policies. Instead, public policies would maintain the current configuration and focus on incremental innovation within the given framework.

Authors advocating transformation to sustainability mirror this perception of public policies. In their view, states and public policies are part of the incumbents and rather inhibit transformative innovation. Such innovation takes place outside state actors, they emerge from society and are rather constrained by governmental actors than being promoted (e.g. Leggewie and Welzer 2010). In this view, public policies and related institutions are an object of transformation rather than causal for transformative change. Public policies constrain niche innovation rather than promoting and enabling them and niche actors rather prefer being independent from governmental regulation and subsidies.

While there is certainly evidence for such view, there are at the same time governmental actors, programs and institutions that are actually interested in promoting transformation. This reflects the (often criticized) fragmentation of governments. While there are institutions that promote and protect the different sectors of the economy, infrastructures, housing, mobility, consumption etc. there are at the same time institutions that question the current configuration. This is in line with the above-mentioned theories of incremental policy-making. Policy entrepreneurs continuously seek for policy innovation and seek in competition for attention and approval. This process can be understood as an evolutionary model of public policy (Cairney 2013). Policy-making can be understood as a creative process when policy entrepreneurs search for the right moment in time for their adoption (Kingdon 1995). Evolutionary theory would also imply that rapid and substantive policy change is possible as a result of a punctuated equilibrium (Baumgartner and Jones 1993). This can be understood as a third order change that goes beyond incremental improvements (Hall 1993). Cairney (2013) suggests combining theories of evolutionary policy-making with complexity theory to explain the possibility of far reaching changes. Although this has not yet been fully exploited in empirical studies, it seems to be a promising starting point for developing policy options for shaping transformation.

The typical process of transformations, and their evolutionary character opens up some options for action for actors that enable or actually shape socio-technical and—economic systems even within the constraints of public policy-making. An evolutionary understanding of the policy process does not demand a fully integrated, holistic and reflexive long-term political strategy for transformation, which seems rather unrealistic. Instead, a policy approach based on incremental trial and error has a high affinity with the evolutionary process of transformation as depicted above.

4.4 Shaping Transformation: Possible Policy Approaches

Given the processes of transformation and the role of system innovation (including social and institutional innovation), the complexity of socio-technical and—economic systems, and the constraints and opportunities of public policy-making, the following spheres of activity could be exploited in order to shape transformations to sustainability.

Against the importance of social and institutional innovation in niches for transformation, public policies could firstly conduct systematic surveys on social innovation and trends. There is social innovation and changes in values that have considerable potential to contribute to transformation and sustainability. For example, vegetarianism and the increasing demand for healthy nutrition, the demand for car sharing rather than car-ownership, or changes in the valuation of status from work vis-a-vis spare time are examples for relevant trends. There are, however, also countervailing social innovation and trends: fast fashion, growing demand for electronic gadgets, increasing space used for living, etc. From the point of view of shaping transformation, it would be important to survey and assess such innovation and trends. This would be a prerequisite for scaling up and diffusing social practices that are favorable from the point of view of sustainability. It would be also necessary to enable a "greening of social innovation" (Jacob 2015). Based on surveys and assessments, public policies could aim to promote options that would meet the demands expressed in societal trends but that are beneficial from the point of view of the environment. Such integration would complement environmental policy integration and technology assessments by a social dimension. Given the importance of social innovation for transformation (as well as for the stabilization of the existing systems) surveys and assessments could be envisaged as a standard practice of transformative policies.

In the same vein, programs to support social innovation could be justified. The need for innovation policies is widely accepted, as economic actors have little incentives to invest in research and development and bearing high risks of failure. Against the payoffs from innovation for society, it can be justified to invest public funds in subsidies for innovation. So far, innovation policies largely focus on technologies. Social innovation has similar potential for payoffs for society, and similar costs and risks for individuals. From the viewpoint of their importance for transformation, in particular those (social and technological) innovations that entail an alternative vision for a system configuration could be promoted. As the success (in terms of technical feasibility and acceptance) of such innovation cannot be predicted, a policy that promotes transformative social and technical innovation would rather aim for developing a pool of innovation instead of focusing on a single solution and picking a winner. Such a program could be coined as a transformative innovation policy (Chataway et al. 2017). A policy to promote social innovation would complement existing innovation policies.

From the viewpoint of the complexity of socio-technical and—economic systems, and the impossibility to predict their development, policies and regulation could be developed and applied in an experimental mode. Institutional innovations are key for transformation, however, in the context of incremental policy-making it is unlikely to adopt policies that would challenge the current system configuration. Actors from within the political system that are committed to preserve the status quo would object such approaches. In the framework of simulation games (e.g. www.flaechenhandel.de for a simulation with tradeable permits for land use among decision makers from communities), policy labs (e.g. open policy-making in the UK, EU Policy lab, etc.) or regulatory innovation zones (e.g. Bauknecht et al. 2015) experimental policy-making can be justified. Very much as it is the case for social and technological innovation policies, a pool of innovation seems desirable rather than betting on a single policy. This implies that policy coherence is not a necessity for a transformative policy.

Another policy approach is the integrated assessment of a given system and the exploration of potential scenarios for an alternative configuration. Such integrated assessments would collect the best available knowledge about how relevant systems operate and about their impacts on the different aspects of sustainable development. In order to develop socially robust knowledge that would be accepted as a basis for decision-making and that would provide legitimacy for policies, it is necessary to involve stakeholders into the process of the assessment. This is a good practice for international assessments and involvement of policy makers (Kowarsch 2016).

Besides knowledge about current system configurations and their impacts on sustainability, visions of alternative configurations are powerful to legitimize niche innovation and public policies. Visions can be understood as desirable future states of the society (John et al. 2015; Wiek and Iwaniec 2014). They are key for processes of transformations (Loorbach 2010; Jacob et al. 2018). The development of visions is a well established part of strategies and in the context of planning processes (Wiek and Iwaniec 2014; Jacob et al. 2015). However, it has been pointed out that there are limitations to the actual shaping of visions and related processes (SRU 2016), not least due to the multiple actors, norms, discourses, technologies, institutions (Brand 2016). This entails not only competing visions between old and new system configurations, but also different variants from new systems (Nill 2009). Besides the function to provide legitimacy for objectives and policy instruments, it has been argued that visions and the process of their definition have an impact of their own: It provides security and coordination for innovation and investments (Jacob 1998; Jacob and Jänicke 1998; Jänicke 2012; Berkhout 2006; Sondeijker et al. 2006; Voß et al. 2006). In order to have such impacts, visions need to be concrete and specific enough to provide guidance and at the same time sufficiently abstract to be applicable to a variety of actors and activities (Wiek and Iwaniec 2014; John et al. 2015; McPhearson et al. 2016). They should frame discourses and knowledge in order to enable learning (Voß et al. 2006; Vulturius and Swartling 2015). Last but not least, visions frame expectations and thereby mobilize actors within systems of innovation (van Lente 1993; Konrad 2006). It has been demonstrated that shared expectations contribute to development of new technological regimes respectively innovation system (Bakker et al. 2011; Bergek et al. 2008). It seems likely that this is applicable also for social and institutional innovation.

However, the content and (potential) impact decreases if visions are contested and diverse. This is the case with current debates on sustainable development: is this to be achieved by post-growth or by green economy? By efficiency or sufficiency? Centralized or decentralized? Exactly because visions should point out problems, responsibilities, and provide legitimization and mobilization, they are contested and subject to competition among actors. They result from a process during which problems, possible solutions and expectations of various actors are exchanged (Ingeborgrud 2017). While governmental actors are not able to prescribe societal visions, they are certainly relevant actors in their development and—perhaps even more important—can provide venues and processes during which actors meet.

Yet another action area is the development and enabling of new actors. Transformation research suggests that it is typically innovators from outside the existing systems that establish niches and initiate transformative processes. The existing actors (incumbents) tend to prefer incremental improvements and avoid a devaluation of their investments. Therefore, a transformative policy would enable new actors to provide services, if not create them. The feed-in tariffs can be used as an example: by means of this regulation, the providers of renewable energy were granted access to the power grid. Beyond this, the law incentivized the formation of new power suppliers. This came along with the establishment of interest groups that supported the respective policies. Creating new actors thereby improves the conditions for scaling up of environmental policies—and vice versa.

Finally, and certainly most demanding are policies that would organize the phase out of existing structures. This has been coined as exnovation (Heyen 2016). The concept aims to highlight that beyond innovation there is a need to replace unsustainable structures that remain in operation despite the demonstrated feasibility of alternative configurations. There are many examples of innovation that are taken up by economic structures without altering the core practices. Innovations are taken up by mainstream business through diversification: e.g. electricity from renewable energy is provided alongside electricity from coal and nuclear; organic food is produced alongside conventional food; battery powered vehicles are produced together with resource intensive SUVs. In many cases, conventional products are maintained by subsidies or by increasing exports. Other forms of adapting to system innovation are hybrid forms: Old and new technologies, practices and institutions are mixed. Both old and new systems operating at the same time as well as hybrid forms may lead to the situation that environmental performance during a transformation is even worse than in the phases of relative stability before and after a transformation. Based on an analysis of previous policies of phase out and structural change, Heyen (2016) proposes to organize exnovation as a long term and time-sensitive process that is accompanied by compensatory measures for affected enterprises, regions and employees.

To summarize, a transformative environmental policy that aims to shape transformation towards sustainability in a context of countervailing actors, interests, institutions and power within and beyond government has a number of options. These options mirror typical phases and causalities of past transformations:

- Surveying and assessing social and institutional innovation and trends: To inform policy-making, provide legitimacy for policies that built on these innovations and to enable a greening of social innovation and trends that would be unsustainable,
- Development and implementation of innovation policy that promotes social innovation,
- Regulatory experimentation in simulation games, policy labs and regulatory innovation zones,
- Integrated participatory assessments, including development of scenarios for socio-technical and—economic systems at stake,
- Initiating and moderating the development of visions for alternative system configuration,
- Enabling or establishment of actors that provide services to society in an alternative configuration,
- Organizing the exnovation out of incumbent systems.

The proposed options for action for shaping system innovation and transformation towards sustainability follow the different phases of a transformation. They increase in regards of capacity requirement: observation and analysis is least demanding, innovation programs and experimentation would require funds. Initiating the development of a vision establishing and enabling actors is likely to meet resistance by the established actors and their corresponding departments and organizing exnovation is most demanding in terms of legitimacy, power and funds. However, as transformations are progressing, a built up in capacities for environmental actors can be expected.

4.5 Conclusions

Although innovation oriented environmental policies have considerable potentials in reducing emissions and resource depletion that are yet underexploited, a strategy based on efficiency improvements is not sufficient for staying in the planetary boundaries. The de-carbonization of the economy and society, the effective protection of biodiversity, a circular economy based on renewable raw materials and without waste, the limitation of nitrogen emissions from agriculture, and other pressing environmental problems certainly cannot be effectively solved by end of the pipe measures, but also efficiency improvements are unlikely to achieve what is needed. The achievements of eco-innovation are often compensated, if not overcompensated by shifting of problems, rebound effects and growth effects.

Unsustainable production and consumption patterns are deeply rooted in complex socio-technical and socio-economic systems. Addressing the systemic nature, including relevant institutions, social practices, infrastructures, cultural embedded values and norms imply a fundamental transformation of such systems. Social practices, cultural norms, the institutional and infrastructural embedding of resource depletion and emissions have not yet been sufficiently subject of environmental policies. Given the autonomous, co-evolutionary and emergent character of such systems on the one hand, and the limitations of environmental policies on the other hand, a top down planning and strategic steering of such systems is unrealistic.

The evolutionary character of transformation processes, the affinity to experimentation also in regards of social and institutional innovation, the role of soft factors, like the coordinating and legitimizing role of visions and the provision of the necessary evidence base for possible future transformative pathways open up opportunities for public policymaking even within the constraints of fragmentation and incrementalism. Policymaking could make use of societal trends and innovation to a much greater extent than it is the case today. For doing so, policy options are available that seem to be consistent with the constraints of environmental policymaking and the respective capacities.

Such policies should not, however, replace the current environmental policies, but should rather be developed and applied as complementary approaches: In view of urgently pressing environmental problems, it would be a risk to wait for transformative innovation and change. Processes of transformation have an enormous potential to reshape societies. The limitations for actually initiating, steering and re-directing societal systems impose a risk for policymaking. At the same time, conventional environmental policymaking has still an agenda as potentials of eco-innovation and efficiency improvements are yet not fully exploited.

References

Bakker, S., van Lente, H., & Meeus, M. (2011). Arenas of expectations for hydrogen technologies. *Technological Forecasting and Social Change, 78*, 152–162.

Bauknecht, D., Heinemann, C., Stronzik, M., & Schmitt, S. (2015). Austesten von regulatorischen Innovationen – das Instrument der Regulatorischen Innovationszone. *Energiewirtschaftliche Tagesfragen, 65*(7), 61–64.

Baumgartner, F., & Jones, B. (1993). *Agendas and instability in American politics*. Chicago, IL: Chicago UP.

Bergek, A., Jacobsson, S., Carlsson, B., Lindmark, S., & Rickne, A. (2008). Analyzing the functional dynamics of technological innovation systems: A scheme of analysis. *Research Policy, 37*(3), 407–429.

Berkhout, F. (2006). Normative expectations in systems innovation. *Technology Analysis & Strategic Management, 18*(3/4), 299–311.

Bleischwitz, R., & Jacob, K. (2011). Innovative Ressourcenpolitikansätze zur Gestaltung der Rahmenbedingungen – Ein Überblick. In P. Hennicke, K. Kristof, & T. Götz (Eds.), *Aus weniger mehr machen – Strategien für eine nachhaltige Ressourcenpolitik in Deutschland*. München: Oekom.

Brand, U. (2016). "Transformation" as a new critical orthodoxy. The strategic use of the term "transformation" does not prevent multiple crisis. *GAIA, 25*(1), 23–27.

Cairney, P. (2013). What is evolutionary theory and how does it inform policy studies? *Policy & Politics, 41*(2), 279–298.

Chataway, J., Daniels, C., Kanger, L., Ramirez, M., Schot, J., & Steinmueller, E. (2017). *Developing and enacting transformative innovation policy. A comparative study*. Retrieved from http://www.johanschot.com/wordpress/wp-content/uploads/2017/06/Developing-and-enacting-Transformative-Innovation-Policy-A-Comparative-Study-FINAL-DRAFT.pdf

Dimitropoulos, A., Oueslati, W., & Sintek, C. (2016). *The rebound effect in road transport*. Paris: OECD.

Geels, F. W. (2002). Technological transitions as evolutionary reconfiguration processes: A multi-level perspective and a case-study. *Research Policy, 31*(8–9), 1257–1274. https://doi.org/10.1016/S0048-7333(02)00062-8

Geels, F. W., & Kemp, R. (2012). The multi-level perspective as a new perspective for studying sociotechnical transitions. In F. W. Geels, R. Kemp, G. Dudley, & G. Lyons (Eds.), *Automobility in transition? A socio-technical analysis of sustainable transport*. London: Routledge.

Grießhammer, R., & Brohmann, B. (2016). *Wie Transformationen und gesellschaftliche Innovationen gelingen können. Transformationsstrategien und models of change für nachhaltigen gesellschaftlichen Wandel*. Baden-Baden: Nomos.

Hall, P. (1993). Policy paradigms, social learning, and the state. *Comparative Politics, 25*(3), 275–293.

Haum, R., Jacob, K., & Hertin, J. (2010). Integration von Umwelt-, Innovations- und Industriepolitiken: Eine evaluation von Europäischen Strategien und Maßnahmen. In R. Steurer & R. Trattnigg (Eds.), *Nachhaltigkeit regieren. Eine Bilanz zu Governance-Prinzipien und-Praktiken* (pp. 97–120). München: Oekom.

Heyen, D. A. (2016). *Exnovation: Herausforderungen und politische Gestaltungsansätze für den Ausstieg aus nicht-nachhaltigen Strukturen*. Berlin. https://www.oeko.de/fileadmin/oekodoc/WP-Exnovation.pdf

Hinterberger, F., & Schmidt-Bleek, F. (1999). Dematerialization, MIPS and factor 10 physical sustainability indicators as a social device. *Ecological Economics, 29*(1), 53–56. https://doi.org/10.1016/S0921-8009(98)00080-9

Ingeborgrud, L. (2017). Visions as trading zones: National and local approaches to improving urban sustainability. *Futures, 96*, 57–67.

Jacob, K. (1998). Umweltpolitik by objectives? Ergebnisse und Überlegungen aus einer Evaluationsstudie der Chemikalienpolitik. *Ökologisches Wirtschaften, 5*, 27–29.

Jacob, K. (2015). Greening des gesellschaftlichen Wandels. *Ökologisches Wirtschaften*, (3), 30.

Jacob, K., & Jänicke, M. (1998). Ökologische Innovationen in der chemischen Industrie: Umweltentlastung ohne Staat? Eine Untersuchung und Kommentierung zu 182 Gefahrenstoffen. *Zeitschrift für Umweltpolitik & Umweltrecht, 21*(4), 519–547.

Jacob, K., Kannen, H., & Niestroy, I. (2015). An international comparison of sustainability strategies. In B. Stiftung (Ed.), *Developing successful sustainability strategies: Strategies for a sustainable future in Germany, Europe and Worldwide*. Gütersloh: Bertelsmann Stiftung.

Jacob, K., Graaf, L., Wolff, F., & Heyen, D. A. (2018). *Transformative Umweltpolitik: Ansätze zur Förderung gesellschaftlichen Wandels*. Berlin: BMUB.

Jänicke, M. (2012). Dynamic governance of clean-energy markets: How technical innovation could accelerate climate policies. *Journal of Cleaner Production, 22*, 50–59.

Jänicke, M., & Jacob, K. (2013). A third industrial revolution? In B. Siebenhüner, M. Arnold, K. Eisenack, & K. Jacob (Eds.), *Long term governance for social ecological change* (pp. 47–70). New York: Routledge.

Jänicke, M., & Rennings, K. (2011). Ecosystem dynamics: The principle of co-evolution and success stories from climate policy. *International Journal of Technology Policy and Management, 11*(3/4), 198–219.

John, B., Withycombe Keeler, L., Wiek, A., & Lang, D. (2015). How much sustainability substance is in urban visions? An analysis of visioning projects in urban planning. *Cities, 48*, 86–98.

Kingdon, J. W. (1995). *Agendas, alternatives and public policies* (2nd ed.). New York: Harper & Collins.

Knaggård, Å. (2014). What do policy-makers do with scientific uncertainty? The incremental character of Swedish climate change policy-making. *Policy Studies, 35*(1), 22–39. https://doi.org/10.1080/01442872.2013.804175

Kondratieff, N. D. (1926). Die langen Wellen der Konjunktur. *Archiv für Sozialwissenschaften und Sozialpolitik, 56*, 573–609.

Konrad, K. (2006). The social dynamics of expectations: The interaction of collective and actor-specific expectations on electronic commerce and interactive television. *Technology Analysis & Strategic Management, 18*, 429–444.

Kowarsch, M. (2016). *A pragmatist orientation for the social sciences in climate policy. How to make integrated economic assessments serve society*. Berlin: Springer.

Leggewie, C., & Welzer, H. (2010). Cultural and societal requirements of The Great Transformations. *Journal of Renewable and Sustainable Energy, 2*(3), 1–13.

Lindberg, H. (2013). *Knowledge and policy change* (pp. 62–88). Newcastle upon Tyne: Cambridge Scholars Publishing.

Lindblom, C. E. (1959). The science of "muddling through". *Public Administration Review, 19*(2), 79–88. https://doi.org/10.2307/973677

Loorbach, D. (2010). Transition management for sustainable development: A prescriptive, complexity-based governance framework. *Governance: An International Journal of Policy, Administration, and Institutions, 23*(1), 161–183.

Maeße, J. (2013). Ökonomie, Diskurs, Regierung. *Globale Politische Ökonomie*. https://doi.org/10.1007/978-3-658-01294-6

Mayntz, R., & Scharpf, F. W. (1995). Der Ansatz des akteurzentrierten Institutionalismus. In R. Mayntz, F. W. Scharpf, & Max-Planck-Institut für Gesellschaftsforschung (Eds.), *Gesellschaftliche Selbstregelung und Politische Steuerung, Schriften des Max-Planck-Instituts für Gesellschaftsforschung* (Vol. 23, pp. 39–72). Frankfurt a. M.: Campus Verlag.

McPhearson, T., Iwaniec, D. M., & Bai, X. (2016). Positive visions for guiding urban transformations toward sustainable futures. *Current Opinion in Environmental Sustainability, 22*, 33–40.

Nill, J. (2009). *Ökologische Innovationspolitik. Eine evolutorisch-ökonomische Perspektive*. Marburg: Metropolis.

Rennings, K. (2000). Redefining innovation – eco-innovation research and the contribution from ecological economics. *Ecological Economics, 32*(2), 319–332. https://doi.org/10.1016/S0921-8009(99)00112-3

Rennings, K. (2014). Der mikro-ökonomische Rebound Effekt. Ergebnisse eines inter- und transdisziplinären Forschungsprojekts. Vortrag auf der Green Economy Konferenz des BMBF 18.11.2014. https://www.fona.de/mediathek/event/ge2014/beitrag/c1_rennings_klaus_01_presentation_ge2014.pdf

Rennings, K., Rammer, C., Oberndorfer, U., Jacob, K., & Boie, G. (2008). Instrumente zur Förderung von Umweltinnovationen. Bestandsaufnahme. In *Bewertung und Defizitanalysen*. Berlin: UBA und BMU.

Sachverständigenrat für Umweltfragen (SRU). (2016). Umweltgutachten 2016. Impulse für eine integrative Umweltpolitik. Hg. v. SRU. Online verfügbar unter http://www.umweltrat.de/SharedDocs/Downloads/DE/01_Umweltgutachten/2016_Umweltgutachten_HD.html. Zuletzt geprüft am 27.08.2017.

Schot, J., & Geel, F. W. (2008). Strategic niche management and sustainable innovation journeys: theory, findings, research agenda, and policy. *Technology Analysis & Strategic Management, 20* (5), 537–554.

Schumpeter, J. A. (1942). *Capitalism, socialism, and democracy*. London: Harper & Brothers.

Semmling, E., Peters, A., Marth, H., Kahlenborn, W., & de Haan, P. (2015). *Rebound-Effekte: Ihre Bedeutung für die Umweltpolitik*. Dessau: Adelphi-Consult.

Sondeijker, S., Geurts, J., Rotmans, J., & Tukker, A. (2006). Imagining sustainability: the added value of transition scenarios in transition management. *Foresight, 8*(5), 15–30.

van Lente, H. (1993). *Promising technology. The dynamics of expectations in technological developments*. PhD Thesis, Enschede.

Voß, J.-P., Bauknecht, D., & Kemp, R. (Eds.). (2006). *Reflexive governance for sustainable development*. Cheltenham: Edward Elgar.

Vulturius, G., & Swartling, Å. (2015). Overcoming social barriers to learning and engagement with climate change adaptation: Experiences with Swedish forestry stakeholders. *Scandinavian Journal of Forest Research, 30*, 217–225.

WBGU. (2011). *Welt im Wandel Gesellschaftsvertrag für eine Große Transformation.* Berlin: WBGU.

Weizsäcker, E. U., Lovins, A. B., & Lovins, H. L. (1995). *Faktor Vier. Doppelter Wohlstand – halbierter Naturverbrauch* (Der neue Bericht an den Club of Rome). München.

Wiek, A., & Iwaniec, D. (2014). Quality criteria for visions and visioning in sustainability science. *Sustainability Science, 9*, 497–512.

Chapter 5
Outlook: Can Environmental Product Standards Enable Eco-Innovation?

Albert Roger

5.1 Sustainability Challenge

Environmental labels have become increasingly present in our daily lives. According to a report from the OECD, the total number of environmental labeling schemes quintupled from 1988 to 2009 (Gruère 2013).

Environmental standards aim at providing information to consumers on the environmental quality of a given good (product standards) or process (process standard). When related to products, as is the case with environmental labels, the so-called Environmental Product Standards (EPS)[1] are aimed at rectifying the asymmetry between consumers and firms regarding the information about the environmental effects of good's consumption (Fuerst and McAllister 2011; Galarraga 2002). The labeled products are called "credence goods" since the consumer cannot evaluate the quality signalled by the label, not even after its purchase. Therefore, trustfulness is key for the effectiveness of EPS (Prag et al. 2016).

Despite the variety of EPS, we can distinguish between *mandatory* and *voluntary* product standards. As shown in Fig. 5.1, we can find different types of *mandatory/voluntary* information on a product's label.

[1]Environmental product standards can be also referred to as Environmental Labeling and Information Schemes (ELIS) (Gruère 2013; Klintman 2016; Prag et al. 2016).

A. Roger (✉)
Centre for European Economic Research (ZEW), Mannheim, Germany
e-mail: roger@zew.de

J. Horbach, C. Reif (eds.), *New Developments in Eco-Innovation Research*, Sustainability and Innovation, https://doi.org/10.1007/978-3-319-93019-0_5

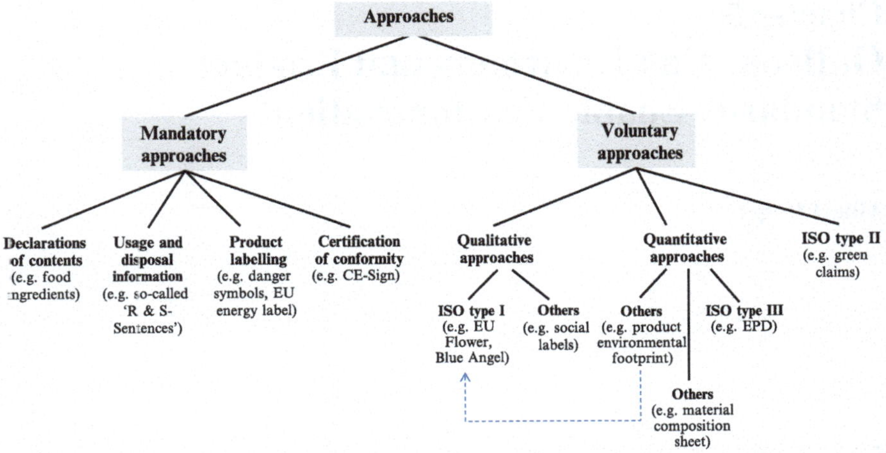

Fig. 5.1 Classification of different information transmission approaches in EPS. Source: Rubik (2015)

5.1.1 Mandatory Environmental Product Standards (MEPS)

Mandatory environmental product standards require every firm that is willing to introduce products in the market to provide certain information about the characteristics of the product (e.g.: information on health impacts, safety, etc.).

Two prominent examples of MEPS are the European Energy Labeling Directive, epitomized the EU "Energy Label", and the Ecodesign Directive. The former focusses on the consumption of energy and other resources (e.g.: water) during the usage of the product. The latter, instead, extends specific "Ecodesign requirements" to the whole life-cycle of the product (i.e. including recycling or disposal of the product after the usage phase). These environmental requirements concern one of the following five aspects, namely: resource consumption, waste, emissions, hazardous substances and physical impacts in the use phase (Molenbroek et al. 2014).

5.1.2 Voluntary Environmental Product Standards (VEPS)

Voluntary environmental product standards (VEPS) constitute a different approach, leaving the decision whether to perform more environmentally-friendly (and signal it with a label) or not (Rubik 2015) to market players. The International Organization for Standardization (ISO) has developed an own taxonomy to classify VEPS. This classification follows the ISO 14020 series and distinguishes between three types of VEPS (Gruère 2013):

- **Type I (ISO 14024)—Eco-labels:** "Voluntary, multiple criteria-based third party programs that award a license authorizing the use of environmental labels on products. These labels provide qualitative environmental information" (ISO 2000: 1). They are covered by ISO 14024 published in April 1999, last reviewed and confirmed in 2009 (Rubik 2015). Examples of Type I labels are among others: German "Blue Angel", "Nordic Swan", European "EcoLabel" and Canadian Environmental Choice.

- **Type II (ISO 14021)—Self-declared Environmental Claims:** "Self-declared environmental claim made by manufacturers, importers, distributors, retailers, or anyone else likely to benefit from such a claim without independent third-party certification (ISO 1999: 3). They are covered by ISO 14021 published in 1999." (Rubik 2015) Examples of Type II labels are among others: Recycled content and Biodegradable.
- **Type III (ISO 14025)—Environmental Declarations:** "Quantified environmental data using predetermined parameters and, where relevant, additional environmental information. (ISO 2006: 2). They are covered by ISO 14025 published in 2006." (Rubik 2015) Examples of Type II labels are among others: "Eco-Leaf" and Korean Environmental Declaration of Products.

The main differences between these three types are:

1. Type I and III cover multiple criteria whereas Type II covers only a single area.
2. Similarly Type I and III are life-cycle based, which is not the case for Type II.
3. Type II labels do not need to be third-party certified whereas it is compulsory for Type I and III.
4. Type I labels are selective, namely the symbol of the label allows to differentiate between products with and without that label.

This taxonomy, however, does not gather the full diversity of VEPS. It fails to include mixed labels (i.e. quantitative or qualitative labels that don't fall into the ISO categories), such as the "Fairtrade" label or some other quantitative labels like the carbon footprint. Therefore, we will complete this classification with the one proposed by Rubik (2015), which also includes the three ISO types (see Fig. 5.1). This classification distinguishes between quantitative and qualitative labels, and Type II labels.

5.1.3 *Qualitative Labels*

- **Type I ISO Labels:** Eco-labels are voluntary product standards that consider the entire life-cycle of the product. Their approach is to label the products with the best above-average environmental performance to set them apart. The first eco-label was introduced in Germany in 1978, the German "Blue Angel". It was followed by the "Nordic Swan" and the Japanese "Eco-Mark" in 1989 (Rubik 2015).
- **Social Labels and Standards:** These product standards aim at covering social features such as social rights, child labor or minimum wages, e.g.: the former

"Rugmark" label now "GoodWeave International" or the "Fairtrade" label among others (Potts et al. 2014; Rubik 2015).
- **Certificates of Conformity:** These certificates might address diverse issues (i.e. not only one single environmental characteristic), but they generally certify the fulfilment of certain environmental requirements and are often concerned with upstream resource extraction. Three well-known examples are the "FSC" (Forest Stewardship Council) label, the "MSC" (Marine Stewardship Council), and the "Rainforest Alliance" label. The FSC is a scheme created in 1992 under an NGO (Forest Stewardship Council), which certifies that companies fulfil a number of forestry requirements (Prag et al. 2016; Rubik 2015).

5.1.4 *Quantitative Labels*

- **Type III ISO Labels:** Also referred to as "Environmental product declarations", are a type of standards mainly oriented towards business partners (e.g. public procurers or retailers). They provide quantified environmental data for a product, given certain parameters. The data provided should be based on life-cycle assessment tools and calculations should consider supply chains (Rubik 2015). e.g.: Japanese "Eco-Leaf" and "International EPD® System".
- **Product Footprint:** The environmental issues addressed depend on the type of footprint (e.g.: ecological, water, carbon, land, etc.). An example of one of these footprints is the "Product Environmental Footprint" (PEF) created by the European Commission under the "Single Market for Green Products Initiative". The PEF measures the environmental performance of products throughout their entire life-cycle (i.e. including recycling and disposal after usage phase), considering relevant environmental impacts of all steps needed to get the product to the consumer. The PEF has been tested from 2013 to 2017 with the collaboration of more than 280 companies and organizations (Rubik 2015).
- **Material Composition:** Without any reference to ISO standards, suppliers might be willing to give consumers information on the composition of their products. Two prominent examples are the electronics and the car industry, where global players ask their suppliers to deliver information on the composition of the products and pre-products. E.g.: "Material Composition Declaration for Electrotechnical Products" of the Consumer Electronics Association (CEA) (Rubik 2015).

5.1.5 *Type II Labels*

This type of labels refer to self-declared environmental claims, which do not undergo an audit process. These kind of labels have raised some issues referring to their trustfulness, which might affect other EPS. A recent study of the OECD gives a first insight on the different types of environmental claims and the possibilities to punish misleading claims (Klintman 2016).

5.1.6 Others

Some of the most extended standards, which are third-party certified but neither life-cycle-based nor multi-criteria standards do not fall within these three types. Prominent examples are organic certified products, the "Energy Star" label or third-party certified labels, which are not life-cycle based, such as energy performance or fuel efficiency labels (Gruère 2013).

Classification by Sector

Besides the above mentioned extended-ISO classification, it is possible to take a sectoral approach to study EPS. In a study published in 2016, the OECD differentiates between four main sectors in which environmental labels are used, namely seafood; coffee, fruits and vegetables; forest products; and appliances (Prag et al. 2016).

5.1.7 Seafood

This sector includes one of the most well-known environmental labels, the "dolphin safe" tuna. The consequences of the implementation of this label have been broadly analyzed among others in Teisl et al. (2002). Introduced in the 1990s, its aim is to reduce the dolphin mortality rate. By focusing on a single issue, it contrasts with other labels that address the sustainability of capture fisheries, such as the "Marine Stewardship Council" (MSC) or "Friends of the Sea" (FOS) (Fig. 5.2).

Besides these kind of labels, an OECD study (OECD 2011) noted that many retailers are introducing self-declared environmental claims. In 2011, Client Earth (a non-profit environmental law organization) found out that 32 out of 100 products examined in UK supermarkets carried misleading, unverified or unverifiable claims (Client Earth 2011). According to Prag et al. (2016) the introduction of Type II labels has contributed to increase consumer confusion in this sector. One of the reasons stated is that Type I labels find low recognition among consumers compared to self-declared claims.

5.1.8 Coffee, Fruits and Vegetables

This sector is characterized not only by the variety of labels but also by their co-existence (e.g.: multiple certification). The variety of EPS makes it possible to observe both a horizontal and vertical differentiation. The horizontal differentiation occurs in terms of the variety of environmental attributes certified (e.g.: organic, fair

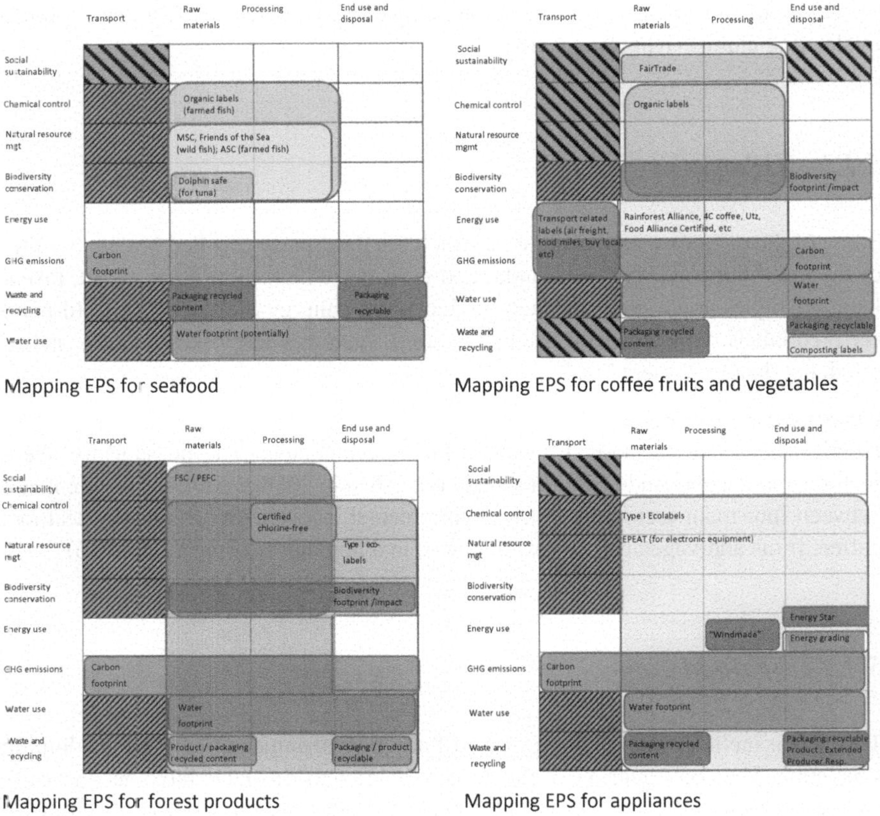

Fig. 5.2 Types of EPS by Sector. Source: Prag et al. (2016)

trade, bird-friendly, etc.) whereas the vertical differentiation takes place on the quality ladder in terms of environmental stringency (Prag et al. 2016). An empirical field study, realized in 12 countries[2] based on a control group, found evidence that on average certified farms had higher yields and that double certification significantly increased yields by a substantial amount (COSA 2013).

5.1.9 *Forest Products*

There are two main consolidated and internationally recognized EPS, namely the "Forest Stewardship Council "(FSC) and the "Programme for the Endorsement of Forest Certification" (PEFC). Both were created in the early 1990s. They primarily focus on

[2]Mexico, Guatemala, Nicaragua, Costa Rica, Colombia, Peru, Côte d'lvoire, Ghana, Tanzania, Vietnam, Indonesia, Papua New Guinea.

environmental performance, but they also address other social issues like workers' safety or community relationships. Nevertheless, one of the main criticisms of these labels is that while they are present in 80 countries, they heavily focus on OECD countries. Together they achieve a coverage in Europe and North America of 88% of the forested area, globally, however, this percentage shrinks to a 9.1% (Prag et al. 2016).

5.1.10 Appliances

In this sector, two well-known labels related to products' energy consumption are competing with other EPS. These two labels use different incentive mechanisms. The "Energy Star", a U.S. government-backed label, is a seal-type certification attributed to the top performers within a product category. The European Union "Energy Label", in contrast, is a type of mandatory grading scheme that covers large household appliances (the so-called white goods e.g.: washing machines, refrigerators, etc.). Going from A (most efficient) to G (least efficient), it assigns every device a corresponding energy efficiency level (see Box 5.1 for more details).

Grading schemes have become mandatory in OECD countries and often include minimum performance standards. The main asymmetry between the U.S. and EU for mandatory energy efficiency grading schemes lies in their coverage. Whilst both are mandatory for white appliances, it is only the U.S. scheme that is mandatory for office equipment. These two labels coexist with other EPS like the multi-attribute label on electronics (EPEAT), the "Windmade" label (certifies manufacturers purchasing renewable energy) or "Extended Producer Responsibility" (EPR) (a waste-related label) (Prag et al. 2016).

Although the revision of the Energy Labeling regulation to replace Directive 2010/30/EU has taken into account the critique concerning the existence of the A+,++,+++ classes, it still lacks information on absolute energy use. Furthermore, label enforcement (market surveillance) is still weak under the new revision (ECOS 2017).

Box 5.1 Energy Labels in the Appliances' Sector: The Case of the European Energy Efficiency Label and the "Energy Star"

The EU "Energy Label" has received much criticism since its modification in 2010 (Arditi et al. 2013). Most of it was related to the introduction of the additional A+, ++ and +++ categories, claiming that providing the A class too easily would undermine consumers' incentives to purchase high-efficient devices. Nevertheless, a recent study has added new critique to the list. The report, signed by four NGOs, intended to scrutinize the testing procedures for the EU "Energy Label" categories.

One of the main findings of the study is that there are discrepancies between the class declared by the appliances' producers and the class measured in a test.

(continued)

Box 5.1 (continued)
In the case of the fridges, 50% of the appliances were found to be one class less efficient than the class reported. Furthermore, they criticize the way the Energy Efficiency Index (EEI) is computed. The aim of this index is to compare the measured energy consumption of the model (kWh/year) to the standard energy consumption depending on the volume. The main drawback of this formula is that it relies heavily on the reference volume, which itself depends, among others, on the type of fridge (e.g. with or without freezer) or the climate. Since the absolute energy use is not clearly communicated to the consumer through the label, he might purchase a new appliance rated A+++, which might consume more energy than his former A+ device (CLASP et al. 2017).

5.1.11 Classification by Impact on the Market

EPS can be also classified depending on the market impact they want to generate. Often Type I labels (e.g. "Energy Star") target only the products with the most environmental-friendly performance (i.e. representing 15 to 30% of the market), see Fig. 5.3a. NGO-backed voluntary EPS tend to differ more in quality; thereby some of them focus on the best performers and others aim at increasing average energy performance (e.g. food labels), see Fig. 5.3b. Finally, the third type of EPS rates the actual environmental performance of the product, this is the case for footprint labels (e.g. water footprint or carbon footprint) or grading schemes (e.g. EU "Energy Label"). The drawback of this label type is that if certification is voluntary, only good performers are willing to be certified, see Fig. 5.3c (Prag et al. 2016).

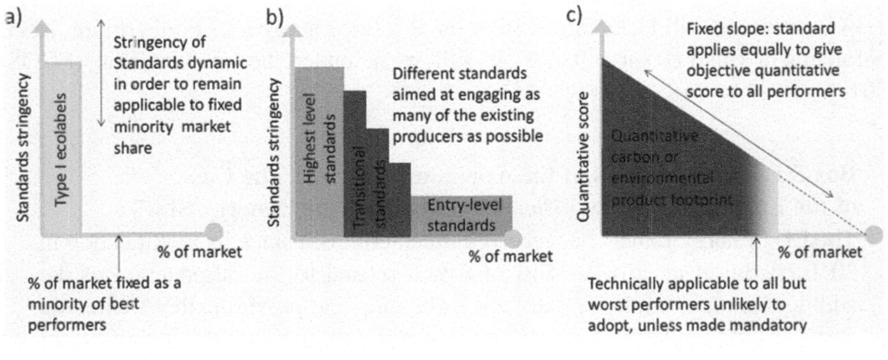

Fig. 5.3 Comparing market objectives of different types of EPS. Source: Prag et al. (2016)

5.2 Eco-Innovation in Practice

Environmental Product Standards can become a cornerstone on the journey towards achieving the Sustainable Development Goals (SDG). The SDGs were defined by the United Nations as part of the new sustainable agenda. As we have seen in the previous section, EPS can go beyond pure environmental issues. Many NGO-backed product standards involve some kind of social criteria for their certification besides the environmental attributes. Well-known examples can be found among food labels such as the "Fairtrade" or the "UTZ" label. Therefore, the connection between EPS and the SDGs reaches beyond solely environmentally-related SDGs like n°6 (Clean water and sanitation), n°7 (Affordable and Clean Energy), n°9 (Industry, innovation and infrastructure), n°12 (Responsible Consumption and Production), n°13 (Climate action) or n°14 (Life Below Water). EPS also contribute to SDGs n°8 (Decent Work and Economic Growth), n°10 (Reduced Inequalities) and n°15 (Life on Land).

5.2.1 Eco-Innovation Impact

The effectiveness of EPS heavily relies on consumers' trust in the environmental impact of the given labels. Nevertheless, as claimed by Cohen and Vandenbergh (2012), only few studies on the effectiveness of EPS generally satisfy the standards of rigorous empirical research since they lack random assignment or quasi-experimental design.

Box 5.2 Compiling Evidence on the Impacts of EPS is Challenging
As already mentioned, one of the cornerstones for the success of environmental labels is consumers' trust, which depends on the evaluation of a label's effectiveness. Thus providing solid evidence of the economic and environmental impact of labels should be a priority. In the RESOLVE (2012) study, a Steering Committee assessed the impacts of environmental labels in four sectors: agriculture, forestry, fisheries and aquaculture. The "impacts" they defined were the changes in the quality and resilience of ecosystems, changes in resource efficiency and livelihoods, and changes in social welfare within the workplace and wider community. In their study, they acknowledge that the main problem in the evaluation of EPS is the identification of an appropriate counterfactual, i.e. finding an answer to the question of what would have happened in absence of the certification scheme. Nevertheless, the main barrier to providing evidence based on counterfactual settings consists in the costs and logistical challenges of experimental and quasi-experimental designs.

Environmental and Economic Impacts

One recent piece of evidence has been provided by Asensio and Delmas (2017). They scrutinize the effectiveness of U.S. energy efficiency building

(continued)

Box 5.2 (continued)
labels induced by three main labels, namely: the U.S. Department of Energy's Better Building Challenger, the U.S. EPA "Energy Star" program and the U.S. Green Building Council's "Leadership in Energy and Environmental Design" (LEED) program. In this case, in order to cope with the mentioned non-randomness problem, they use matching techniques to compare the performance of participating buildings with the one of similar buildings that are not part of these programs. They find energy savings of about 18% to 30%, depending on the program. Nevertheless, these programs do not substantially reduce emissions in small and medium sized buildings, which represent about two-thirds of commercial sector building emissions.

Also, using propensity score matching to control for self-selection bias, Blackman and Naranjo (2012) find that eco-certification of coffee improves growers' environmental performance. Furthermore, they find that it significantly reduces chemical input use and increases the adoption of some environmentally friendly practices.

In the RESOLVE (2012) study as well as in the Kjeldsen et al. (2014) study, further evidence, very often survey- or case-study-based, can be found on the environmental and economic impacts of environmental labels. Additionally, the meta-study of Carlson and Palmer (2016) compiles the main case studies on the impact of two eco-labeling schemes in developing countries, namely the "Forest Stewardship Council" (FSC) and "Marine Stewardship Council" (MSC). From the case studies, they conclude that producers benefit in varied ways from certification. They do not seem to receive benefits in the form of price premiums or market access, but mainly intangible benefits, i.e. learning, governance, community empowerment, and reputational benefits.

Cohen and Vandenbergh (2012) review evidence on the role of product labeling and its influence on consumer and firm behavior. In their paper, they classify the provided types of evidence (non-quasi-experimental) into two types: industry and market studies of product sales, and consumer surveys of label awareness, use and stated preferences. In the following, I present a summary of their main analyses.

5.2.1.1 Industry and Market Studies

They use the environmental impact evaluation of the U.S. Energy Star program as an example. According to them, the estimated emission reduction benefits from the program (U.S. EPA 2008; Brown et al. 2002, 515) cannot be solely attributed to the EPS' implementation. They claim that the estimation of the emission reduction is based on the market penetration of the "Energy Star" and engineering estimates. Nevertheless they claim that, as it is not possible to know whether these products would have been manufactured and purchased in absence of the label, it is hard to attribute all the estimated energy-efficiency benefits to the program itself.

Later on they analyze the market impact of the introduction of an EPS, by considering different possible impacts such as: demand rebound, substitution effects or price-band specific impact. In their paper, they discuss first an example of demand rebound, namely a case study analyzing the impact of EPS, and more specifically of the "dolphin-safe tuna" label on the market. In the case of this label, public environmental concerns about dolphin killings spurred a drop in tuna demand. Interestingly, the introduction of the dolphin-safe label increased tuna demand (Teisl et al. 2002).

Then they analyze further market impacts of EPS such as substitution effects. They discuss a study of Bjørner et al. (2004) on the "Nordic Swan". The study sheds light on the consequences of the introduction of this EPS on the willingness to pay for certified toilet paper, which they find it increased from 13% to 18%. In the case of paper towels, however, they found few evidence. In order to explain the results, they argue that most environmentally friendly consumers are more likely to avoid buying any paper towels and would rather substitute them by cloth.

Finally they claim that the impact of EPS can also depend on the price of the good. They discuss an experiment done in an Australian grocery shop (Vanclay et al. 2011), where the introduction of an EPS increased the demand for the most environment-friendly products on average by 4%. Furthermore if those products had also been the cheapest, the demand would have increased by 20%. Part of the evidence on the higher willingness to pay for environmentally-certified goods comes from participation in green electricity programs in the U.S. (Bird and Sumner 2010; Kotchen and Moore 2007).

5.2.1.2 Consumer Surveys

Cohen and Vandenbergh (2012) review evidence on surveys used to uncover consumers' preferences. They discuss several papers starting with a paper by Borchers et al. (2007) where they found a positive willingness to pay for green electricity among consumers, especially for solar energy, using hypothetical purchases. Then they review a work by Clark et al. (2003), which is in the same line but surveying real purchases. There they found that altruism towards the environment followed by altruism towards regional residents to be the most important factors for purchasing green electricity. They argue that there is, however, few evidence on higher willingness to pay for carbon emission reductions besides energy saving motivations or other personal benefits. They suggest that willingness to pay for green goods might depend on consumer preferences, income, taste and product category (Jacobsen et al. 2012; Wiser et al. 2000; Michaud et al. 2013). For further information on studies related to the willingness to pay see Box 5.3.

Finally they discuss evidence on consumers' awareness of environmental labels. They argue that while purchasers seem to know about the existence of green labels in some sectors, like energy-efficient labels in the appliances sector (Ottman 2011), other sectoral labels attract little attention (e.g.: seafood). According to a U.S. survey, only 18% of the respondents were aware of the "Marine Stewardship Council" label on sustainable fish (Ottman 2011). Furthermore, as shown in Murray and Mills (2011), awareness might be different across household income levels.

5.3 Eco-Innovation Challenge

5.3.1 Drivers, Benefits and Barriers to Eco-Innovation

In my review of the drivers, benefits and barriers of adoption of EPS by firms, I will build upon two main pieces of evidence namely the EVER study (2005), which evaluated the "EU Ecolabel" and a study evaluating the "Nordic Swan" (Kjeldsen et al. 2014).

5.3.1.1 Drivers of Adopting EPS

Survey results from the EVER study (2005) suggest that one of the main motivations for companies to apply for the "EU Ecolabel" scheme is to exploit business opportunities offered by higher consumer awareness of environmental issues. Different studies (Horbach et al. 2012; Wagner 2008; Demirel and Kesidou 2012) suggest that firms' decision to certify their products might be due to societal pressures and market requirements, e.g.: gain access to certain markets, green procurement, green demand, etc. (Iraldo and Barberio 2017).

A study evaluating the "Nordic Swan" (Kjeldsen et al. 2014) uses a survey to analyse the motivations behind firms' decision to obtain certification. The main reasons found were to obtain or to sustain a green profile and to increase sales. Interestingly, the source of motivation was top-managements' idea of using the "Nordic Swan" label as part of a strategical environmental focus. Besides this, other companies claim that their decision to obtain the certification was driven by the will to be at the forefront of upcoming changes in environmental regulation.

From these two studies we can conclude that motivations behind the decision to adopt an EPS are diverse and might depend on the context of the firm. Nevertheless market-related issues, e.g.: competitive pressure and demand pull, seem to have been two major drivers of this decision.

5.3.1.2 Benefits of EPS

There is few evidence from quantitative or qualitative studies on economic or environmental benefits obtained by firms due to the implementation of the "EU Ecolabel". In the case of the "EU Ecolabel", most of the evidence comes from the EVER study (2005), in which some surveyed firms state having received economic and environmental benefits. Other companies, for instance, state having experienced a modest increase in market share and sales. Besides the economic benefits, the surveyed firms also state an improvement in their environmental performance. One of the reasons mentioned for this improvement was environmental knowledge acquisition during the Ecolabel implementation process, which for some firms induced them to set environmental targets (Iraldo and Barberio 2017).

Evidence from the "Nordic Swan" by Kjeldsen et al. (2014) suggests that some companies have gained a competitive advantage by being recognized as market leaders. Furthermore, even companies that did notice an increase in sales recognize that they would have lost their market share if they had not adopted the "Nordic Swan" label. Firms participating in the "Nordic Swan" labeling scheme declared significant gains in resource efficiency (Iraldo and Barberio 2017).

Box 5.3 Impact of EPS: Consumers' Willingness to Pay and Bunching Effects

Beyond the before mentioned market effects, environmental labels might not only impact consumers' willingness to pay, and thus have an effect on price premiums, but also influence firms' strategy to develop "green" product characteristics, by steering them towards a label's rating scheme.

Willingness to Pay

Evidence on price premiums and EPS is not clear. There are some studies that find higher prices for labelled products but other studies discuss whether the price premium can really be only attributed to the label.

Examples of studies finding a positive impact on price premiums are Ward et al. (2011) and Fuerst and McAllister (2011). Using survey results, Ward et al. (2011) find that consumers are on average willing to pay an extra $249.82-$349.30 for a fridge with the "Energy Star" label. Fuerst and McAllister 2011) find that eco-labels in commercial offices (LEED and "Energy Star" labels) obtain higher rental rates and an average sales premium of 18% for "Energy Star" and 25% for LEED labelled office buildings.

More recent studies come to different results for other EPS. Namely, Kortelainen et al. (2016) find no evidence that carbon reduction labels have an impact on detergent prices or demand. Furthermore, scrutinizing the reasons for such results, they find the specific design of the label to be responsible for the lack of success. Park (2017) suggests in his findings that the price premium in the Korean television market does not result from the energy efficiency label itself. Energy-efficient products already had a higher price before the introduction of the label.

Bunching Effects

As previously mentioned, EPS might not only affect firms in their pricing strategy, they can also have an influence on the "environmental" quality of the goods they provide. Shewmake and Viscusi (2015) find that firms respond to

(continued)

Box 5.3 (continued)
environmental label stringency by strategically incorporating green features to achieve higher ratings. Firms incorporate green attributes to the offered goods such that product bunch around notches. This appears to be a consequence of producers strategically building homes to achieve ratings, which is consistent with the absence of a price premium for points beyond rating cut-offs. Recent results by Houde (2017) for the refrigerator market are in line with those of Shewmake and Viscusi (2015).

5.3.1.3 Barriers to EPS

The EVER study (2005) on the "EU Ecolabel" found that label holders considered the so-called "red tape/documentation"[3] and the costs of compliance with label's criteria to be the main barriers. In the same study, the non-label holders explain the main reasons why they abandoned the certification scheme or decided not to enter it. Those can be summed up in four main types: lack of recognition by future demand, high costs of implementation, high costs of license, and lack of economic incentives.

Similarly, a study evaluating the "Nordic Swan" found that the main barriers were the overall cost of implementation and the application procedure (Kjeldsen et al. 2014). With regard to the implementation costs, the indirect costs appeared to be even higher than the direct ones (e.g.: cost of application procedure, changes in the production process, consulting costs or human capital training costs). Besides the costs, time spent in the overall application process appeared to be another barrier, namely time used to understand environmental criteria and time spent collecting documentation.

5.3.2 *Trade and EPS*

A UNEP report (UNEP 2005) analyses the possible Technical Barriers to Trade (TBT) of five eco-labeling programs ("Blue Angel" label, IFOAM accreditation, FSC and MSC, and "Fairtrade" label). Analyzing the possible impact of eco-labeling on trade, it outlines some of the requirements in the TBT Agreement (particularly the Annex 3 of the TBT Agreement: Code of Good Practice "Standards Code") and the challenges associated with their application in the case of EPS. The following list subsumes the main trade-related challenges:

- Article F of the Standards Code calls on standard bodies to base their work on relevant existing international standards. This presents problems in the case of

[3]Red tape is the term used to define the bureaucratic process that companies need to fulfil including documentation of the compliance with the criteria to adopt the label.

ecolabels. Besides the generic ISO 14020 series of eco-labeling template standards and the generic ISO 14040 life-cycle assessment standards, there are very few international labels. Ecolabels are generally developed based on national environmental priorities and preferences.

- Since developing countries are standard takers, some in the trading community (e.g.: developing country exporters) argue that a proliferation of ecolabels can greatly increase the cost to these countries of accessing different markets.

  They argue that since most ecolabels are developed by non-governmental bodies outside the traditional standards networks, it is likely that practitioners are unaware of the procedurals provisions. Thus, it is often difficult for producers in one country to obtain information on the existence or specific requirements of an ecolabel in another country.

  Besides these points, they mention that it is generally accepted that conditions in developing countries are such that the certification costs are higher than in developed countries. This is mainly due to the lack of availability of domestic certification services, the size of the facilities and the gap between existing practices and the requirements of the ecolabel.
- Many ecolabels maintain a monopoly over the accreditation of conformity assessment service providers and therefore do not enter into mutual recognition agreements with other competent bodies, e.g.: FSC auditors are also forbidden from certifying to any other sustainable forest management standards.

Finally, the lack of data makes it impossible to quantify barriers to market access arising from environmental requirements. Ecolabels may impose additional burdens on companies from developing countries, but they do not necessarily impose a greater burden than any other kind of standard. Nevertheless, if ecolabel's requirements are not designed with a clear understanding of the domestic environmental, social and economic context of the developing country, adoption of the ecolabel could impose inappropriate requirements.

Box 5.4 Impact of EU Ecodesign and Energy Labeling on R&D and Technological Innovation

A study ordered by the European Commission in 2014 (Braungardt et al. 2014), evaluating the impact of the EU Ecodesign and the EU "Energy Label" on R&D and technological innovation, found that the directives seemed to have stimulated innovation in some of the studied sectors. They identify a list of factors contributing to the "innovation friendliness" of policy instruments, they find that the Ecodesign and Labeling directives fulfil a number of these criteria.

Using patent data, they show that the Ecodesign directive did typically not have a significant effect on the patenting activities of the affected companies. They argue that the firms already had the necessary technologies to meet the directive's requirements but they lacked the incentives to bring them to the

(continued)

Box 5.4 (continued)
market. Thus, they recognize the role of the directives in the promotion of the diffusion of high-efficiency technologies.

In their case-study analysis, they observe that for the consumer market, information-related barriers to the adoption of energy efficiency innovation are predominant and are adequately addressed by the labeling legislation for the high-efficiency segment and by the Ecodesign directive for the low-cost segment.

Policy Recommendations

As a result of their analysis, they developed a list of policy recommendations for policymakers in order to enhance the positive impact of regulations on R&D and innovation. These could be divided into six categories:

Increasing stringency of regulatory requirements: focus on engaging innovative manufacturers, including a stage in the "Methodology for the Ecodesign of Energy-related Products" (MEERP) to investigate innovation (best not available technology).

Market surveillance and control: long-term impact of the regulation can only be assured if the legislation is enforced.

Recasting of the labeling classes: stakeholders highlighted that incentives to innovate are limited when the top of the classes are reached too early.

Sector specific innovation dynamics: the innovation dynamics might vary from sector to sector. Therefore, in order to enhance the impact of regulation on innovation, this effect should be taken into account.

Consumer response to Labeling: the impact of labeling on consumers' decision varies between different products, sales structure and member states. In order to enhance the impact of labeling, these kind of effects need to be taken into account.

Complementary measures: they recommended to use green public procurement to identify the best performing class of products and thereby incentivize energy efficient innovations.

5.3.3 Eco-Innovation Gap: New Insights from Behavioral Economics

The emergence of a new field studying reaction of human behavior to different incentives is helping to shed some light on key issues related with EPS. Behavioral economics can provide a new glance to understand the gap between provided information and the way consumers might react to it. This is key for EPS since the information on the environmental attributes is mainly proportioned through a label. Thereby the way consumers will react to a certain label will depend mainly on *what* kind of information is provided (i.e.: sustainability information) and *how* the information is provided. Behavioral economics can help enhancing the design of labels such that consumers react purchasing the most sustainable good. Hereby I summarize key insights from a literature review by Gerarden et al. (2017) as well as some main experimental results from an OECD report (OECD 2017).

5.3.3.1 Impact of Cognitive Biases on Information Perception

A major problem in the design of labels is to take into account how the consumer might react to the different types of information provided. Namely, if he will finally purchase the most sustainable product he is willing to pay for. At this stage different perception biases might prevent him from doing the right purchase. Indeed there might be some **imperfect information** problems happening, i.e. consumers might not be provided the right information on the potential economic savings of their energy-efficient purchase. Consumers might also be **myopic**, i.e. they might under-value discount rates on their energy-efficient investments thus revealing them their discount rates might help them do more sustainable purchases. Furthermore, consumers might have **cognitive limitations,** that is, if they are exposed to a variety of complex information they might struggle to disentangle the right one for doing the most sustainable purchase (e.g.: energy metrics are often hard to interpret in terms of economic savings). Finally, consumers might have **loss aversion**, i.e. they might react differently depending on how the message is formulated (e.g.: people strongly prefer avoiding losses rather than acquiring gains). Further references can be found in Table 5.1. From these different examples we can see that the design of the label can be a key factor in the orientation of the consumer towards the most sustainable choice.

5.3.3.2 Insights from Experimental Evidence

A recent study from the OECD (2017) gathered the results of some behavioral experiments on EPS (an extended summary can be found in Tables 5.2 and 5.3). Some of the main findings of these studies are that consumers do not give attention to the actual sustainability quantifier (e.g.: energy consumption) but rather to a provided label (e.g.: energy efficiency letter A, B, etc.). Furthermore, different energy label designs were compared to the actual design of the EU "Energy Label", they find that all alternative designs actually outperformed the current label in physical stores. Thus showing that there is still room for improvement of it. Finally, information of product sustainability was compared among different food products. The results showed that in foods products more attention was given to price and nutritional information rather than to sustainability information.

Table 5.1 Overview of behavioral biases

Problem	Effect	Solution	Reference
Imperfect information: Consumers might not be provided with enough information on a product, or they may not pay attention to the available information or have difficulties using it.	E1. Leads to significant undervaluation of energy efficiency by the consumers E2. Firms as consumers of energy-efficiency technologies may underinvest in profitable energy-efficiency technologies E3. Providing information to consumers may lead some consumers to consume more energy. If they are informed about their own and their neighbors' energy consumption, those who are consuming below the average tend to consume more energy (Schultz et al. 2007).	S1. Providing simple information on the economic value of saving energy leads to an increase in cost-effective energy-efficiency decisions. S2. Presenting a cost and benefits analysis and additional information on projects. S3. Designing the information provision carefully, providing peer comparisons, and changing reference points. S4. Informed third parties (governments and private labeling programs) can fill the information gap.	E1-S1. Allcott et al. (2015) E1-S1. Newell and Siikamäki (2014) E2-S2. Anderson and Newell (2004) 2.Bloom et al. (2013) E3-S3. Schultz et al. (2007) E3-S3. Allcott and Sweeney (2014) E3-S3. Allcott (2011) E3-S3. Ayres et al. (2013) E3-S3. Allcott and Rogers (2014) S4. Davis and Metcalf (2014) (effects of state-specific Energy Guide labels) S4. Sallee (2014) (effects of coarse energy-efficiency certifications) S4. Houde (2018) (positive effects of certification program) S4. Houde (2017) (crowding out effect of certification) S4. Eichholtz et al. (2010, 2013), Brounen and Kok (2011), Kahn and Kok (2014), Wallls et al. (2013)
Myopia: Consumers tend to minimize their costs but there might be an inconsistency between cost-minimizing behavior and the discount rates that consumers use. They may consider the upfront investment costs, and not be aware of or not pay attention to operating costs.	E4. Consumers may undervalue discount rates and energy efficiency. E5. Consumers have different individual discount rates and individual time preferences.	S5. Revealing the discount rates helps consumers with making their decisions rationally on average.	E4. Allcott and Wozny (2014) (fuel economy) E5. Newell and Siikamäki (2015) E5. Bradford et al. (2014) S5. Newell and Siikamäki (2014)

(continued)

Table 5.1 (continued)

Problem	Effect	Solution	Reference
Cognitive limitations: Heuristics and bounded rationality.	E6. Heuristics and bounded rationality prevents consumers from analyzing benefits and costs of investing in energy-consuming goods. E7. Consumers misperceive the information provided by fuel economy ratings (the MPG illusion). E8. Consumers' preferences for cars depend on the metric and scale of information on energy labels.	S6. Consumers' preferences are improved by redesigning the energy labels by adjusting the scale of energy labels based on expected lifetime, providing multiple translations of energy-efficiency metrics, comparing products, or providing environmental ratings.	E6. Gillingham et al. (2009) E6. Sanstad and Howarth (1994) E7. Larrick and Soll (2008) E7. Allcott (2013) E8-S6 .Camilleri and Larrick (2014) E8-S6. Ungemach et al. (2017) E8-S6. US Environmental Protection Agency (2015)
Loss aversion and reference points: People strongly prefer avoiding losses to acquiring gains.	E9. Consumers and firms investing in energy-efficiency are sensitive to reference points and loss aversion. E10. Every consumer has different opinions on the right level of energy efficiency (heterogeneity problem).	S7. Encouraging goal-setting programs makes consumers reduce their energy consumption to meet their own goal. S8. Different types of information on energy labels may affect individuals' reference points.	E9-S7. Harding and Hsiaw (2014) E9-S7. Abrahamse et al. (2007) E9-S7. Carrico and Riemer (2011) E10-S8. No reference on this point. It is stated that existing research does not provide enough evidence on it

5.4 What Can Policy Makers Do?

In this policy outlook I aimed at providing a broad and depth picture of the landscape in which EPS and eco-innovation interact. Even if there are some clear messages that can be taken from this outlook, I would like to emphasize that often lessons can only be drawn on a case-by-case basis. Thus the main lesson to be taken is the need for quasi-experimental policy evaluation. Besides this main point I would like to list some additional messages:

- Consumers need to be better informed to be able to disentangle between Type I and Type II labels. Furthermore information about "Self-declared environmental claims" (Type II labels) having legal implications needs to be clearly communicated to consumers.

Table 5.2 ***Behavioral Lever:*** Simplification and framing. ***Environmental Objective (A & B):*** A: Promote private investment in more efficient technologies; B: Incentivize environmentally sustainable consumption patterns. ***Behavioral Issues (C, D, E & F):*** C: Attitude-behavior gap; status-quo bias; myopic preferences. A relatively small number of purchases of energy efficient appliances underline a discrepancy between consumers' stated intentions to reduce expenditures on energy and their behavior at the moment of the purchase, where energy efficiency is only one among various product attributes under scrutiny

		Env. Obj.	Beh. Iss.	Ref.
Energy Efficiency Labeling and Consumer Behavior	One of the main factors that affects the impact of energy labels is their design and consumers' comprehension of the information provided. The experiment supported by the European Commission on televisions, washing machines and light bulbs with different energy efficiency labels reveals that letter-based scales are better understood by consumers than numerical scales. They also find that, among letter-base labels, consumers tend to choose products with labels scaled from A to D, rather than those ranged from A+++ to D.	A	C	London Economics and IPSOS (2014)
Drivers of the purchase of energy efficient durables	Consumers make their purchasing decisions considering upfront capital costs and operating costs. However, it is not easy to calculate the accurate discount rate and this creates an energy efficiency gap. The study conducted in Switzerland aimed to observe whether and to what extend there is an energy efficiency gap using two different labels: the EU energy label and a new monetary costs and lifetime-oriented label displaying the information on annual electricity costs and lifetime energy costs of a product. Their sample includes freezers, vacuum cleaners, tumble dryers and televisions. Both labels increase the share of energy efficient products sold. In terms of reduction in average annual energy consumption, the new label leads to a higher decrease (9.6%) in tumble dryers' energy consumption compared to the EU energy label (8%). Both labels are inefficient in increasing the sales of energy efficient freezers. Besides this, the new label is less effective for products with low annual energy costs (vacuum cleaners). Since those products already consume low	A	C	Schubert and Stadelmann (2016)

(continued)

Table 5.2 (continued)

		Env. Obj.	Beh. Iss.	Ref.
	level of energy, the improvement in their energy efficiency displayed by the new label can be neglected by the consumers.			
Energy Efficiency Labeling for Online Retail	The aim of this study is to measure what is the most effective way to provide information on energy efficiency labels to increase sales of energy efficient products on online retail platforms. They use four labels with different designs in different appliances: refrigerators, televisions, washing machines and light bulbs. They found out that Label 3 is the most effective one and that even the least effective label is better than the no information scenario to increase online sales. In physical stores, all four labels outperformed the standard EU energy label. 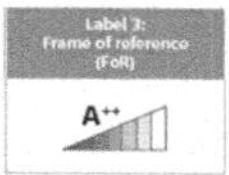	A	C	ECORYS et al. (2014)
Detection of an Energy Efficiency Fallacy	In this experiment they aim at analysing how consumers interpret information provided on energy labels. They observed that consumers tend to estimate rather lower energy consumption for high energy efficiency labels (A-label) than for low energy efficiency labels (B-label) although both labels state the same level of energy consumption. In other words, consumers make their decision based on energy efficiency information rather than on annual electricity consumption information.	A	C	Wächter et al. (2015, 2016)

- Policymakers should analyze the consequences of the multiplication of EPS and the possible loss in trustfulness that this might induce in already existing labels. Policymakers can play a key role for building consumer's trust on EPS.
- Providing consumers with sectoral information on the different types of labels available, the type of environmental quality that they award and their scope would help consumers to do a more rational choice according to their preferences.

Table 5.3 ***Behavioral Lever:*** Simplification and framing. ***Environmental Objective (A & B):*** A: Promote private investment in more efficient technologies; B: Incentivize environmentally sustainable consumption patterns. ***Behavioral Issues (C, D, E & F):*** C: Attitude-behavior gap; status-quo bias; myopic preferences. A relatively small number of purchases of energy efficient appliances underline a discrepancy between consumers' stated intentions to reduce expenditures on energy and their behavior at the moment of the purchase, where energy efficiency is only one among various product attributes under scrutiny

		Env. Obj.	Beh. Iss.	Ref.
Understanding Consumer Perception of Energy Labels	While designing a label it is important to measure how consumers react to the information provided on it. This experiment emphasizes the importance of the effect of the EU Energy Label on consumers' purchasing decision. According to the result, consumers pay more attention on energy-related information if they are provided with energy labels. Energy efficiency ratings are processed faster via labels. However, the current format of the label makes consumers pay more attention on energy-efficiency information rather than the actual energy consumption. Therefore, it might increase the energy consumption as high energy efficiency does not directly imply low energy consumption.	A	D	Wächter et al. (2015, 2016)
Environmental Footprint Labeling and Consumer Behavior	While designing a label it is important to provide informative and simple information. In this study, consumers are exposed to two different designs of labels, environmental footprint label and carbon footprint label, while their willingness to pay for environmentally friendly products is measured. According to the results, consumers are willing to pay more for environmentally friendly products (washing machines and televisions; not light bulbs) if these are provided with labels regardless of the design. Besides this, they also found out that if consumers are exposed to explanatory information on the label, their understanding of the label is higher and they tend to purchase the better performing product. In addition, consumers focus more on standard indicators than environmental or	A + B	C	Ipsos MORI et al. (2012)

(continued)

Table 5.3 (continued)

		Env. Obj.	Beh. Iss.	Ref.
	carbon footprint indicators. In other words, they pay more attention on the simple information.			
Testing CO2 / Car Labeling Options and Consumer Information	The objective of this study is to observe the effectiveness of different car eco-labels and that of mandatory information on fuel efficiency and CO2 emissions. They found out that it is more easily processed if a car is rated compared with cars from all classes (absolute format) rather than if it is rated compared with cars in the same class (relative format) or than the combination of both formats. Besides, it is revealed that providing information on fuel economy is more effective than providing information on CO2 emissions.	A	E	Codagnone et al. (2013)
Consumer Use of Sustainability Information	In this study it is aimed to observe consumers' reactions to sustainability-related information on the food they purchase. According to the results, consumers pay more attention on information about price, nutritional value and origin of raw materials than sustainability information and they think that sustainability is less important than the other attributes. However, they also found out that consumers pay more attention to sustainability information than information on the other components in the future.	B	F	ECORYS et al. (2015)

- In the design or modification of future EPS, it is important to account for firms' strategic behavior (e.g.: bunching effects, price premiums, etc.). Furthermore it is important to make sure that the sustainability index allowing to differentiate between goods is really aligned with environmental goals (e.g.: avoid that consumers might purchase a new fridge attributed a higher energy efficiency rate than their previous one but actually consuming more energy).
- Since there are some case studies indicating that some EPS have rather enhanced the diffusion of already existing innovations rather than incentivized the creation of new innovations, this should be taken into account for the design of future EPS.

- In this outlook I have made the difference between so-called *voluntary* and *mandatory* labels. Nevertheless, the mandatory character of EPS might be misleading since without proper market surveillance mechanisms it is impossible to enforce *mandatory* labels.
- From a more global perspective there is still a need to understand the different implications of EPS in developing and developed countries. Thereby, it is important to understand to which extent these do not represent a barrier to access markets for certain countries.
- From the evaluation of two voluntary EPS we have learnt that implementation costs together with indirect adoption costs (i.e.: human capital training) are the main barriers for the EPS's adoption. Therefore, providing support for the implementation might be a good policy to foster adoption of environmental labels.
- New insights from behavioral economics have taught us highly valuable lessons on not only *how* to design new labels but also on *what* kind of information we should provide through them. Future policymakers should take into account cognitive biases and consumer myopia in their design of new labels.

5.5 Summary

Environmental product standards (EPS) certifying environmental product attributes are key for fostering sustainable consumption, which is an essential measure for achieving the Sustainable Development Goals (SDGs) adopted by the United Nations. EPS, also called environmental labels or eco-labels, are intended to describe environmental features of consumer goods and raise consumers' awareness about sustainability. By fostering sustainable consumption they can become one of the main policy instruments for tackling climate change. They can be mandatory, where the provision of information is compulsory, or voluntary. In both cases EPS aim at correcting the information asymmetry between consumers and providers. Evidence shows that demand-pull is a decisive factor for firms' to voluntarily provide environmental quality. Thus, by enhancing consumers' awareness, it can spur eco-innovation. Nevertheless, EPS have also raised some concerns about barriers to trade and "greenwashing". Furthermore, the recent multiplication of EPS has fostered a label competition, confusing prospective consumers, and thus endangering potential sustainability benefits resulting from EPS. The aim of this chapter is to provide policymakers with an overview on how EPS can support eco-innovation. For this purpose I first describe the different types of labels and review evidence on the different impacts of EPS. Later on, I analyze drivers, benefits and barriers of adoption of EPS and their relation to eco-innovation and environmental performance. Finally, I provide an overview on new behavioral insights to EPS.

References

Abrahamse, W., Steg, L., Vlek, C., & Rothengatter, T. (2007). The effect of tailored information, goal setting, and tailored feedback on household energy use, energy-related behaviors, and behavioral antecedents. *Journal of Environmental Psychology, 27*(4), 265–276.

Allcott, H. (2011). Social norms and energy conservation. *Journal of Public Economics, 95*(9–10), 1082–1095.

Allcott, H. (2013). The welfare effects of misperceived product costs: Data and calibrations from the automobile market. *American Economic Journal: Economic Policy, 5*(3), 30–66.

Allcott, H., Knittel, C., & Taubinsky, D. (2015). Tagging and targeting of energy efficiency subsidies. *American Economic Review, 105*(5), 187–191.

Allcott, H., & Rogers, T. (2014). The short-run and long-run effects of behavioral interventions: Experimental evidence from energy conservation. *American Economic Review, 104*(10), 3003–3037.

Allcott, H., & Sweeney, R. (2014). The role of sales agents in information disclosure: Evidence from a field experiment. *Management Science, 63*(1), 21–39.

Allcott, H., & Wozny, N. (2014). Gasoline prices, fuel economy, and the energy paradox. *Review of Economics and Statistics, 96*(5), 779–795.

Anderson, S. T., & Newell, R. G. (2004). Information programs for technology adoption: The case of energy-efficiency audits. *Resource and Energy Economics, 26*(1), 27–50.

Arditi, S, Meli, L., & Toulouse, E. (2013). Revising EU energy label: Evolution or revolution?. *ECEEE summer study proceedings. ECEEE 2013 summer study – Rethink, Review, Restart.*

Asensio, O. I., & Delmas, M. A. (2017). The effectiveness of U.S. energy efficiency building labels. *Nature Energy, 2*(4), 17033.

Ayres, I., Raseman, S., & Shih, A. (2013). Evidence from two large field experiments that peer comparison feedback can reduce residential energy usage. *Journal of Law, Economics, and Organization, 29*(5), 992–1022.

Bird, L., & Sumner, J. (2010). *Green power marketing in the United States: A status report* (Technical Report NREL/TP-6A20-4903). Washington, DC: National Renewable Energy Laboratory.

Bjørner, T. B., Hansen, L. G., & Russell, C. S. (2004). Environmental labeling and consumers' choice—an empirical analysis of the effect of the Nordic Swan. *Journal of Environmental Economics and Management, 47*(3), 411–434.

Blackman, A., & Naranjo, M. A. (2012). Does eco-certification have environmental benefits? Organic coffee in Costa Rica. *Ecological Economics, 83*, 58–66.

Bloom, N., Eifert, B., Mahajan, A., McKenzie, D., & Roberts, J. (2013). Does management matter? Evidence from India. *Quarterly Journal of Economics, 128*(1), 1–51.

Borchers, A. M., Duke, J. M., & Parsons, G. R. (2007). Does willingness to pay for green energy differ by source? *Energy Policy, 35*(6), 3327–3334.

Bradford, D., Courtemanche, C., Heutel, G., McAlvanah, P., & Ruhm, C. (2014). *Time preferences and consumer behavior* (National Bureau of Economic Research Working Paper 20320).

Braungardt, S., Molenbroek, E., Smith, M., Williams, R., Attali, S., & McAlister, C. (2014). *Impact of eco-design and energy/tyre labelling on R&D and technological innovation* (ECOFYS, Project Number: DESNL13606).

Brounen, D., & Kok, N. (2011). On the Economics of Energy Labels in the Housing Market. *Journal of Environmental Economics and Management, 62*(2), 166–179.

Brown, R., Webber, C., & Koomey, J. (2002). Status and future directions of the ENERGY STAR program. *Energy, 27*(5), 505–520.

Camilleri, A. R., & Larrick, R. P. (2014). Metric and scale design as choice architecture tools. *Journal of Public Policy and Marketing, 33*(1), 108–125.

Carlson, A., & Palmer, C. (2016). A qualitative meta-synthesis of the benefits of eco-labeling in developing countries. *Ecological Economics, 127*, 129–145.

Carrico, A. R., & Riemer, M. (2011). Motivating energy conservation in the workplace: An evaluation of the use of group-level feedback and peer education. *Journal of Environmental Psychology, 31*(1), 1–13.

Clark, C. F., Kotchen, M. J., & Moore, M. R. (2003). Internal and external influences on pro-environmental behavior: Participation in a green electricity program. *Journal of Environmental Psychology, 23*(3), 237–246.

CLASP., et al. (2017). *Closing the "reality gap" – Ensuring a fair energy label for consumers.*

Client Earth. (2011). *Environmental claims on supermarket seafood: Improving product labelling & consumer protection*. London: ClientEarth.

Codagnone, C., Bogliacino, F., & Veltri, G. (2013). *Testing CO2/Car labelling options and consumer information*. Report for the European Commission, Brussels.

Cohen, M. A., & Vandenbergh, M. P. (2012). The potential role of carbon labeling in a green economy. *Energy Economics, 34*, S53–S63.

COSA. (2013). The COSA measuring sustainability report: Coffee and Cocoa in 12 countries. In *The Committee on Sustainability Assessment*. Philadelphia, PA.

Davis, L. W., & Metcalf, G. E. (2014). *Does better information lead to better choices? Evidence from energy-efficiency labels* (National Bureau of Economic Research Working Paper 20720).

Demirel, P., & Kesidou, E. (2012). *Stimulating different types of eco-innovation in the UK: Government policies and firm motivations* (STPS Working Papers 1203). Ankara, Turkey: STPS—Science and Technology Policy Studies Center, Middle East Technical University.

ECORYS, Tilburg University, & GfK. (2014). *Study on the effects on consumer behaviour of online sustainability information displays*. Report for the European Commission, Brussels. https://doi.org/10.2759/52063

ECORYS, Tilburg University, & GfK. (2015). *Milan BExpo 2015: A behavioural study on food choices and eating habits*. Report for the European Commission, Brussels. https://doi.org/10.2838/537411

ECOS. (2017). The revised energy labelling regulation. Brussels.

Eichholtz, P., Kok, N., & Quigley, J. M. (2010). Doing well by doing good? Green office buildings. *American Economic Review, 100*(5), 2492–2509.

Eichholtz, P., Kok, N., & Quigley, J. M. (2013). The economics of green building. *Review of Economics and Statistics, 95*(1), 50–63.

EVER. (2005). *Evaluation of EMAS and eco-label for their revision* (Report 1: Options and recommendations for the revision process. 26.12.2005 Part B: The EU Eco-label).

Fuerst, F., & McAllister, P. (2011). Eco-labeling in commercial office markets: Do LEED and energy star offices obtain multiple premiums? *Ecological Economics, 70*(6), 1220–1230.

Galarraga, I. (2002). The use of eco-labels: A review of the literature. *Environmental Policy and Governance, 12*(6), 316–331.

Gerarden, T. D., Newell, R. G., & Stavins, R. N. (2017). Assessing the energy-efficiency gap. *Journal of Economic Literature, 55*(4), 1486–1525.

Gillingham, K., Newell, R. G., & Palmer, K. (2009). Energy efficiency economics and policy. *Annual Review of Resource Economics, 1*, 597–620.

Gruère, G. (2013). *A characterisation of environmental labelling and information schemes* (OECD environment working papers, No. 62). Paris: OECD Publishing

Harding, M., & Hsiaw, A. (2014). Goal setting and energy conservation. *Journal of Economic Behavior and Organization, 107*(Part A), 209–227.

Horbach, J., Rammer, C., & Rennings, K. (2012). Determinants of eco-innovations by type of environmental impact—The role of regulatory push/pull, technology push and market pull. *Ecological Economics, 78*, 112–122.

Houde, S. (2017). Bunching with the stars: How firms respond to environmental certification. *American Economic Journal: Economic Policy* (Revise and Resubmit).

Houde, S. (2018). How consumers respond to product certification and the value of energy information. *The Rand Journal of Economics, 49*, 453–477. https://doi.org/10.1111/1756-2171.12231.

Ipsos MORI, London Economics, & AEA. (2012). *Research on EU product label options*. Report for the European Commission, Brussels.

Iraldo, F., & Barberio, M. (2017). Drivers, barriers and benefits of the EU ecolabel in European companies' perception. *Sustainability, 9*(5), 751.

ISO. (1999). *ISO 14021. Environmental labels and declarations – Self-declared environmental claims (Type II environmental labelling)*. Geneva: International Organization for Standardization.

ISO. (2000). *ISO 14020. Environmental labels and declarations – General principles*. Geneva: International Organization for Standardization.

ISO. (2006). *ISO 14025. Environmental labels and declarations – Type III environmental declarations – Principles and procedures*. Geneva: International Organization for Standardization.

Jacobsen, G., Kotchen, M. J., & Vandenbergh, M. P. (2012). The behavioral response to voluntary provision of an environmental public good: Evidence from residential electricity demand. *European Economic Review, 56*(5), 946–960.

Kahn, M. E., & Kok, N. (2014). The capitalization of green labels in the California housing market. *Regional Science and Urban Economics, 47*, 25–34.

Kjeldsen, U. B., Wied, M., Lange, P., Tofteng, M., & Lindgaard, K. (2014). *The Nordic Swan and companies. It is worthwhile to acquire the swan label?* Copenhagen, Denmark: Nordic Council of Ministers.

Klintman, M. (2016). *A review of public policies relating to the use of Environmental Labelling and Information Schemes (ELIS)* (OECD Environment Working Papers, No. 105). Paris: OECD Publishing.

Kortelainen, M., Raychaudhuri, J., & Roussillon, B. (2016). Effects of carbon reduction labels: Evidence from scanner data. *Economic Inquiry, 54*(2), 1167–1187.

Kotchen, M. J., & Moore, M. R. (2007). Private provision of environmental public goods: Household participation in green-electricity programs. *Journal of Environmental Economics and Management, 53*(1), 1–16.

Larrick, R. P., & Soll, J. B. (2008). The MPG Illusion. *Science, 320*(5883), 1593–1594.

London Economics, & IPSOS. (2014). *Study on the impact of the energy label and potential changes to it – On consumer understanding and on purchase decisions*. Report for the European Commission, Brussels.

Michaud, C., Llerena, D., & Joly, I. (2013). Willingness to pay for environmental attributes of non-food agricultural products: A real choice experiment. *European Review of Agricultural Economics, 40*(2), 313–329.

Molenbroek, E., Smith, M., Groenenberg, H., Waide, P., Attali, S., Fischer, C., et al. (2014). *Final technical report: Evaluation of the energy labelling directive and specific aspects of the eco-design directive* (Project Number: BUINL13345), ECOFYS.

Murray, A. G., & Mills, B. F. (2011). Read the label! energy star appliance label awareness and uptake among U.S. consumers. *Energy Economics, 33*(6), 1103–1110.

Newell, R. G., & Siikamäki, J. (2014). Nudging energy efficiency behavior: The role of information labels. *Journal of the Association of Environmental and Resource Economists, 1*(4), 555–598.

Newell, R. G., & Siikamäki, J. (2015). Individual time preferences and energy efficiency. *American Economic Review, 105*(5), 196–200.

OECD. (2011). *Fisheries and aquaculture certification*. Paris: OECD Publishing. https://doi.org/10.1787/9789264119680-en

OECD. (2017). *Tackling environmental problems with the help of behavioral insights*. Paris: OECD Publishing. https://doi.org/10.1787/9789264273887-en

Ottman, J. (2011). *The new rules of green marketing*. San Francisco: Barrett-Koehler.

Park, J. Y. (2017). Is there a price premium for energy efficiency labels? Evidence from the Introduction of a Label in Korea. *Energy Economics, 62*, 240–247.

Potts, J., Lynch, M., Wilkings, A., Huppé, G. A., Cunningham, M., & Voora, V. A. (2014). *The state of sustainability initiatives review 2014: Standards and the green economy*. Winnipeg, MB: International Institute for Sustainable Development.

Prag, A., Lyon, T., & Russillo, A. (2016). Multiplication of Environmental Labelling and Information Schemes (ELIS): Implications for environment and trade (OECD Environment Working Papers, No. 106). Paris: OECD Publishing.

RESOLVE. (2012). *Steering committee of the state-of-knowledge assessment of standards and certification. Toward sustainability: The roles and limitations of certification*. Washington, DC: RESOLVE.

Rubik, F. (2015). Life cycle management: Labelling, declarations and certifications at the product level–different approaches. In *Life cycle management* (pp. 65–77). Dordrecht: Springer.

Sallee, J. M. (2014). Rational inattention and energy efficiency. *Journal of Law and Economics, 57*(3), 781–820.

Sanstad, A. H., & Howarth, R. B. (1994). Consumer rationality and energy efficiency. *Proceedings of the ACEEE, 1*, 175–183.

Schubert, R. & Stadelmann, M. (2016). *Energy-using durables: Driving forces of purchase decisions*. Report for the Swiss Federal Office of Energy, Bern.

Schultz, P. W., Nolan, J. M., Cialdini, R. B., Goldstein, N. J., & Griskevicius, V. (2007). The constructive, destructive, and reconstructive power of social norms. *Psychological Science, 18*(5), 429–434.

Shewmake, S., & Viscusi, W. K. (2015). Producer and consumer responses to green housing labels. *Economic Inquiry, 53*(1), 681–699.

Teisl, M. F., Roe, B., & Hicks, R. L. (2002). Can eco-labels tune a market? Evidence from dolphin-safe labeling. *Journal of Environmental Economics and Management, 43*(3), 339–359.

U.S. Environmental Protection Agency. (2015). *Learn about the Fuel Economy Label*. https://www.epa.gov/greenvehicles/learn-about-fuel-economy-label

U.S. EPA (Environmental Protection Agency), Office of Inspector General. (2008). *Improvements needed to validate reported ENERGY STAR benefits* (Report No. 09-P-0061). Washington, DC.

UNEP. (2005). *The trade and the environmental effects of ecolabels: Assessment and response*. http://wedocs.unep.org/handle/20.500.11822/8313

Ungemach, C., Camilleri, A. R., Johnson, E. J., Larrick, R. P., & Weber, E. U. (2017). Translated attributes as choice architecture: Aligning objectives and choices through decision signposts. *Management Science, 64*(5), 2445–2459.

Vanclay, J. K., Shortiss, J., Auselbrook, S., Gillespie, A. M., Howell, B. C., Johanni, R., Maher, M. J., Mitchell, K. M., Stewart, M. D., & Yates, J. (2011). Customer response to carbon labelling of groceries. *Journal of Consumer Policy, 34*(1), 153–160.

Wagner, M. (2008). Empirical influence of environmental management on innovation: Evidence from Europe. *Ecological Economics, 66*, 392–402.

Walls, M., Palmer, K., Gerarden, T., & Bak, X. (2013). *Is energy efficiency capitalized into home prices? Evidence from three U.S. cities* (pp. 13–18). Resources for the Future Discussion Paper.

Wächter, S. M., Sütterlin, B., & Siegrist, M. (2015). The misleading effect of energy efficiency information on perceived energy friendliness of electric goods. *Journal of Cleaner Production, 93*, 193–202. https://doi.org/10.1016/j.jclepro.2015.01.011.

Wächter, S. M., Sütterlin, B. & Siegrist, M. (2016). *Investigating energy-friendly consumer behavior: The role of labels, information, and decision-making strategies in the context of energy consumption*. Report for the Swiss Federal Office of Energy, Bern.

Ward, D. O., Clark, C. D., Jensen, K. L., Yen, S. T., & Russell, C. S. (2011). Factors influencing willingness-to-pay for the ENERGY STAR® label. *Energy Policy, 39*(3), 1450–1458.

Wiser, R., Bolinger, M., & Holt, E. (2000). Customer choice and green power marketing: A critical review and analysis of experience to date. *Proceedings of the ACEEE 2000 Summer study on energy efficiency in buildings*.

Chapter 6
Disentangling Technological Innovations: A Micro-Econometric Analysis of their Determinants

Andreas Ziegler

6.1 Introduction

Corporate innovations are an important component in the process of technological change and economic growth. However, innovations lead to positive externalities since the society benefits from them, while the costs have to be predominantly borne by a single firm. As a consequence, these externalities could result in a suboptimal number of innovations, which makes them interesting for policy makers and scholars. The determinants of innovation activities have therefore been econometrically analyzed since appropriate disaggregated firm-level data have been available (e.g. Lunn 1986; Brouwer and Kleinknecht 1996; Galende and de la Fuente 2003; Czarnitzki and Kraft 2004). Furthermore, one specific subset of innovations, namely environmental innovations, has recently been focused on in environmental and industrial economics (e.g. Rehfeld et al. 2007; Carrión-Flores and Innes 2010; Horbach et al. 2012). This type of innovations particularly receives increasing attention from (environmental) policy because it does not only produce the knowledge spillovers of innovations in general, but additionally limits the environmental burden and thus leads to further positive externalities.

Many econometric studies in the field of innovation activities consider R&D measures or (environmental) patents. However, these indicators have several shortcomings (e.g. Brouwer and Kleinknecht 1999). For example, R&D is obviously only an input at the beginning of the total innovation process. In this respect, patents are certainly a better indicator. In contrast to innovation indicators that are built on the basis of surveys, the main advantage of the use of (environmental) patents is that

---

With the permission of the publisher Taylor & Francis, this chapter is a reprint of the corresponding article in *Journal of Environmental Planning and Management, 58*(2), 315–335.

A. Ziegler (✉)
Department of Economics, University of Kassel, Kassel, Germany
e-mail: andreas.ziegler@uni-kassel.de

J. Horbach, C. Reif (eds.), *New Developments in Eco-Innovation Research*, Sustainability and Innovation, https://doi.org/10.1007/978-3-319-93019-0_6

they are more objective and reliable with respect to innovation activities (e.g. Rave et al. 2011). However, (environmental) patents only measure inventions which need not necessarily be translated into new products or processes and thus are also rather inputs of (marketed) innovations. Furthermore, (environmental) patents do not capture all relevant innovation activities such as new production processes. Against this background, output indicators for innovations have been developed since the beginning of the 1990s, for example, in the Community Innovation Surveys (CIS), which are conducted in several European countries.

These indicators particularly refer to technological innovations, i.e. product and process innovations, besides organizational innovations. The latter innovation type is not a direct component in the process of technological change and economic growth although it can support technological innovations. With respect to product and process innovations, several studies show that their distinction is important for the impact on economic performance (e.g. Cabagnols and Le Bas 2002). While product innovations are rather directed towards product differentiation, process innovations are more a means of reducing costs, so that it is likely that the determinants of these technological innovation types differ (e.g. Lunn 1986). As a consequence, the determinants of general product and process innovations are often separately econometrically analyzed (e.g. Flaig and Stadler 1994; Baldwin et al. 2002; Cabagnols and Le Bas 2002; Martínez-Ros and Labeaga 2002). This separate analysis is also transferred to a few econometric studies of environmental technological innovations, which thus distinguish between environmental product and process innovations (e.g. Rehfeld et al. 2007; Wagner 2007, 2008; Rave et al. 2011).

One important stimulus of studies considering the determinants of environmental product and/or process innovations is the empirical testing of the well-known Porter hypothesis (Porter and van der Linde 1995). This hypothesis suggests that environmental regulation can have a positive impact on (environmental and non-environmental) technological innovations and even on economic performance. Against this background, the effect of pollution abatement expenditures or toxic air releases (as environmental regulation indicators) is not only analyzed for specific environmental technological innovations (e.g. Brunnermeier and Cohen 2003; Carrión-Flores and Innes 2010), but also for general technological innovations (e.g. Jaffe and Palmer 1997). However, these studies only refer to the industry level since corresponding firm-level data (e.g. for pollution abatement expenditures) are not available so far.

A further stimulus of empirical studies of determinants of environmental product and/or process innovations is the discussion of non-mandatory approaches in environmental policy (e.g. Khanna 2001). It is, for example, argued that public voluntary programs encouraging proactive environmental management such as 33/50 (which was initiated by the U.S. Environmental Protection Agency) or negotiated agreements between business and government (e.g. Koehler 2007) may be useful supplements to traditional mandatory command and control regulations (e.g. Arora and Cason 1995; Khanna and Damon 1999). Other initiatives refer to environmental management systems (EMS), which can be considered as unilateral agreements by firms. For these initiatives it is argued that they are able to promote both environmental product and process innovations (e.g. Rennings et al. 2006). The most important programs in this respect are ISO 14001 (developed by the International Organization for Standardization) and EMAS (Eco-Management and Audit Scheme, introduced by the European Commission).

Based on a unique firm-level data set from the German manufacturing sector, this paper complements former empirical studies of the determinants of environmental product and process innovations. Its contribution is three-fold: First, our micro-econometric analysis does not only include environmental, but also non-environmental technological innovations, and particularly disentangles environmental and non-environmental product and process innovations. This disentanglement seems to be important since reliable (environmental) policy conclusions require knowledge of the driving factors for environmental technological innovations, but also the corresponding relationship to the driving factors for non-environmental technological innovations (e.g. Rave et al. 2011). Second, our data allow the analysis of both innovating and non-innovating firms. In contrast, most former econometric studies in this field are only based on the subgroup of innovating or even environmentally innovating firms (e.g. Rennings and Rammer 2009; Rave et al. 2011; Belin et al. 2011; Horbach et al. 2012).[1] While these former studies are certainly very valuable for the understanding of driving factors for environmental relative to non-environmental technological innovations, the analysis of the determinants of environmental technological innovations obviously requires the inclusion of the main comparison group, namely non-innovating firms. Third, our disaggregated technological innovation data allow the application of multivariate (binary) probit models. Compared to univariate (binary) probit models, this approach is able to jointly analyze several dependent dummy variables and to estimate correlations coefficients in the underlying latent variables.

The remainder of the paper is organized as follows: The second section reviews the relevant empirical literature and develops a conceptual framework for the determinants of environmental and non-environmental product and process innovations. The third section explains the data and variables of the empirical analysis. The fourth section discusses the econometric approaches and the estimation results, while the final section draws some conclusions.

6.2 Conceptual Framework and Literature Review

6.2.1 Definition of Innovations

For the definition of (environmental or non-environmental) innovations, we (in line with Rehfeld et al. 2007; Wagner 2007, 2008; Horbach 2008; Frondel et al. 2008;

[1]An exception in this respect is the study of Horbach (2008), which also includes both environmental and non-environmental technological innovations. However, the corresponding data do not allow the important disentanglement of (environmental and non-environmental) product and process innovations.

Rennings and Rammer 2009; Belin et al. 2011; Horbach et al. 2012) conceptually refer to the Oslo Manual of OECD and Eurostat (2005). According to this, an innovation is the implementation of a new or significantly improved product or production process, a new marketing method, or a new organizational method. With respect to technological innovations, the distinction between product and process innovations is therefore crucial. A specific product innovation is the introduction of a good that is new (e.g. the introduction of specific digital cameras) or significantly improved with respect to its characteristics or intended uses (e.g. the implementation of GPS navigational systems in cars). This includes improvements in technical specifications, components, materials, incorporated software, user friendliness, or other functional characteristics.

In contrast, a process innovation is the implementation of a new or significantly improved production method (e.g. a new automation equipment on a production line) or delivery method with respect to the logistics of a firm (e.g. a bar-coded or active Radio Frequency Identification goods-tracking system). This includes changes in techniques, equipment, and software. Two common features of product and process innovations are that they must have already been implemented (i.e. a new or significantly improved product must have been introduced to the market and a new or significantly improved process must have been brought into actual use in the operations of a firm, typically within the last 3 years) and that they only need to be new to the firm itself, not necessarily to the market (i.e. products or processes may already have been implemented by other firms). Due to these definitions, we consider the output of the total innovation process and do not use patents as a proxy for product and process innovations as discussed above.

Environmental product and process innovations as specific types of technological innovations consist of new or significantly improved products and production processes, which additionally avoid or reduce environmental pollution (e.g. Rennings and Zwick 2002). According to this, an environmental product innovation means the introduction of a new environmentally friendly product or an environmentally improved product, for example, a product with environmentally less harmful components (e.g. solvent-free paints) or a product with less energy use (e.g. more energy efficient cars or washing machines). An environmental process innovation means the implementation of an environmentally more friendly composition of production processes comprising process integrated measures (e.g. the increase of material or energy efficiency), process-internal recycling (e.g. water recycling), measures with respect to the own energy production (e.g. concerning less carbon dioxide), or end-of-pipe measures (e.g. filters for desulphurization).

According to the Oslo-Manual of OECD and Eurostat (2005), organizational innovations refer to the implementation of new organizational methods in business practices, in the workplace organization, or in external relations of the firm. Examples for organizational innovations in business practices are the recent introduction of management systems for general production or supply operations, such as supply chain management systems, business reengineering, lean production, and quality management systems. In this respect, the recent certification of a quality management system according to ISO 9001, but also the recent certification of an EMS according to ISO 14001 or EMAS can be considered as organizational innovations. Due to the environmental focus, the latter certifications can furthermore specifically be considered as an environmental organizational innovation. As aforementioned, organizational innovations are not a direct component in the process of technological change and economic growth, but can strongly support technological innovations.

6.2.2 *Determinants of Technological Innovations*

Environmental organizational innovations are therefore an important first group of determinants of environmental product and process innovations. However, we do not consider (environmental) organizational innovations which are new organizational methods that have not been used before in the firm. Instead we more generally examine organizational measures that need not have necessarily been implemented recently. The most important environmental organizational measures are certified EMS according to ISO 14001 or EMAS, as aforementioned. The world-wide ISO 14001 initiative is sponsored by the International Organization for Standardization (e.g. Potoski and Prakash 2005a). This organization has established standards that must be adopted for a certification according to ISO 14001, so that a facility must undertake an initial comprehensive review of its environmental practices and systems, formulate and implement an action plan for environmental management, identify internal governance responsibilities for environmental issues, and have a plan to correct environmental problems (e.g. Potoski and Prakash 2005b). Furthermore, the certification requires third-party audits. EMAS has been introduced in 1993 by the European Commission (e.g. Wätzold and Bültmann 2001). The adoption of EMAS requires facilities (besides third-party audits with independent environmental verifiers and registration bodies) to publish an environmental statement.

According to several econometric studies (e.g. Rennings et al. 2006; Rehfeld et al. 2007), certified EMS play an important role for environmental product and process innovations. Rennings and Rammer (2009) argue that EMS represent important internal capabilities for successful environmental technological innovations. Their argument is therefore based on the resource-based view of the firm (e.g. Wernerfelt 1984; Barney 1991), which emphasizes the importance of internal capabilities or resources that are valuable, rare, and difficult to imitate or substitute and therefore fundamental for innovation activities (e.g. Galende and de la Fuente 2003). Due to their environmental focus, it is obvious that EMS are mainly directed towards environmental product and process innovations and less towards non-environmental technological innovations. According to Horbach et al. (2012), EMS help to overcome incomplete information within a firm so that they are particularly important for the implementation of cleaner technologies. Therefore, it can be hypothesized that EMS have stronger impacts on environmental process innovations compared to environmental product innovations.

The discussion so far refers to certified EMS. However, Wagner (2008) notes that in some cases the certification of EMS is only a symbolic gesture, whereas the environmental performance and thus the propensity to environmental technological innovations of the corresponding firm is rather weak. Therefore, it seems to be important to additionally consider further specific environmental organizational measures such as waste disposal or take-back systems of the products or life cycle assessment activities. While these measures are product-related, it seems to be plausible that a life cycle consideration of products also generates information for improving production processes. Furthermore, life cycle assessment activities can also have an impact on non-environmental and not only on environmental technological innovations. Finally, Horbach et al. (2012) argue that not only environmental but also general organizational measures can be relevant for environmental technological innovations.[2] With respect to such general organizational measures, a quality management system, for example, certified according to ISO 9001 can play an important role (e.g. Wagner 2008). Similar to the effect of EMS, however, general organizational measures seem to have stronger impacts on environmental process innovations if they are, for example, targeted at increasing efficiencies in the environmentally more friendly layout of production processes.

Due to their contribution to internal technological capabilities, environmental and general organizational measures can be more generally considered as specific supply or technology push factors, which are widely assumed as important driving forces of technological progress (e.g. Rehfeld et al. 2007).[3] In this respect, the stimulus through technological capabilities seems to be particularly relevant for the first stages within the total innovation process. Against this background, a key factor for technological capabilities and technological opportunities are investments in R&D, which are a major input into the innovation process (e.g. Baldwin et al. 2002). Such investments are certainly important for the development of environmental and non-environmental product and process innovations (e.g. Rave et al. 2011) so that it can be hypothesized that R&D positively affect all types of technological innovations. Furthermore, it can be hypothesized that firms with only one facility have lower technological opportunities for the development of environmental and non-environmental product and process innovations than firms that consist of more than one facility since the latter are able to use intra-firm spillover effects.

Another main group of determinants are demand or market pull factors (e.g. Rennings and Rammer 2009), which emphasize the role of pressure from consumers and competing firms. One environmentally relevant market pull factor refers to activities on environmental markets, i.e. a firm that sells products on such markets has obviously a stronger propensity for the development of new products which avoid or reduce environmental pollution. Similarly, it can be hypothesized that environmental product innovations are stimulated if environmental issues are an

---

[2]This hypothesis is empirically confirmed in the study of Horbach (2008).

[3]For the discussion of technology push factors, particularly in comparison to market pull factors, see also Horbach (2008), Rennings and Rammer (2009), Belin et al. (2011), Rave et al. (2011), Horbach et al. (2012).

important competition factor. In contrast, the competition factors prices, quality, and customers seem to not only be important for environmental, but also for non-environmental technological innovations. While prices and quality can be assumed to positively affect cost-saving production processes and production processes with a higher quality, customers seem to be more relevant for product and particularly environmental product innovations (e.g. Kammerer 2009). An additional general market pull factor is competition intensity, which can be expected to positively affect technological and particularly product innovations (e.g. Cabagnols and Le Bas 2002). In contrast, the relevance of stronger consumer as opposed to industrial customer relationships for technological innovations is a priori ambiguous. A final relevant market pull factor refers to competitive pressure from international markets (e.g. Czarnitzki and Kraft 2004), although it is not clear a priori whether it has a higher impact on environmental or non-environmental product or process innovations.

Finally, several studies discuss the important role of (environmental) regulation, which is obviously able to stimulate environmental technological innovations (e.g. Rehfeld et al. 2007; Horbach 2008; Rennings and Rammer 2009; Belin et al. 2011; Rave et al. 2011; Horbach et al. 2012). However, it should be noted that the effect of regulation can hardly be identified with common econometric analyses, at least on the basis of cross-sectional data. As also mentioned by Kammerer (2009), former empirical studies mostly use compliance with regulation as innovation goal or the perception of regulatory stringency as regulation indicators. However, such variables are only subjective indicators, which need not necessarily be strongly correlated with real stringency of environmental regulations. A main drawback of these former studies is that these regulation indicators are only available for innovating firms so that directly drawing conclusions from their effects is difficult since this analysis requires the inclusion of non-innovating firms as a comparison group, as discussed above. A reliable empirical analysis of the effects of regulation would require firm-level panel data over a long time period or at least indirect indicators such as pollution abatement and control or environmental expenditures (e.g. Brunnermeier and Cohen 2003; Böhringer et al. 2012). Unfortunately, however, such firm-level panel data are not available so far and data on pollution abatement and control or environmental expenditures are only available at the sector level. As a consequence, we do not analyze the effect of regulation, but focus on the role of organizational measures, technology push factors, as well as market pull factors.

6.2.3 Data and Variables

The data for our empirical analysis were collected by means of a telephone survey at the Centre for European Economic Research (ZEW) in Mannheim, Germany, in 2003. The stratified representative sample considering two firm size classes (less than 200 and at least 200 employees), two regions (Western and Eastern Germany), and eleven industries was drawn from the population of all (innovating and non-innovating) German manufacturing firms with 50 or more employees. The interviewees were the responsible production managers (R&D manager, environmental manager, general manager) since previous case studies showed that they were the most competent respondents for this kind of survey. 588 or 24.5% of the 2399 firms that were reached participated in the survey. With respect to the aforementioned two firm size classes, two regions, and eleven industries, it was tested whether the percentages in the sample comply with the percentages in the population. The appropriate two-tailed tests show that the underlying null hypotheses can never be rejected at the 10% level of significance. Therefore, sample selection is not a strong problem. We exclude firms founded in the years 2002 or 2003 because several variables refer to the period between 2001 and 2003.

With respect to the different technological innovation types, the firms were asked whether they were planning to implement environmental and non-environmental product and process innovations by the end of 2005, irrespective of any innovation activity in the past. The inclusion of future technological innovations and thus lagged explanatory variables helps to weaken potential endogeneity problems in our econometric analysis. For the corresponding binary dependent variables, we therefore distinguish between "product innovations" and "process innovations" that take the value one if the firm was planning to implement such innovations (which include both environmental and non-environmental innovations). Furthermore, we distinguish between "environmental technological innovations" and "non-environmental technological innovations" that take the value one if the firm was planning to implement such innovations (which include both product and process innovations). In order to further disentangle these types of technological innovation and consistent with our conceptual framework, we finally distinguish between the dummy variables "environmental product innovations", "environmental process innovations", "non-environmental product innovations", and "non-environmental process innovations".

With respect to organizational measures as the first main group of explanatory variables in the econometric analysis, we particularly consider environmentally oriented organizational measures. According to the conceptual framework, the certification of EMS plays a crucial role in this respect. The corresponding dummy variable "EMS" takes the value one if at least one facility was certified according to ISO 14001 or EMAS. For specific environmental organizational measures we additionally examine the two dummy variables "life cycle" and "disposal", which take the value one if the firm performed environmental life cycle assessment activities and if it carried out measures with respect to waste disposal or redemption of own products, respectively. In line with the discussion in the conceptual framework, we finally consider an indicator for general organizational measures. The corresponding dummy variable "ISO 9001" takes the value one if at least one facility had a certified quality management system according to ISO 9001.

According to the conceptual framework, R&D is a key technology push factors. The corresponding dummy variable "R&D" takes the value one if a firm carried out R&D activities in the previous year. Furthermore, the dummy variable "one facility" takes the value one if the firm consisted of only one facility. With respect to market pull factors, we examine nine explanatory variables in total. The dummy variable "environmental market" takes the value one if a firm sold products on the environmental market. The dummy variables "environment important", "prices important", "quality important", and "customers important" take the value one if the firm stated that environmental issues, prices, quality, and customers, respectively, were an important (and not less or average important) factor in delivering competitive advantages in the last 3 years. With respect to competition intensity, the dummy variable "increased competition" takes the value one if the firm stated that competitive pressure would increase in the next 3 years. As indicator for consumer or industrial customer relationships, we consider "sales consumers", which is the percentage of sales to consumers (divided by 100) relative to the sales to industrial consumers in the previous year. As indicators for competitive pressure from international markets, we examine the two dummy variables "main market abroad" and "exports", which take the value one if the main market was abroad (and not national or regional) in the last 3 years and if the firm exported in the previous year.

Finally, we include some firm-specific control variables such as firm size. Its effect on technological innovations is ambiguous a priori since positive impacts (e.g. due to scale advantages of large firms) and negative impacts (e.g. due to a higher flexibility of small firms) are possible (e.g. Brouwer and Kleinknecht 1996). The corresponding variable "employees" is the natural logarithm of the number of employees. Firm age as a further indicator for organizational resources (e.g. Galende and de la Fuente 2003) can be expected to positively affect technological innovations, although younger firms can also be more innovative in order to increase their market share (e.g. Askildsen et al. 2006). The corresponding variable "age" is measured as the natural logarithm of age in years. Moreover, we incorporate the regional dummy variable "Western Germany", which takes the value one if the firm operated in the Western part ("alte Bundesländer" excluding Berlin) of Germany, as well as eight industry dummy variables. These variables aim at controlling for differences, for example, in competition (e.g. Cabagnols and Le Bas 2002) and particularly (environmental) policy as discussed above.[4]

Table 6.1 reports the descriptive statistics (i.e. the means and standard deviations) for the dependent and explanatory variables in the econometric analysis. In addition, Table 6.2 reports the pairwise (Pearson) correlation coefficients between the main explanatory variables, which refer to organizational measures as well as to technology push and market pull factors. The latter table reveals, for example, relatively strong correlations between single environmental and general organizational measures as well as between "R&D" and the two variables for competitive pressure from international markets. In the following econometric analysis, the number of firms decreases to values between 372 and 386 due to incomplete data for some examined variables.

[4]The estimates of the parameters for the sector dummy variables are not reported for reasons of brevity, although they were included in all econometric models.

Table 6.1 Descriptive statistics

Variables	Number of firms	Mean	Standard deviation
Product innovations	452	0.75	0.43
Process innovations	463	0.80	0.40
Environmental technological innovations	453	0.73	0.44
Non-environmental technological innovations	461	0.81	0.39
Environmental product innovations	447	0.38	0.49
Environmental process innovations	462	0.64	0.48
Non-environmental product innovations	457	0.68	0.47
Non-environmental process innovations	461	0.67	0.47
EMS	467	0.27	0.44
Life cycle	466	0.16	0.36
Disposal	470	0.37	0.48
ISO 9001	475	0.65	0.48
R&D	473	0.75	0.43
One facility	473	0.54	0.50
Environmental market	470	0.16	0.37
Environment important	473	0.21	0.40
Prices important	476	0.86	0.35
Quality important	475	0.86	0.35
Customers important	475	0.90	0.30
Increased competition	473	0.70	0.46
Sales consumers	450	0.39	0.44
Main market abroad	474	0.37	0.48
Exports	474	0.84	0.37
Employees	467	4.92	1.03
Age	476	2.96	1.34
Western Germany	475	0.88	0.32

Note: The statistics refer to firms that were founded previous to the year 2002

6.3 Econometric Analysis

6.3.1 Econometric Approaches

While the determinants of the pairs "product innovations" and "process innovations" as well as "environmental technological innovations" and "non-environmental technological innovations" are separately analyzed jointly in bivariate (binary) probit models (e.g. Flaig and Stadler 1994), the determinants of "environmental product innovations", "environmental process innovations", "non-environmental product innovations", and "non-environmental process innovations" are jointly examined in multivariate (binary) probit models with four equations (see Appendix). Besides parameters of the explanatory variables, these models incorporate correlation coefficients between the two or four dependent dummy variables in the corresponding stochastic components of the underlying latent variables. If these correlations are not considered, biased and inconsistent

Table 6.2 Pairwise (Pearson) correlation coefficients, number of firms varies between 441 and 475

	EMS	Life cycle	Disposal	ISO 9001	R&D	One facility	Environmental market	Environment important	Prices important	Quality important	Customers important	Increased competition	Sales consumers	Main market abroad	Exports
EMS	1														
Life cycle	0.29	1													
Disposal	0.06	0.14	1												
ISO 9001	0.26	0.10	0.03	1											
R&D	0.11	0.13	0.05	0.27	1										
One facility	−0.14	−0.14	−0.06	−0.17	−0.24	1									
Environmental market	0.03	0.13	0.11	0.10	0.12	−0.03	1								
Environment important	0.12	0.06	0.07	0.04	0.06	−0.05	0.10	1							
Prices important	0.01	−0.01	0.06	0.03	−0.08	0.02	−0.07	0.04	1						
Quality important	0.04	0.08	0.06	0.05	0.04	−0.05	−0.00	0.08	−0.09	1					
Customers important	−0.02	0.10	−0.01	−0.00	0.15	−0.06	0.05	0.16	−0.07	0.24	1				
Increased competition	−0.01	0.07	0.05	−0.04	−0.07	−0.04	0.02	−0.07	0.17	−0.06	−0.05	1			
Sales consumers	−0.11	−0.04	0.08	−0.28	−0.05	0.06	−0.06	−0.01	−0.05	−0.05	−0.05	−0.02	1		
Main market abroad	0.09	−0.05	−0.10	0.12	0.28	−0.06	−0.00	0.05	−0.02	−0.02	0.01	−0.01	−0.13	1	
Exports	0.06	0.05	0.01	0.17	0.34	−0.19	0.01	0.04	0.03	0.02	0.05	0.01	−0.11	0.31	1

parameter estimations are possible. While multivariate probit models in the bivariate case are straightforwardly estimated with the maximum likelihood method, we have to apply the simulated counterpart of this method that incorporates the Geweke-Hajivassiliou-Keane (GHK) simulator (Börsch-Supan and Hajivassiliou 1993; Geweke et al. 1994; Keane 1994) for the estimation in the four equation case. In this respect, we use 50 random draws in the GHK simulator. Furthermore, we always consider the robust estimations of the standard deviation of the parameter estimates (e.g. White 1982).

6.3.2 Estimation Results

Table 6.3 reports the estimation results in the two bivariate probit models. The two dependent variables in the first model approach are "product innovations" and "process innovations". In contrast, the two dependent variables in the second model approach are "environmental technological innovations" and "non-environmental technological innovations". Table 6.4 reports the corresponding estimation results in the multivariate probit model with the four dependent variables "environmental product innovations", "environmental process innovations", "non-environmental product innovations", and "non-environmental process innovations".[5] Both tables show that the null hypothesis that no explanatory variable is related with the dependent variables is strongly rejected at extremely low significance levels on the basis of the underlying Wald or simulated Wald tests. Furthermore, the tables underpin the importance of applying multivariate instead of univariate probit models due to the significantly positive correlations in the stochastic components of the underlying latent variables, particularly between environmental and non-environmental process innovations as well as between both types of non-environmental technological innovations.[6]

According to Table 6.3, one main estimation result is the very strong significant correlation of "R&D" with all types of technological innovation, which points to the expected crucial role of this technology push factor. Furthermore, several market pull factors also play an important role. As expected, the second model approach reveals that activities on environmental markets are strongly connected with environmental technological innovations. In addition, increased competitive pressure in the next years seems to be a particular stimulus for product innovations and non-environmental technological innovations. According to the second model approach, firm size is significantly positively related with environmental technological innovations, but not significantly related with non-environmental technological

[5]Both tables report whether a parameter of the explanatory variables is different from zero at the 1%, 5%, or even 10% significance level. However, the following discussion of the estimation results focuses on corresponding correlations at the 1% or 5% significance level since a rejection of the underlying null hypothesis that the parameter is zero at the 10% significance level seems to be insufficient for robust conclusions, particularly if the relatively small sample sizes are considered.

[6]The only exception is the insignificant correlation between environmental process and non-environmental product innovations in the stochastic components.

Table 6.3 Maximum likelihood estimates in bivariate probit models, number of firms = 382 in model 1, number of firms = 386 in model 2

	Dependent variables (model 1)		Dependent variables (model 2)	
Explanatory variables	Product innovations	Process innovations	Environmental technological innovations	Non-environmental technological innovations
EMS	0.02	0.27	0.27	0.05
Life cycle	−0.03	0.82**	0.19	0.54**
Disposal	0.30*	0.14	0.30*	0.13
ISO 9001	0.25	0.19	0.35**	0.25
R&D	1.27***	0.62***	0.46**	0.74***
One facility	−0.22	−0.26	0.14	0.11
Environmental market	0.38	−0.41*	0.66***	0.01
Environment important	0.33	−0.15	0.30	−0.20
Prices important	−0.29	−0.05	0.04	−0.62**
Quality important	−0.13	0.56**	0.14	0.39*
Customers important	−0.00	−0.28	0.31	−0.57*
Increased competition	0.47**	−0.01	−0.32*	0.39**
Sales consumers	−0.21	−0.41**	−0.26	−0.19
Main market abroad	−0.07	−0.11	0.22	−0.00
Exports	0.13	0.14	0.26	0.41*
Employees	0.12	0.17*	0.25***	0.02
Age	−0.08	0.04	0.04	−0.00
Western Germany	−0.52*	−0.00	−0.13	−0.29
Constant	−0.31	−1.02	−2.18***	0.48
Wald test statistics	171.56***		146.61***	
Correlation coefficients of stochastic components	0.38***		0.39***	

Note: *** (**, *) means that the appropriate parameter or (with respect to the Wald test) that at least one parameter of the explanatory variables is different from zero at the 1% (5%, 10%) significance level, respectively

innovations, which suggests that scale advantages of large firms are more relevant for environmental technological innovations. With respect to organizational measures, Table 6.3 reveals that life cycle assessment activities are significantly positively correlated with process innovations and with non-environmental technological innovations, whereas the corresponding relation with environmental technological innovations is insignificant. Similarly, "EMS" is insignificantly related with any technological innovation type (including environmental technological innovations), whereas the certification according to ISO 9001 is significantly positively correlated with environmental technological innovations.

The estimation results in Table 6.3 are certainly a first basis for the understanding of driving factors for the implementation of new or significantly improved environmental

Table 6.4 Simulated maximum likelihood estimates in the multivariate probit model, number of firms = 372

Explanatory variables	Dependent variables: Environmental product innovations	Environmental process innovations	Non-environmental product innovations	Non-environmental process innovations
EMS	0.14	0.36^{**}	−0.11	0.17
Life cycle	0.35^{*}	0.45^{**}	0.03	0.87^{***}
Disposal	0.46^{***}	0.12	0.16	0.14
ISO 9001	0.05	0.43^{**}	0.26	0.05
R&D	0.44^{**}	0.41^{**}	1.22^{***}	0.51^{***}
One facility	0.04	0.05	−0.05	−0.15
Environmental market	0.95^{***}	0.12	−0.06	−0.00
Environment important	0.63^{***}	0.01	−0.18	−0.14
Prices important	0.13	−0.08	−0.32	−0.30
Quality important	−0.00	0.14	−0.12	0.56^{***}
Customers important	0.53^{*}	0.36	−0.08	-0.41^{*}
Increased competition	0.24	−0.21	0.40^{**}	0.19
Sales consumers	-0.41^{**}	−0.27	−0.04	-0.40^{**}
Main market abroad	0.30^{*}	−0.03	−0.13	−0.16
Exports	0.28	0.06	0.22	0.38^{*}
Employees	0.20^{**}	0.14^{*}	0.05	−0.01
Age	−0.03	−0.01	−0.02	0.08
Western Germany	−0.25	−0.13	−0.43	−0.13
Constant	-2.96^{***}	-1.34^{**}	−0.37	−0.36
Simulated Wald test statistic	316.17^{***}			

Simulated correlation coefficients of stochastic components		Environmental process innovations	Non-environmental product innovations	Non-environmental process innovations
	Environmental product innovations	0.34^{***}	0.34^{***}	0.24^{**}
	Environmental process innovations		0.05	0.54^{***}
	Non-environmental product innovations			0.60^{***}

Note: *** (**, *) means that the appropriate parameter or (with respect to the simulated Wald test) that at least one parameter of the explanatory variables is different from zero at the 1% (5%, 10%) significance level, respectively

or non-environmental products or processes. However, our conceptual approach suggests several stimuli of organizational measures as well as technology push and market pull factors for specific environmental or non-environmental product or process innovations. These stimuli cannot be completely captured by bivariate probit models as basis of the estimation results in Table 6.3 since they still comprise relatively aggregated dependent variables. Therefore, a further disentanglement of the dependent variables and thus of the different types of technological innovations is certainly very valuable. As a consequence, we focus on the discussion of the estimation results in Table 6.4. These results are based on multivariate probit models and thus on four technological innovation types that are more disaggregated so that a more precise analysis of the role of organizational measures as well as of technology push and market pull factors is possible.

One important modification of the estimation results refers to the relevance of the certification of EMS according to ISO 14001 or EMAS, which is not significantly correlated with environmental technological innovations according to Table 6.3. In contrast, Table 6.4 reveals a significantly positive relation of "EMS" with environmental process innovations, which is completely consistent with the expectations from the conceptual approach (e.g. Horbach et al. 2012). In line with Wagner (2008), who emphasizes the relevance of environmental organizational measures besides the certification of EMS, life cycle assessment activities are also significantly positively connected with environmental process innovations. Besides these environmental organizational measures, general organizational measures are obviously also an important driving factor for environmental process innovations since the parameter for "ISO 9001" is significantly different from zero. This estimation result suggests that (in line with Horbach 2008; Wagner 2008; Horbach et al. 2012) general organizational measures like the certification of a quality management system according to ISO 9001 are able to additionally support the increase of efficiencies in the environmentally more friendly composition of production processes.

Interestingly, life cycle assessment activities are also significantly positively correlated with non-environmental process innovations. This suggests that (in line with the corresponding estimation result in Table 6.3) these activities are generally a relevant stimulus for process innovations.[7] In contrast, measures with respect to waste disposal or redemption of own products are clearly environmentally and product-related since "disposal" is strongly significantly positively correlated with environmental product innovations. As expected from the conceptual approach, the parameters for "environmental market" and "environment important" are additionally strongly significantly different from zero with respect to environmental product innovations. Therefore, it can be concluded that activities on environmental markets and environmental issues as an important competition factor are crucial market pull factors for the development of new products which avoid or reduce environmental pollution. Another significant market pull factor for environmental product innovations (just as for non-environmental process innovations) refers to "sales consumers". The corresponding negative parameter estimate implies that industrial

[7]The corresponding correlation of "life cycle" with environmental product innovations is only different from zero at the 10% significance level, which seems to be too high to draw robust conclusions from this result, as discussed above.

customer relationships are obviously a more relevant stimulus for technological innovations than consumer relationships. Finally, the hypothesis of a higher flexibility of small firms for technological innovations cannot be confirmed. In contrast, firm size is significantly positively correlated with environmental product innovations, possibly due to scale advantages of large firms.

Overall, Table 6.4 reveals that the market pull factors are only relevant for non-environmental technological innovations in a few cases. In line with the expectations from the conceptual approach, "increased competition" is significantly positively connected with non-environmental product innovations and quality as an important competition factor is significantly positively related with non-environmental process innovations.[8] Furthermore, no single market pull factor is significantly connected with environmental process innovations. In line with Table 6.3, however, "R&D" is most relevant since this technology push factor is strongly positively correlated with all types of technological innovations. R&D activities are therefore obviously a key factor for technological capabilities and technological opportunities and thus not only crucial for non-environmental product innovations, but also for non-environmental process innovations as well as for environmental product and process innovations (even though the correlation with non-environmental technological innovations is slightly stronger).

6.3.3 *Alternative Model Specifications and Robustness Checks*

In order to test the robustness of the estimation results in Table 6.4, several additional model specifications were examined.[9] For example, we experimented with higher or lower numbers of random draws in the GHK simulator for the simulated maximum likelihood estimation of the multivariate probit model with four equations. The corresponding estimation results are qualitatively nearly identical with those in Table 6.4, even with only ten random draws. Moreover, we substituted "R&D" by a dummy variable for the existence of an R&D department (e.g. Brouwer et al. 1999) as a stronger indicator for technological capabilities. However, the positive correlations of this technology push factor with technological innovations are weaker than the corresponding correlations of "R&D" and even not significant with environmental process innovations. Furthermore, we examined the inclusion of an indicator for external financial constraints (e.g. Czarnitzki 2006), measured by a dummy variable for a credit rating index[10] and an indicator for skill structure, measured by the ratio of

[8]The corresponding correlations with environmental product innovations (for "increased competition") or environmental process innovations (for "quality important") are not significant, although the respective parameter estimates are (in line with the corresponding estimation results in Table 6.3) also positive.

[9]The corresponding estimation results are not reported for reasons of brevity, but are available upon request.

[10]This variable is based on evaluations of "Creditreform", the largest German credit rating agency.

the number of salaried employees with a university or college degree relative to the number of all salaried employees.[11] While the credit rating indicator is not significantly related with any technological innovation type, employee skills are significantly positively related with non-environmental product innovations. However, these estimation results should be treated with caution since the corresponding model specifications comprise much lower numbers of observations due to many missing values for these two explanatory variables.

With respect to environmental organizational measures, we also examined two separate dummy variables for the certification according to ISO 14001 or EMAS (e.g. Ziegler and Seijas Nogareda 2009) instead of the joint variable "EMS". In contrast to the significant correlation of "EMS" with environmental process innovations according to Table 6.4, the separate variables are not significantly related with any technological innovation type. This result is possibly due to multicollinearity problems since the two certifications according to ISO 14001 and EMAS are highly correlated, i.e. many firms stated that they either had no EMS certification or a double ISO 14001 and EMAS certification. This strengthens our use of the variable "EMS". However, the estimation results in Table 6.4 could be influenced by additional multicollinearity problems. According to Table 6.2, "EMS" and "life cycle" are also highly positively correlated so that it can be difficult to identify parameters that are significantly different from zero. Therefore, we experimented with the exclusion of "life cycle". In this case, the correlation of the certification of EMS with environmental process innovations is slightly stronger and particularly more robust than the corresponding correlation of the ISO 9001 certification. This estimation result still suggests that general organizational measures are an important stimulus of environmental process innovations, but strengthens the high relevance of environmental organizational measures.[12]

Table 6.2 also reveals very strong correlations between our two variables of competitive pressure from international markets and R&D activities. The corresponding multicollinearity problems could explain the insignificant or at least less robust correlations of "main market abroad" and "exports" with the different types of environmental and non-environmental product and process innovations according to Table 6.4. Therefore, we experimented with the exclusion of "R&D". In this case, "main market abroad" is significantly positively related with environmental product innovations and "exports" is significantly positively related with non-environmental product innovations and non-environmental process innovations. In line with Czarnitzki and Kraft (2004), these results suggest the relevance of competitive pressure from international markets as a market pull factor. However, it should be noted that the exclusion of "R&D" as a main technology push factor can lead to omitted variable biases.

---

[11]In line with, for example, Horbach (2008) and Horbach et al. (2012), the skill structure in a firm can be considered as an additional technology push factor.

[12]The corresponding estimation results additionally reveal a weak correlation of "EMS" with non-environmental process innovations, however, only at the 10% significance level.

Finally, we also analyzed the already implemented environmental and non-environmental product and process innovations in the last 3 years between 2001 and 2003 as dependent variables. In this case, several parameter estimates differ from the corresponding parameter estimates in Table 6.4, which are based on multivariate probit models with planned environmental and non-environmental product and process innovations by the end of 2005 as dependent variables. However, the analysis of implemented technological innovations between 2001 and 2003 should also be treated with caution since the interpretation of causal effects (for a discussion of the causality problem, see also Ziegler and Seijas Nogareda 2009) is ambiguous due to the time structure of the explanatory and dependent variables. Furthermore, it should be mentioned that a significantly positive correlation never switches to a significantly negative correlation. Overall, it should be additionally noted that the different technological innovation types are highly persistent over time due to the extremely high correlations between already implemented and planned environmental and non-environmental product and process innovations, which vary between 0.6 and 0.8 over both 3-year periods.

6.4 Conclusions

Product and process innovations are an important component in the process of technological change and economic growth, which makes them very interesting for scholars and particularly policy makers. In this respect, the subgroup of environmental technological innovations receives specific attention since they are expected to generate a double dividend, i.e. limit the environmental burden and contribute to the technological modernization of the economy. However, the incentives to produce such a double dividend are weak due to market failures. The costs for an environmental technological innovation have to be borne by a single firm, although the whole society benefits from such a measure due to positive spillovers. Even if an environmental product or process innovation can be successfully marketed, it is difficult for a firm to receive the profits of this technological innovation if the corresponding knowledge is easily accessible for imitators and the environmental benefits have a public good character. On the basis of a unique firm-level data set from the German manufacturing sector, this paper therefore empirically examines the determinants of different technological innovation types. In contrast to former studies, it disentangles environmental and non-environmental product and process innovations and includes non-innovating firms as a relevant comparison group.

The micro-econometric analysis with multivariate probit models points to the extremely high relevance of R&D activities as a key factor for technological capabilities and technological opportunities and thus for technological innovations. While this estimation result is not surprising and in line with former empirical studies in industrial and environmental economics (e.g. Baldwin et al. 2002; Rave et al. 2011), our study implies that R&D is not only a crucial stimulus for non-environmental technological innovations, but also for specific environmental product and process innovations. In order to generate the aforementioned double dividend of environmental technological innovations, firm-internal R&D activities should certainly be encouraged by (environmental) policy. Specific possible policy measures refer to the corresponding relief of administrative barriers, the advancement of patent protection, the support of cooperation of public and firm-internal R&D, or the direct financial promotion of R&D activities by subsidies.

Our empirical analysis also points to the relevance of a few market pull factors, particularly for non-environmental product and process innovations, such as the importance of the competition factor quality, a generally high competition intensity, or industrial customer relationships. Such industrial customer relationships as well as activities on environmental markets are obviously also important stimuli for environmental product innovations. This specific type of environmental technological innovations is additionally strongly stimulated if environmental issues are an important competition factor. The estimation result for this market pull factor can be an additional interesting direction for (environmental) policy, which can, for example, promote the consideration of environmental issues (e.g. by specific information and advertisement campaigns) when consumers purchase (particularly durable) goods. In contrast, our empirical analysis implies no significant relevance of any market pull factor for environmental process innovations.

According to our estimation results, this specific type of environmental technological innovations is particularly stimulated by organizational measures. In this respect, the certification of EMS and (consistent with Wagner 2008) specific environmental organizational measures such as life cycle assessment activities (which are also very relevant for non-environmental process innovations) play an important role. Furthermore, specific product-related environmental organizational measures like waste disposal or take-back systems of the products seem to stimulate environmental product innovations. In line with Horbach et al. (2012), our empirical analysis interestingly points to the additional relevance of general organizational measures for environmental process innovations such as the certification of a quality management system according to ISO 9001. With respect to the aforementioned double dividend of environmental technological innovations, our estimation results therefore support the public encouragement of certified EMS, but also of specific environmental and general organizational measures. Possible directions for corresponding policy measures are public information campaigns on the benefits of such organizational measures, the privilege of ISO 14001, EMAS, and ISO 9001 certified firms in the public procurement, or direct subsidies for such organizational measures, which are often (e.g. in the case of an EMS certification) rather expensive.

However, it should be mentioned that our conclusions (like the corresponding conclusions from former empirical studies in this field) for the stimulation of environmental technological innovations are necessarily somewhat preliminary. By showing that, for example, certified EMS could also reversely be affected by environmental product or process innovations, Ziegler and Seijas Nogareda (2009) argue that the causal relationship between environmental organizational measures and environmental technological innovations is ambiguous. If such organizational measures are really more likely to be realized by already environmentally active firms, they would not need separate public support. In this case, corresponding policy measures can even lead to windfall profits for these firms. Reliable conclusions with respect to the causality of the relationship between organizational measures and environmental technological innovations would require the inclusion of valid instruments into the econometric analysis. However, such instruments are not available and could be constructed within panel data analyses at the best. Therefore, the main reason for this causality question still being open is that necessary firm-level panel data over a long time period are not available so far.

As a consequence, the collection of such panel data comprising environmental technological innovations seems to be a main direction for further research in this field. As discussed above, such panel data over a long time period would also allow a reliable empirical analysis of the direct effects of regulation on environmental product and process innovations. Furthermore, these data would enable the reliable control for unobserved firm characteristics as well as the analysis of dynamic effects. The inclusion of corresponding lagged dependent variables, i.e. lagged environmental and non-environmental product and process innovations, would be the basis for the analysis whether or particularly which specific technological innovation types breed which technological innovation types (e.g. Flaig and Stadler 1994). A final direction for further research is an analysis of the determinants of even more disaggregated (environmental and/or non-environmental) product and process innovations. While this paper disentangles environmental and non-environmental product and process innovations at the firm level for the first time, these four technological innovation types can still be considered somewhat aggregated because they consist of different kinds of product and process innovations. For example, our definition of technological innovations comprises new or significantly improved products and processes that could indeed be new for the market, but essentially only have to be new for the firm itself.

Appendix: Multivariate (Binary) Probit Models

We assume that a firm i ($i = 1,\ldots,N$) implements a specific type j ($j = 1,\ldots,J$) of technological innovations if the expected profit from implementation is greater than the expected profit from not implementing it. The underlying latent unobservable variables are as follows:

$$U_{ij} = \beta'_j x_{ij} + \varepsilon_{ij}$$

The U_{ij} can be interpreted as an attraction measure for the profit with respect to a specific type of technological innovations. We assume that a firm i implements a technological innovation type j if $U_{ij} > 0$. Based on this, we define the observable indicator variables:

$$Y_{ij} = \begin{cases} 1 & \text{if } U_{ij} > 0 \\ 0 & \text{otherwise} \end{cases}$$

The $j = 1,\ldots,J$ vectors x_{ij} of the respective K_j explanatory variables are $x_{ij} = (x_{i1},\ldots,x_{iKj})'$ and the corresponding unknown parameter vectors are $\beta_j = (\beta_1,\ldots,\beta_{Kj})'$. P($Y_{ij} = 1$) denotes the probability for the implementation of a technological innovation type j. Since we consider probit models, the stochastic components ε_{ij} are standard normally distributed. The assumption that the ε_{ij} are mutually independent over all $j = 1,\ldots,J$ leads to simple univariate (binary) probit models. Flexible multivariate (binary) probit models are based on the assumption that the ε_{ij} are jointly normally distributed with:

$$\varepsilon_i = (\varepsilon_{i1}, \ldots, \varepsilon_{iJ}) \sim \mathrm{N}_J(0; \Sigma_J)$$

Instead of the one-dimensional probabilities P($Y_{ij} = 1$) and P($Y_{ij} = 0$), these models comprise for each firm i the following J-dimensional probabilities in the maximum likelihood estimation:

$$\begin{aligned} &P(Y_{i1} = y_{i1}, \ldots, Y_{iJ} = y_{iJ}) \\ &= \Phi_J\left(x_{i1}, \ldots, x_{iJ}; \beta_1, \ldots, \beta_J; \rho_{12}, \ldots, \rho_{1J}, \rho_{23}, \ldots, \rho_{2J}, \ldots, \rho_{J-1,J}\right) \end{aligned}$$

These probabilities depend on the realized values $y_{i1},\ldots,y_{iJ}$ that equal to one or zero (e.g. Greene 2003). The log-likelihood function therefore has the following structure:

$$\ln L = \sum_{i=1}^{N} \ln \Phi_J\left(x_{i1}, \ldots, x_{iJ}; \beta_1, \ldots, \beta_J; \rho_{12}, \ldots, \rho_{1J}, \rho_{23}, \ldots, \rho_{2J}, \ldots, \rho_{J-1,J}\right)$$

According to this, the probabilities in multivariate probit models comprise $J(J-1)/2$ correlation coefficients in Σ_J that can be estimated besides the parameters of the explanatory variables (the corresponding variance parameters in Σ_J are restricted due to model identification). In the case of the bivariate probit model with $J = 2$ the maximum likelihood estimation (which only comprises one correlation coefficient in the stochastic components) is straightforward. In contrast, the estimation of multivariate probit models with sizeable J is more complex due to the arising multiple integrals in the probabilities P($Y_{i1} = y_{i1},\ldots,Y_{iJ} = y_{iJ}$) and thus in the distribution functions $\Phi_J(\cdot)$ of the J-dimensional normal distribution. The computation of these J-dimensional integrals (in our multivariate probit model with $J = 4$ types of technological innovations) is not feasible with deterministic numerical integration methods.

But the probabilities can be accurately approximated with (unbiased) stochastic simulation methods, i.e. with repeated transformed draws of pseudo-random numbers. By incorporating a simulator, one obtains the simulated counterpart of P ($Y_{i1} = y_{i1},\ldots,Y_{iJ} = y_{iJ}$). In comparative Monte Carlo experiments, it has been shown that the Geweke-Hajivas-siliou-Keane (GHK) simulator (Börsch-Supan and Hajivassiliou 1993; Keane 1994; Geweke et al. 1994) outperforms other simulation methods with respect to the approximation of the true probability. The incorporation of these simulators in the maximum likelihood estimation leads to the simulated maximum likelihood estimation of multivariate probit models. Based on these estimates of the unknown parameters, simulated counterparts of classical test statistics can additionally be applied (e.g. Ziegler 2007).

References

Arora, S., & Cason, T. N. (1995). An experiment in voluntary environmental regulation: Participation in EPA's 33/50 program. *Journal of Environmental Economics and Management, 28*, 271–286.

Askildsen, J. E., Jirjahn, U., & Smith, S. C. (2006). Works councils and environmental investment: Theory and evidence from German panel data. *Journal of Economic Behavior and Organization, 60*, 346–372.

Baldwin, J., Hanel, P., & Sabourin, D. (2002). Determinants of innovative activity in Canadian manufacturing firms. In A. Kleinknecht & P. Mohnen (Eds.), *Innovation and firm performance: Econometric explorations of survey data* (pp. 86–111). New York, NY: Palgrave.

Barney, J. B. (1991). Firm resources and sustained competitive advantage. *Journal of Management, 17*, 99–120.

Belin, J., J. Horbach, & Oltra, V. (2011). *Determinants and specificities of eco-innovations – An econometric analysis for the French and German industry based on the community innovation survey* (Cahiers du GREThA n°2011-17).

Böhringer, C., Moslener, U., Oberndorfer, U., & Ziegler, A. (2012). Clean and productive? Empirical evidence from the German manufacturing industry. *Research Policy, 41*, 442–451.

Börsch-Supan, A., & Hajivassiliou, V. A. (1993). Smooth unbiased multivariate probability simulators for maximum likelihood estimation of limited dependent variable models. *Journal of Econometrics, 58*, 347–368.

Brouwer, E., & Kleinknecht, A. (1996). Firm size, small business presence and sales of innovative products: A micro-econometric analysis. *Small Business Economics, 8*, 189–201.

Brouwer, E., & Kleinknecht, A. (1999). Innovative output, and a firm's propensity to patent. An exploration of CIS micro data. *Research Policy, 28*, 615–624.

Brouwer, E., Budil-Nadvornikova, H., & Kleinknecht, A. (1999). Are urban agglomerations a better breeding place for product innovation? An analysis of new product announcements. *Regional Studies, 33*, 541–549.

Brunnermeier, S. B., & Cohen, M. A. (2003). Determinants of environmental innovation in US manufacturing industries. *Journal of Environmental Economics and Management, 45*, 278–293.

Cabagnols, A., & Le Bas, C. (2002). Differences in the determinants of product and process innovations: The French case. In A. Kleinknecht & P. Mohnen (Eds.), *Innovation and firm performance: Econometric explorations of survey data* (pp. 112–149). New York: Palgrave.

Carrión-Flores, C. E., & Innes, R. (2010). Environmental innovations and environmental performance. *Journal of Environmental Economics and Management, 59*, 27–42.

Czarnitzki, D. (2006). Research and development in small and medium-sized enterprises: The role of financial constraints and public funding. *Scottish Journal of Political Economy, 53*, 335–357.

Czarnitzki, D., & Kraft, K. (2004). An empirical test of the asymmetric models on innovative activity: Who invests more into R&D, the incumbent or the challenger? *Journal of Economic Behavior and Organization, 54*, 153–173.

Flaig, G., & Stadler, M. (1994). Success breeds success. The dynamics of the innovation process. *Empirical Economics, 19*, 55–68.

Frondel, M., Horbach, J., & Rennings, K. (2008). What triggers environmental management and innovation? Empirical evidence for Germany. *Ecological Economics, 66*, 153–160.

Galende, J., & de la Fuente, J. M. (2003). Internal factors determining a firm's innovative behaviour. *Research Policy, 32*, 715–736.

Geweke, J., Keane, M., & Runkle, D. (1994). Alternative computational approaches to inference in the multinomial model. *The Review of Economics and Statistics, 76*, 609–632.

Greene, W. H. (2003). *Econometric analysis*. New York: Prentice Hall.

Horbach, J. (2008). Determinants of environmental innovation – New evidence from German panel data sources. *Research Policy, 37*, 163–173.

Horbach, J., Rammer, C., & Rennings, K. (2012). Determinants of eco-innovations by type of environmental impact – The role of regulatory push/pull, technology push and market pull. *Ecological Economics, 78*, 112–122.

Jaffe, A. B., & Palmer, K. (1997). Environmental regulation and innovation: A panel data study. *The Review of Economics and Statistics, 79*, 610–619.

Kammerer, D. (2009). The effects of customer benefit and regulation on environmental product innovation: Empirical evidence from appliance manufacturers in Germany. *Ecological Economics, 68*, 2285–2295.

Keane, M. (1994). A computationally practical simulation estimator for panel data. *Econometrica, 62*, 95–116.

Khanna, M. (2001). Non-mandatory approaches to environmental protection. *Journal of Economic Surveys, 15*, 291–324.

Khanna, M., & Damon, L. A. (1999). EPA's voluntary 33/50 program: Impact on toxic releases and economic performance of firms. *Journal of Environmental Economics and Management, 37*, 1–25.

Koehler, D. A. (2007). The effectiveness of voluntary environmental programs – A policy at a crossroads? *The Policy Studies Journal, 35*, 689–722.

Lunn, J. (1986). An empirical analysis of process and product patenting: A simultaneous equation framework. *The Journal of Industrial Economics, 34*, 319–330.

Martínez-Ros, E., & Labeaga, J. M. (2002). Modelling innovation activities using discrete choice panel data models. In A. Kleinknecht & P. Mohnen (Eds.), *Innovation and firm performance: Econometric explorations of survey data* (pp. 150–171). New York: Palgrave.

OECD and Eurostat. (2005). *Oslo manual: Guidelines for collecting and interpreting innovation data*. Paris: OECD Publishing.

Porter, M. E., & van der Linde, C. (1995). Toward a new conception of the environment-competitiveness relationship. *Journal of Economic Perspectives, 9*, 97–118.

Potoski, M., & Prakash, A. (2005a). Covenants with weak swords: ISO 14001 and facilities' environmental performance. *Journal of Policy Analysis and Management, 24*, 745–769.

Potoski, M., & Prakash, A. (2005b). Green clubs and voluntary governance: ISO 14001 and firms' regulatory compliance. *American Journal of Political Science, 49*, 235–248.

Rave, T., Goetzke, F., & Larch, M. (2011). *The determinants of environmental innovations and patenting* (Ifo Working Paper No. 97).

Rehfeld, K.-M., Rennings, K., & Ziegler, A. (2007). Integrated product policy and environmental product innovations – An empirical analysis. *Ecological Economics, 61*, 91–100.

Rennings, K., & Rammer, C. (2009). Increasing energy and resource efficiency through innovation – An explorative analysis using innovation survey data. *Czech Journal of Economics and Finance, 59*, 442–459.

Rennings, K., & Zwick, T. (2002). The employment impact of cleaner production on the firm level – Empirical evidence from a survey in five European countries. *International Journal of Innovation Management, 6*, 319–342.

Rennings, K., Ziegler, A., Ankele, K., & Hoffmann, E. (2006). The influence of different characteristics of the EU environmental management and auditing scheme on technical environmental innovations and economic performance. *Ecological Economics, 57*, 45–59.

Wagner, M. (2007). On the relationship between environmental management, environmental innovation and patenting: Evidence from German manufacturing firms. *Research Policy, 36*, 1587–1602.

Wagner, M. (2008). Empirical influence of environmental management on innovation: Evidence from Europe. *Ecological Economics, 66*, 392–402.

Wätzold, F., & Bültmann, A. (2001). The implementation of EMAS in Europe: A case of competition between standards for environmental management systems. In M. Glachant (Ed.), *Implementing European environmental policy – The impacts of directives in the member states* (pp. 134–177). Northampton, MA: Edward Elgar.

Wernerfelt, B. (1984). A resource-based view of the firm. *Strategic Management Journal, 5*, 171–180.

White, H. (1982). Maximum likelihood estimation of misspecified models. *Econometrica, 50*, 1–25.

Ziegler, A. (2007). Simulated classical tests in multinomial probit models. *Statistical Papers, 48*, 655–681.

Ziegler, A., & Seijas Nogareda, J. (2009). Environmental management systems and technological environmental innovations: Exploring the causal relationship. *Research Policy, 38*(5), 885–893.

Chapter 7
The Impact of Resource Efficiency Measures on the Performance of Small and Medium-Sized Enterprises

Jens Horbach

7.1 Introduction

The effects of environmental measures on firm growth and productivity increasingly receive attention because a growing number of studies show positive effects contrary to the "traditional" view that environmental activities only raise production costs. The profitability of green investment is crucial for the diffusion of the resulting technologies but the knowledge about these performance effects is still limited. Positive performance effects may be based on cost savings stemming from the introduction of cleaner production processes connected with lower material and/or energy use. Porter and van der Linde (1995) find another positive effect. Following the Porter hypothesis, regulation-induced early introduction of environmental products may lead to first-mover advantages and to the improvement of a firm's competitiveness thus leading to better performance.

The present paper empirically analyzes the effects of environmentally active behavior on the performance of a firm. The analysis is based on the 2013 wave of the Eurobarometer data for small and medium-sized firms (SMEs). Looking at SMEs is particularly interesting since small firms might be especially affected by the financial burden of introducing resource efficiency measures, which are costly in the short run. On the one hand, these short run costs may constitute an important barrier to invest in cleaner technologies for SMEs with limited financial possibilities despite considerable cost saving effects in the long run. On the other hand, it might be possible that especially "young pioneers" confirm the validity of the Porter hypothesis. There are many examples from the past which show that young and small firms are often more likely to develop totally new ideas and products whereas big and established firms are not able to change their innovation paths. Thus, first

J. Horbach (✉)
Faculty of Business, Augsburg University of Applied Sciences, Augsburg, Germany
e-mail: jens.horbach@hs-augsburg.de

J. Horbach, C. Reif (eds.), *New Developments in Eco-Innovation Research*, Sustainability and Innovation, https://doi.org/10.1007/978-3-319-93019-0_7

mover advantages are often based on the activities of small and medium-sized firms. The problem is that small firms might be less able to bear the risks of developing new products. Strict environmental regulation in a country may create opportunities and demand security thereby reducing the risks for SMEs. The Eurobarometer data also allows answering the question of which environmental and resource-related measures lead to positive performance effects. Typical end-of-pipe measures, such as water purification by sewage treatment plants, might lead to negative performance effects whereas the introduction of energy saving measures might increase a firm's performance.

The chapter is organized as follows. Section 7.2 summarizes the main theoretical considerations on the relationship of eco-innovation and performance and the empirical literature. In Sect. 7.3.1, the Eurobarometer data basis and main descriptive results are presented. Section 7.3.2 shows the results of bivariate probit models analyzing the effects of resource efficiency measures on turnover development and self-perceived profitability.

7.2 The Effects of Eco-Innovations on Performance: Theoretical Considerations and Literature Overview

Whereas the implementation of end-of pipe technologies, such as additional filters, may raise production costs thus reducing productivity and international competitiveness, the introduction of cleaner technologies may lead to the opposite result because of reduced use of materials and energy savings. This leads to Hypothesis 1:

H1 Only resource efficiency measures leading to cost savings support performance whereas end-of-pipe measures might reduce performance

Furthermore, the development of greener products may create additional market opportunities. The well-known Porter hypothesis (Porter and van der Linde 1995) stresses the point that environmental regulation helps to overcome eco-innovation barriers consisting in imperfect information, organizational problems and market failures [see Horbach (2015) for a more detailed discussion]. Regulation-induced eco-innovations may thus lead to an increase in competitiveness and even first-mover advantages for eco-innovators. Some authors make a distinction between a weak and a strong version of the Porter hypothesis [see Jaffe and Palmer (1997) and Lanoie et al. (2011) for an empirical analysis]. The so-called weak version of the Porter hypothesis postulates that regulation induces eco-innovations without claiming that these innovations are also socially benign. The strong version goes a step further assuming that regulation-induced innovations overcompensate for the cost of compliance thus leading to an increase in the competitiveness of a firm. The existence of these possible extra-returns and first-mover advantages (Gagliardi et al.

2016) show that investment in resource efficiency might be advantageous compared to other investment activities, which leads to Hypothesis 2 and 3:

H2 A high amount of resource efficiency investment increases the performance of a firm

H3 The greenness of a firm is positively correlated to its performance

Specificities of SMEs

Following the definition of the European Commission (2003), SMEs are defined as firms employing less than 250 workers with an annual turnover that does not exceed 50 million euros. On the one hand, small firms might be more flexible and open to new innovation fields but on the other hand, SMEs might be specifically affected by innovation barriers (Tiwari and Buse 2007; Belitz and Lejpras 2014; Marin et al. 2015a; Ghisetti et al. 2017): high fixed innovation costs could reduce the availability of external financing because of high economic risks whereas big firms may finance the failure of an innovation project with the success of other projects. SMEs may be more affected by labor shortage because bigger firms are more attractive for applicants. For instance, the results of a survey in Hamburg show that "financing" and "finding suitable human resources" were the top innovation barriers for SMEs (Tiwari and Buse 2007). Furthermore, limited internal know-how and resources, missing possibilities to enter foreign markets because of the lack of an adequate logistic structure may reduce the ability to manage innovation processes. Additionally, bureaucratic hurdles such as long administrative procedures may be more problematic for SMEs due to their limited resources.

Therefore, looking at SMEs is particularly interesting since small firms might be especially affected by the financial burden of introducing resource efficiency measures, which are costly in the short run. On the one hand, these short run costs may constitute an important barrier to invest in cleaner technologies for SMEs with limited financial possibilities despite considerable cost saving effects in the long run (see also Ghisetti et al. 2017). Following Soltmann et al. (2015:460), "The development of green products and processes usually implies investing in technologies that lie beyond the firm's traditional technological scope ..." because the firm's resource base has to be enlarged and adapted and/or business processes and working routines have to be changed, too.

On the other hand, it might be possible that especially "young pioneers" confirm the validity of the Porter hypothesis. Many examples from the past show that young and small firms are often more likely to develop totally new ideas and products whereas big and established firms are not able to change their innovation paths. For instance, at the end of the seventies, small firms (Intel and Microsoft) pushed the development of personal computers instead of firms such as IBM that were specialized in the production of mainframe computers. Thus, first mover advantages are often based on the activities of small and medium-sized firms. The problem is that small firms are less able to bear the risks of developing new products. Strict environmental regulation in a country may create opportunities and demand security consequently reducing the risks for SMEs.

However, bigger, older and more experienced firms (Leoncini et al. 2016) are more likely to manage the higher complexity of eco-innovation and provide the more extensive technological experience these innovations require. Therefore, the role of the size and the age of a firm on the relationship of eco-innovation and performance remains an empirical question. As our sample only contains SMEs and not big firms with a long experience and tradition, the aforementioned argument of the young "pioneers" might be more important but SMEs may need external support to manage the complexity of eco-innovation. This leads to Hypothesis 4 and 5:

H4 Young "pioneers" are more likely to perform well
H5 SMEs using external support show better performance

Literature Overview

In the following, the recent empirical literature on the economic effects of eco-innovation is shortly summarized.[1] One part of the studies focuses on the effects of eco-innovation on productivity, further studies on the analysis of the relationship between eco-innovation and firm growth.

Rennings and Rammer (2011) use data from the German innovation survey of 2003. They detect similar success in terms of sales of new products and cost savings resulting from environmental regulation-driven innovations and other, non-environmentally related innovations. This result does not hold for all environmental innovation fields. Whereas regulations in favor of sustainable mobility, recycling, waste management or resource efficiency lead to higher sales, regulations in the field of water management are connected with a decrease in sales.

Ghisetti and Rennings (2014) use the eco-innovation question of the MIP 2009 and the wave of 2011 to measure the return on sales as outcome variable. They find positive profitability effects of innovations leading to a reduction in the use of energy and resources. They also observe that more end-of-pipe oriented innovations such as harmful materials and air, water, noise and soil pollutions show a negative influence on performance. In a study based on the same data basis, Rexhäuser and Rammer (2014) distinguish between regulation-induced and voluntary environmental innovations. Regulation-driven eco-innovations that improve firms' resource productivity seem to have a stronger effect on profitability compared to voluntarily introduced eco-innovations. Mohnen and van Leeuwen (2013) and Rubashkina et al. (2015) confirm the weak but not the strong version of the Porter hypothesis. Marin et al. (2015b) analyze the effects of the EU ETS on economic performance at the firm level. They use different indicators of performance such as value added, turnover, employment, investment, labor productivity, total factor productivity and markup. "Summing up, our estimates suggest that the EU ETS, despite its negative (but small) impacts on productivity and profitability, has stimulated the growth of firms" (Marin et al. 2015b:15).

[1]See also Barbieri et al. (2016) and the contribution of Ghisetti (Chap. 3) in this book for more comprehensive analyses. Please note that this short overview does not consider the employment effects of eco-innovation. For such an overview see e.g. Horbach and Rennings (2013) or Horbach and Janser (2016).

Franco and Marin (2017) use a panel of 8 European countries and 13 manufacturing sectors over the years 2001–2007. The authors measure the direct effects of environmental taxes on productivity but also the indirect effects of induced innovation in upstream and downstream sectors. They find that "... downstream regulation generates opportunities for innovation and may create markets for new and improved intermediate goods, upstream regulation acts as a constraint which negatively affects innovation and, even more strongly, productivity." (Franco and Marin 2017:289).

Hottenrott et al. (2016) show a complementarity effect of green technology adoption and organizational change. Only those green technologies that are accompanied by organizational changes are connected with constant or higher productivity.

Based on panel data of environmental R&D activities, Reif and Rexhäuser (Chap. 8 in this book) show the positive role of corporate social responsibility (CSR) for better financial performance. Firms signaling their environmental engagement through CSR seem to improve their financial performance. In their recent analysis, Soltmann et al. (2015) exploit new industry-level panel data of 12 OECD countries for a time span of 30 years. "The results show that green inventions are U-shape related to performance. However, the turning point is quite high and hence only relevant for a few industries. This indicates that—given the current level of green promotion—market incentives alone are not sufficient to allow the green invention activities of industries to rise considerably." (Soltmann et al. 2015:457). Based on a patent analysis, Lotti and Marin (2017) find that eco-innovations show a lower return compared to other innovations. This seems to be especially true for polluting firms facing high compliance costs.

In a recent paper (Leoncini et al. 2016) on the effects of eco-innovation on firm growth, researchers use quantile regressions and show that green technologies generate higher growth effects compared to other technologies for moderately growing firms but not for rapidly growing companies. Older and experienced firms profit more from the introduction of green technologies due to the complexity of managing eco-innovation. Based on a patent analysis, Colombelli et al. (2015) find that eco-innovation activities are especially benign for already fast growing firms.

The focus of the empirical analysis in Sect. 7.3 lies on the question which different resource efficiency measures are correlated to better performance and profitability for the rarely analyzed case of SMEs. The unique database of the Eurobarometer 2013 allows exploring the specific situation and constraints of SMEs.

7.3 Empirical Analysis

7.3.1 Data Basis and Descriptive Statistics

The analysis uses data from the Eurobarometer 2013 on resource efficiency and green markets (European Commission 2014). The database includes the 28 Member States of the European Union and furthermore Albania, Israel, Iceland, Liechtenstein, Montenegro, the former Yugoslav Republic of Macedonia, Norway, The

Table 7.1 Distribution of the sample by firm size and sectors

	Number of firms	in %
Number of employees		
1–9 employees	6166	46.0
10–49 employees	4681	34.7
50–249 employees	2660	19.7
Total	13,507	100.0
Sectors by NACE		
B—Mining and quarrying	84	0.6
C—Manufacturing	2890	21.4
D—Electricity, gas, steam and air conditioning	105	0.8
E—Water supply, sewerage, waste management	262	1.9
F—Construction	2216	16.4
G—Wholesale and retail trade, repair	4264	31.6
H—Transportation and storage	737	5.5
I—Accommodation and food service activities	727	5.4
J—Information and communication	464	3.4
K—Financial and insurance activities	231	1.7
L—Real estate activities	276	2.0
M—Professional, scientific and technical services	1253	9.3
Total	13,509	100.0

Source: European Commission (2014), own calculations

Republic of Serbia, Turkey and The United States and focuses on small and medium sized enterprises (up to 249 employees) (see GESIS Data Archive 2014).

The survey covers 13,509 observations (11,207 from the EU) in the manufacturing (NACE category C), services (NACE categories G/H/I/J/K/L/M/N) and industry sector (NACE categories D/E/F). The respondents of the questionnaire had to be a general manager, a financial director or a significant owner [for a more detailed description of the database see European Commission (2014)]. The sample is dominated by very small firms from one to nine employees (see Table 7.1); the most important sectors are "wholesale, retail trade, repair" (32%), manufacturing (21%) and construction (16%). More than one third of the firms offer green products or services; for nearly 20% the turnover share of these products is higher than 75% (see Table 7.2). About 54% of the questioned firms stated that their investments in resource efficiency led to a reduction of production costs (Table 7.2). For most of the firms (85.5%), the turnover share of these investments did not exceed 5%.

7.3.2 *Econometric Model and Estimation Results*

The performance effects of environmental and resource efficiency activities are measured by two different indicators: the turnover development of the preceding 2 years denotes the actual performance development. As this indicator does not

Table 7.2 Resource efficiency measures and green products

	Number of firms[a]	In %
Does your company offer green products or services?		
Yes	3865	30.6
No, but we are planning to do so in the future	1014	8.0
No, and we are not planning to do so	7751	61.4
Total	12,630	100.0
How much do these green products or services represent in your annual turnover?		
1–5%	1279	37.8
6–10%	506	14.9
11–30%	486	14.4
31–50%	269	7.9
51–75%	191	5.6
More than 75%	655	19.3
Total	3386	100.0
Investment to improve resource efficiency (in % of annual turnover)		
Less than 1%	4903	46.6
1–5%	4086	38.9
6–10%	977	9.3
11–30%	373	3.5
More than 30%	174	1.7
Total	10,513	100.0
What impact have the undertaken resource efficiency actions had on the production costs?		
It significantly decreased production costs	720	6.4
It slightly decreased production costs	5291	47.1
It slightly increased production costs	1950	17.4
It significantly increased production costs	558	5.0
It had no impact	2711	24.1
Total	11,230	100.0

Source: European Commission (2014), own calculations
[a]The total numbers of firms differ because of missing values

capture the profitability of the firm, the self-perceived resource investment profitability is used. Due to data restrictions, these two outcome variables are only binary. Firms showing a successful performance in terms of turnover development (*turnoverdev*) might be also more likely to report high self-perceived resource investment profitability (*selfpercprof*). Consequently, the two outcomes may be correlated leading to inconsistent estimates of simple probit models so that a bivariate probit model has to be estimated. This model reads as follows (Greene 2008):

1. $turnoverdev_i = x_i'\alpha + \varepsilon_i$
2. $selfpercprof_i = y_i'\beta + \mu_i$

If the cov $(\varepsilon_i, \mu_i) = \rho$ is zero, "... then the log likelihood for the bivariate probit models is equal to the sum of the log likelihoods of the two univariate probit models.

A likelihood-ratio test may therefore be performed by comparing the likelihood of the full bivariate model with the sum of the log likelihoods for the univariate probit models." (STATACorp 2015:183).

Description of Variables (for a Detailed Description see the Appendix)
The variable *sharegreenprod* gets the value one if the share of green products and services on total turnover is higher than 10%. *Costenv* denotes the situation if the questioned firm had difficulties to introduce resource efficiency measures due to the high costs of these measures. The lack of demand for ecological products and services is denoted by the dummy variable *lackdemand*. If additional profits and an expected increase in competitiveness are main motivations to realize resource efficiency measures the variable *profit* gets the value one. *Highgreenjob* denotes a high share of employees related to environmental issues (e.g. control of environmental regulations, production or marketing of green products etc.). *Investresource* gets the value one if firms spent more than 5% of their yearly turnover on measures improving resource efficiency. *EMS* captures the implementation of an environmental management system. The variables *consumer*, *firm* and *public* describe the different customers of a firm's products and services.

Greenness characterizes firms that declare environmental questions to be core priorities. From their self-perceived perspective, these firms try to go beyond the requirements of environmental regulations. If measures to improve resource efficiency are mainly based on own financial resources and own technical know-how, the variables *ownfinance* and *knowhow* get the value one. On the other side, *extern* denotes the importance of external support for these measures. The variables *measwater*, *measenergy*, *measrenewable*, *measmaterial*, *measwaste*, *measscrap*, *measrecycling* describe measures of the firms to improve resource efficiency in different environmental technology fields. *Size* gets the value one if the number of employees exceeds 50. A firm is defined as *young* if it was founded less than 10 years ago. *Oneperson* denotes firms with only one employee. Sector and country dummies are also included.

For our performance indicator, i.e. the turnover development of the past 2 years, the results of the bivariate probit model show that high investment in resource efficiency measures (*investresource*) increases the performance of a firm supporting H2 (Table 7.3).[2] Self-perceived profitability is also positively correlated to the share of investments in resource efficiency but this may also be due to the fact that high amounts of money spent on resource efficiency measures are motivated by expected positive returns. In contrast to the companies described, firms characterized by low resource investment shares mainly aim at fulfilling regulation requirements. The greenness of a firm measured by a high priority of environmental concerns

[2]It is important to notice that the results of the econometric analysis have to be interpreted as correlations rather than causal effects because of the cross section character of the data. Furthermore, the formulation of some questions does not allow the identification of an exact time structure.

Table 7.3 Performance effects of resource efficiency measures—all sectors

Correlates	Turnover development	Self-perceived profitability
Resource efficiency measures		
EMS	0.01 (0.75)	0.02 (2.73)**
Investresource	0.05 (3.90)**	0.06 (7.08)**
Measenergy	0.01 (0.87)	0.00 (0.05)
Measmaterial	−0.00 (−0.15)	−0.01 (−1.16)
Measrecyc	0.00 (0.39)	0.01 (1.83)+
Measrenewable	0.04 (2.67)**	0.04 (3.32)**
Measscrap	0.02 (1.42)	0.01 (0.82)
Measwater	−0.02 (−1.77)+	0.01 (1.59)
Measwaste	−0.00 (−0.20)	−0.01 (−1.44)
Greenness of the firm		
Greenness	0.03 (2.28)*	0.05 (6.04)**
Highgreenjob	0.05 (3.56)**	−0.00 (−0.09)
Sharegreenprod	0.01 (0.62)	0.01 (1.69)+
Control variables		
Consumer	−0.01 (−1.48)	−0.00 (−0.15)
Extern	0.04 (3.88)**	0.01 (1.04)
Firm	0.06 (6.27)**	−0.00 (−0.70)
Costenv	−0.03 (−2.70)**	−0.03 (−4.73)**
Profit	0.03 (2.17)*	−0.00 (−0.47)
Knowhow	0.01 (1.31)	0.02 (4.38)**
Lackdemand	0.00 (0.07)	−0.00 (−0.64)
Oneperson	−0.12 (−8.25)**	0.01 (0.93)
Ownfinance	0.01 (1.55)	0.04 (6.38)**
Public	0.02 (1.89)+	−0.00 (−0.37)
Size	0.12 (11.0)**	0.00 (0.31)
Young	0.17 (17.8)**	0.01 (1.64)+

Bivariate probit estimation reporting average marginal effects, robust standard errors. Number of observations: 13,376. Concerning dummy variables the marginal effects report changes in probability for discrete changes of these variables from zero to one. Wald χ^2 (144) = 2222. Z-statistics are given in parentheses; +, * and ** denote significance at the 10%, 5% and 1% level, respectively. Rho = 0.12. Likelihood-ratio test of rho = 0: χ^2 (1) = 36.0. Prob > χ^2 = 0.00. Sector/country dummies and constants are included but not reported

(*greenness*) and a high share of green jobs (*highgreenjobs*) is significantly and positively correlated to the turnover development (H3).

Not all measures for improving resource efficiency are connected with a positive performance (H1): an increased use of renewables (*measrenewable*) leads to a higher performance whereas measures to reduce water consumption (*measwater*) are even negatively correlated to the turnover development, corroborating the results of Rennings and Rammer (2011). This result is confirmed for the self-perceived profitability of resource efficiency. Concerning this indicator, recycling related measures (*measrecyc*) also show a weakly significant positive influence.

The introduction of Environmental Management Systems (*EMSs*) seems not to be relevant for the performance indicator but for the self-perceived profitability where *EMS* is highly significant. EMSs help to improve the profitability of resource efficiency related investments by identifying cost or material saving possibilities within a firm. Not surprisingly, firms showing bigger problems to bear the costs of environmental and resource efficiency measures are also low performers documented by the significantly negative influence of *costenv*. This result also holds for the self-perceived resource efficiency indicator.

As concerns further control variables, *young* firms show a better turnover and higher self-perceived profitability of resource related measures (H4). The effect of young firms being more dynamic in developing new ideas and turning them into innovations seems to surpass the fact that old, big and experienced firms are better capable to realize complex eco-innovations. This may be due to the sample exclusively containing SMEs. The *size* of the firm is positively correlated to turnover development, very small firms (*oneperson*) show a significantly weak performance. *External* support significantly helps SMEs to improve performance, which supports H5. Interestingly, this variable is not significant in relation to the self-perceived efficiency indicator. The questioned firms seem to be convinced that their own technical *knowhow* and financial resources (*ownfinance*) are crucial for the success of resource related measures.

In a further step, two separate bi-probit models for production-oriented sectors and the service sector are estimated (Table 7.4). The results show that the cost barrier (*costenv*) is not relevant for the performance in the service sector but only for the production sector. That is not surprising since in the production sector the introduction of resource efficiency related measures often requires high investments in physical capital whereas in the service sector mere organizational and logistic changes are sufficient in many cases. In the production sector, in which the costs of environmental measures seem to play a more important role, firms relying on own financial resources (*ownfinance*) show better performance. Also, resource efficiency measures creating additional market opportunities (*profit*) are more important for services whereas this variable is not significant for the production sector. A high priority for environmental concerns (*greenness*) is positively related to the performance of firms belonging to the service sector but not for manufacturing firms. For the self-perceived profitability, *EMS* as a soft instrument is only significant for services.

7.4 Summary and Conclusions

The paper analyzes the performance effects of different resource efficiency measures. Two indicators to measure performance and the profitability of environmental measures are used: the turnover development of the preceding 2 years denoting the actual performance development and self-perceived resource investment profitability. Due to data restrictions, these two outcome variables are only binary. As

Table 7.4 Performance effects of resource efficiency measures by different sectors

Correlates	Turnover development		Self-perceived profitability	
	Production sectors (NACE B-F)	Service sectors (NACE G-M)	Production sectors (NACE B-F)	Service sectors (NACE G-M)
Resource efficiency measures				
EMS	0.01 (0.76)	0.01 (0.71)	−0.00 (−0.05)	0.04 (3.64)**
Investresource	0.07 (3.60)**	0.03 (1.95)*	0.07 (5.45)**	0.06 (4.81)**
Measenergy	0.02 (0.80)	0.01 (0.68)	0.00 (0.27)	−0.00 (−0.01)
Measmaterial	−0.01 (−0.33)	0.00 (0.28)	−0.01 (−0.88)	−0.01 (−0.79)
Measrecyc	0.01 (0.48)	−0.00 (−0.09)	0.01 (0.58)	0.02 (1.97)*
Measrenewable	0.04 (1.59)	0.04 (1.84)+	0.04 (2.46)**	0.04 (2.34)*
Measscrap	0.03 (1.72)+	0.01 (0.62)	−0.00 (−0.09)	0.01 (0.94)
Measwater	−0.03 (−1.52)	−0.02 (−1.12)	0.04 (2.79)**	−0.00 (−0.33)
Measwaste	−0.03 (−1.32)	0.01 (0.69)	−0.02 (−1.39)	−0.01 (−0.74)
Greenness of the firm				
Greenness	0.01 (0.57)	0.04 (2.55)**	0.03 (2.71)**	0.07 (5.65)**
Highgreenjob	0.05 (2.37)*	0.05 (2.60)**	0.01 (0.55)	−0.01 (−0.63)
Sharegreenprod	0.02 (1.10)	−0.01 (−0.42)	0.03 (2.09)*	0.00 (0.33)
Control variables				
Consumer	0.02 (1.25)	−0.04 (−3.06)**	−0.00 (−0.44)	0.00 (0.30)
Extern	0.04 (2.38)*	0.04 (2.94)**	0.00 (0.20)	0.01 (1.07)
Firm	0.06 (3.91)**	0.06 (4.59)**	−0.00 (−0.29)	−0.01 (−0.59)
Costenv	−0.04 (−2.73)**	−0.01 (−1.04)	−0.03 (−3.25)**	−0.03 (−3.42)**
Profit	−0.00 (−0.06)	0.05 (2.82)**	−0.01 (−0.53)	−0.00 (−0.11)
Knowhow	0.02 (1.80)+	0.00 (0.42)	0.02 (1.90)+	0.03 (3.78)**
Lackdemand	0.01 (0.80)	−0.01 (−0.52)	−0.01 (−1.18)	0.00 (0.31)
Oneperson	−0.12 (−4.68)**	−0.12 (−6.83)**	0.01 (0.56)	0.01 (0.66)
Ownfinance	0.04 (2.74)**	−0.00 (−0.17)	0.02 (2.52)**	0.04 (5.97)**
Public	0.01 (0.82)	0.02 (1.81)+	0.01 (0.54)	−0.01 (−0.87)
Size	0.11 (6.96)**	0.13 (8.44)**	0.03 (2.84)**	−0.02 (−1.83)+
Young	0.19 (12.20)**	0.16 (13.14)**	0.01 (0.76)	0.01 (1.50)

Bivariate probit estimation reporting average marginal effects, robust standard errors. Concerning dummy variables the marginal effects report changes in probability for discrete changes of these variables from zero to one. Z-statistics are given in parentheses; +, * and ** denote significance at the 10%, 5% and 1% level, respectively. Sector/country dummies and constants are included but not reported

Production sectors: number of observations: 5512. Wald χ^2 (130) = 6196. Rho = 0.17. Likelihood-ratio test of rho = 0: χ^2 (1) = 31.2. Prob > χ^2 = 0.00

Service sectors: number of observations: 7864. Wald χ^2 (134) = 1418. Rho = 0.09. Likelihood-ratio test of rho = 0: χ^2 (1) = 11.1. Prob > χ^2 = 0.00

the two outcomes may be correlated, leading to inconsistent estimates of simple probit models, bivariate probit models have been estimated.

The results of these models show that high investment in resource efficiency measures increases the overall performance of a firm. A high self-perceived greenness of a firm and a high share of green employment is positively correlated to performance. In fact, not all measures for improving resource efficiency are connected with a positive performance: increased use of renewables leads to higher performance whereas measures to reduce water consumption are negatively correlated to turnover development. This result shows the need for further research analyzing different resource efficiency technologies. New surveys on eco-innovation and resource efficiency should address this point. Young firms show a better turnover development. The results also show that external financing is significantly important for a good performance. Firms characterized by increased problems to bear the costs of environmental and resource efficiency measures are also low performers. The firm size is positively correlated to performance; significantly negative performance of one person firms has been observed.

The results for the indicator "perceived resource investment profitability" widely confirms the results of the turnover development. Measures introducing renewables are again favorable to improve profitability; furthermore recycling is significantly positively correlated to investment profitability. The significantly positive effect of Environmental Management Systems (EMS) on perceived resource investment profitability is plausible because these systems provide information and thus help to reveal energy or other resource saving potentials in a firm.

All in all, the results show that especially investment in renewable energy technologies is correlated to positive performance effects but the realization of such measures in SMEs is highly dependent on external financing sources.

Appendix: Description of the Variables

Variables	Description	Mean	St. dev.
Dependent variables			
Turnoverdev	1 Increasing turnover during the last 2 years, 0 Constant or decreasing turnover	0.35	0.48
Selfpercprof	1 Highly satisfied with measures to improve resource efficiency, 0 other	0.12	0.32
Resource efficiency measures	Which of the following measures are implemented in your firm (1 yes, 0 no)?		
EMS	Environmental Management System	0.22	0.41
Measenergy	Energy reduction	0.34	0.47
Measmaterial	Material reduction	0.30	0.46
Measrecyc	Recycling	0.24	0.42
Measrenewable	Predominant use of renewable energy	0.08	0.27

(continued)

Variables	Description	Mean	St. dev.
Measscrap	Sale of scrap to other firms	0.18	0.38
Measwater	Reduction of water use	0.25	0.43
Measwaste	Reduction of waste	0.31	0.46
Investresource	1 Resource efficiency investment share on turnover greater than 5%, 0 other	0.11	0.32
Greenness of the firm			
Greenness	1 Environment is a core priority of the firm, firm goes beyond requirements of regulations, 0 other	0.12	0.33
Highgreenjob	1 High share of green jobs, 0 other	0.09	0.28
Sharegreenprod	Share of green products on turnover greater that 10%, 0 other	0.12	0.32
Control variables	1 yes, 0 no		
Consumer	Consumers as end-users	0.63	0.48
Firm	Sales to other firms	0.69	0.46
Public	Sales to public institutions	0.30	0.46
Costenv	Cost of resource efficiency measures as barrier	0.22	0.42
Extern	External support to realize resource efficiency	0.19	0.39
Profit	Improvement of the competition situation as motivation for resource efficiency measures	0.13	0.33
Knowhow	Internal know-how to realize resource efficiency	0.49	0.50
Lackdemand	Lack of demand for eco-products	0.16	0.36
Oneperson	One-person-company	0.08	0.27
Ownfinance	Self-financed resource efficiency measures	0.62	0.48
Size	1 Between 50 and 250 employees, 0 other	0.20	0.40
Young	Age of the firm less than 10 years, 0 other	0.25	0.43
Sector dummies	1 yes, 0 other sector		
Sec1	Mining and quarrying	0.01	0.08
Sec2	Manufacturing	0.21	0.41
Sec3	Electricity, gas, steam and air condition	0.01	0.09
Sec4	Water supply, sewerage, waste management	0.02	0.14
Sec5	Construction	0.16	0.37
Sec6	Wholesale and retail trade, repair	0.32	0.46
Sec7	Transportation and storage	0.05	0.23
Sec8	Accommodation and food service activities	0.05	0.23
Sec9	Information and communication	0.03	0.18
Sec10	Financial and insurance activities	0.02	0.13
Sec11	Real estate activities	0.02	0.14
Sec12	Professional, scientific and technical activities	0.09	0.29
Country dummies	1 yes, 0 other country		
AL	Albania	0.01	0.09
AT	Austria	0.03	0.17
BE	Belgium	0.03	0.17
BG	Bulgaria	0.03	0.17
CY	Cyprus	0.01	0.12

(continued)

Variables	Description	Mean	St. dev.
CZ	Czech Republic	0.03	0.17
DE	Germany	0.04	0.19
HU	Hungary	0.03	0.17
IE	Ireland	0.03	0.17
IL	Israel	0.02	0.15
IS	Island	0.01	0.12
IT	Italy	0.04	0.19
LI	Liechtenstein	0.01	0.09
LT	Lithuania	0.03	0.17
LU	Luxembourg	0.01	0.12
LV	Latvia	0.03	0.17
ME	Montenegro	0.01	0.09
MK	Macedonia	0.01	0.12
MT	Malta	0.01	0.12
NL	Netherlands	0.03	0.17
NO	Norway	0.02	0.15
PL	Poland	0.04	0.19
PT	Portugal	0.03	0.17
RO	Romania	0.03	0.17
RS	Serbia	0.01	0.12
SE	Sweden	0.03	0.17
SI	Slovenia	0.03	0.17
DK	Denmark	0.03	0.17
EE	Estonia	0.03	0.17
ES	Spain	0.04	0.19
FI	Finland	0.03	0.17
FR	France	0.04	0.19
GB	Great Britain	0.04	0.19
GR	Greece	0.03	0.17
HR	Croatia	0.03	0.17
SK	Slovakia	0.03	0.17
TR	Turkey	0.02	0.15
US	United States	0.04	0.19

References

Barbieri, N., Ghisetti, C., Gilli, M., & Nicolli, F. (2016). A survey of the literature on environmental innovation based on main path analysis. *Journal of Economic Surveys, 30*, 596–623. https://doi.org/10.1111/joes.12149.

Belitz, H., & Lejpras, A. (2014). *Financing patterns of innovative SMEs and the perception of innovation barriers in Germany*. DIW discussion paper 1353, Berlin.

Colombelli, A., Krafft, J., & Quatraro, F. (2015). *Eco-innovation and firm growth: Do green gazelles run faster? Microeconometric evidence from a sample of European firms*. GREDEG working papers 2015-12, Nice.

European Commission. (2003). Commission recommendation of 6 May 2003 concerning the definition of micro, small and medium-sized enterprises. *Official Journal of the European Union* (notified under document number C (2003) 1422), Brussels. http://eur-lex.europa.eu/legal-content/EN/TXT/PDF/?uri=CELEX:32003H0361&from=EN.

European Commission. (2014). Flash eurobarometer 381: Small and medium enterprises, resource efficiency and green markets. September 2013. TNS Political & Social [Producer].

Franco, C., & Marin, G. (2017). The effect of within-sector, upstream and downstream environmental taxes on innovation and productivity. *Environmental and Resource Economics, 66*(2), 261–291.

Gagliardi, L., Marin, G., & Miriello, C. (2016). The greener the better? Job creation effects of environmentally-friendly technological change. *Industrial and Corporate Change, 25*(5), 779–807.

GESIS Data Archive: ZA5874. (2014). Dataset version 1.0.0. https://doi.org/10.4232/1.11869.

Ghisetti, C., & Rennings, K. (2014). Environmental innovations and profitability: How does it pay to be green? An empirical analysis on the German innovation survey. *Journal of Cleaner Production, 75*, 106–117.

Ghisetti, C., Mazzanti, M., Mancinelli, S., & Zoli, M. (2017). Financial barriers and environmental innovations: Evidence from EU manufacturing firms. *Climate Policy, 17*(Supplement 1), 131–147.

Greene, W. H. (2008). *Econometric analysis* (6th ed.). Upper Saddle River: Prentice Hall.

Horbach, J. (2015). The role of environmental policy for eco-innovation: Theoretical background and empirical results for different countries. In F. Crespi & F. Quatraro (Eds.), *The economics of knowledge* (pp. 348–360). Routledge: Innovation and Systemic Technology Policy.

Horbach, J., & Janser, M. (2016). The role of innovation and agglomeration for employment growth in the environmental sector. *Industry and Innovation, 23*(6), 488–511.

Horbach, J., & Rennings, K. (2013). Environmental innovation and employment dynamics in different technology fields – An analysis based on the German community innovation survey 2009. *Journal of Cleaner Production, 57*, 158–165.

Hottenrott, H., Rexhäuser, S., & Veugelers, R. (2016). Organisational change and the productivity effects of green technology adoption. *Resource and Energy Economics, 43*, 172–194.

Jaffe, A., & Palmer, K. (1997). Environmental regulation and innovation: A panel study. *The Review of Economics and Statistics, 10*, 610–619.

Lanoie, P., Laurent-Lucchetti, J., Johnstone, N., & Ambec, S. (2011). Environmental policy, innovation and performance: New insights on the Porter hypothesis. *Journal of Economics & Management Strategy, 20*(3), 803–842.

Leoncini, R., Marzucchi, A., Montresor, S., Rentocchini, F., & Rizzo, U. (2016). *Better late than never': A longitudinal quantile regression approach to the interplay between green technology and age for firm growth*. SEEDS Working paper series 16/2016, Ferrara.

Lotti, F., & Marin, G. (2017). Productivity effects of eco-innovations using data on eco-patents. *Industrial and corporate change, 26*(1), 125–148.

Marin, G., Marzucchi, A., & Zoboli, R. (2015a). SMEs and barriers to eco-innovation in the EU: Exploring different firm profiles. *Journal of Evolutionary Economics, 25*(3), 671–705.

Marin, G., Pellegrin, C., & Marino, M. (2015b). *The impact of the European emission trading scheme on multiple measures of economic performance*. SEEDS working paper series 20/2015, Ferrara.

Mohnen P., & Van Leeuwen, G. (2013). *Revisiting the Porter hypothesis: An empirical analysis of green innovation for the Netherlands*. UNU-MERIT working paper series 2, Maastricht.

Porter, M. E., & van der Linde, C. (1995). Toward a new conception of the environment-competitiveness relationship. *Journal of Economic Perspectives, 9*(4), 97–118.

Rennings, K., & Rammer, C. (2011). The impact of regulation-driven environmental innovation on innovation success and firm performance. *Industry and Innovation, 18*(3), 255–283.

Rexhäuser, S., & Rammer, C. (2014). Environmental innovations and firm profitability: Unmasking the Porter hypothesis. *Environmental and Resource Economics, 57*(1), 145–167.

Rubashkina, Y., Galeotti, M., & Verdolini, E. (2015). Environmental regulation and competitiveness: Empirical evidence on the Porter hypothesis from European manufacturing sectors. *Energy Policy, 83*, 288–300.

Soltmann, C., Stucki, T., & Woerter, M. (2015). The impact of environmentally friendly innovations on value added. *Environmental and Resource Economics, 62*, 457–479.

StataCorp. (2015). *The STATA manual 2015*. Texas: Stata.

Tiwari, R., & Buse, S. (2007). *Barriers to innovation in SMEs: Can the internationalization of R&D mitigate their effects?* Hamburg University of Technology working paper no. 50, Hamburg.

Chapter 8
Good Enough! Are Socially Responsible Companies the More Successful Environmental Innovators?

Christiane Reif and Sascha Rexhäuser

8.1 Introduction

In the last years, environmental awareness has been on the rise especially in the societies of industrialized countries. Consumers are willing to pay (WTP) a price mark-up for 'green' products. This provides firms with the opportunity to gain profits and separate themselves from competitors by offering such products (Russo and Fouts 1997). The industry makes use of this trend by supplying adjusted products during whose production process environmental and sustainable aspects are considered. These often called 'green' products can be subsumed as impure public goods in economic literature (see e.g. Kotchen 2006). It is noticeable that more and more companies appear to have become socially and environmentally responsible on a voluntary basis in recent years (Poddi and Vergalli 2009). These voluntary actions of firms are called Corporate Social Responsibility (CSR). Firms can reveal their over-compliance with unobservable attributes through voluntary programs with publicly available information. Bénabou and Tirole (2010) classify this phenomenon as delegated philanthropy of stakeholders. Amongst others, Arora and Cason (1995) provide empirical evidence and Arora and Gangopadhyay (1995) refer to the theoretical background on vertical differentiation. This voluntary social or environmental action may be driven by the demand side of 'green' consumers or 'green' investors, on which we base the considerations of our analysis in this chapter.[1]

We concentrate on voluntary actions of firms in the environmental context. Communicating the environmental performance of a firm can help this firm in its

[1]Lyon and Maxwell (2008) additionally distinguish cost saving considerations and the avoidance of further threats of regulation as further reasons for such self-regulation.

C. Reif (✉) · S. Rexhäuser
Centre for European Economic Research (ZEW), Mannheim, Germany
e-mail: reif@zew.de

J. Horbach, C. Reif (eds.), *New Developments in Eco-Innovation Research*, Sustainability and Innovation, https://doi.org/10.1007/978-3-319-93019-0_8

competition for socially responsible clients (Kitzmueller and Shimshack 2012). The firm knows about its environmentally friendly activities, but for its consumers, employees and investors, it is not easy to obtain, aggregate or compare this information (Bénabou and Tirole 2010). Products can carry information on their environmental performance but not directly on their production process.[2] We argue that CSR can carry unobservable organizational qualities to reduce information asymmetry. We base this on the signaling model explained by Spence (1973) for the job market context and applied to the financial market by e.g. Ross (1977) or Bhattacharya (1979), in which signals serve as information on unobservable attributes (see Spence 2002).[3] We apply the idea of reputational economies of scale by Wernerfelt (1988) to our approach. Note that CSR does not signal the specific attributes of one product, but the firm's environmental R&D activity as a whole. CSR certifications—such as the Global Reporting Initiative (GRI)—are regarded as credible signals to transport firm attributes which cannot be observed easily and build up reputation (see e.g. Akerlof 1970; Toms 2002). Like Terlaak and King (2006), we assume that CSR signals these desired qualifications of firms, in our case information about the environmental attributes of a firm. Thereby it enhances the firm's reputation which is rewarded at the market by consumers, investors or business partners.

The research question is whether a joint strategy of environmental innovation and CSR engagement leads to higher financial performance than either one or none of them. If combined they could signal green attributes of firms. We use the Thompson and Reuters ASSET4 (A4) database of large global companies in panel-structure to analyze the long-term effects of such a joint strategy on firms' market value accounted for by the price-to-sales ratio. The underlying assumption is that CSR signals the environmental performance of a firm or in our approach the environmental R&D commitment. It indicates not only a firm's current activity, but also its long-term commitment to environmental responsibility.

The results support the view that the two strategies (environmental innovation and certified CSR) act as strategic complements in terms of a company's market value. Introducing a single strategy alone did not significantly affect financial performance while companies that adopted both strategies jointly significantly increased their market value compared to the control group. In this sense, the results suggest that a credible signal is needed to successfully disclose otherwise private information on companies' environmental performance.

[2]Akerlof (1970) originally addressed the problem of information asymmetry that arises either from adverse selection or from unobservable attributes (moral hazard).

[3]Please see Riley (2001) for an overview and discussion on the literature of signaling.

8.1.1 Related Literature

In the empirical research on CSR, an overwhelming number of studies addresses the connection of CSR and financial performance (FP), mostly focusing on how CSR influences financial performance.[4] Cochran and Wood (1984), Pava and Krausz (1996), as well as Griffin and Mahon (1997) provide literature reviews of early research. The meta-analyses by Orlitzky et al. (2003) or Margolis et al. (2007) indicate a positive correlation of CSR and FP.

However, empirical analyses hardly take innovation as a factor into account. Hart and Ahuja (1996) include innovation in form of R&D per sales as a control variable in their analyses of how CSR affects different types of operational and financial performances. They observe a positive effect of preventing pollution on financial performance. McWilliams and Siegel (2000) base their empirical research on two main arguments: innovation has a positive effect on FP, and CSR and innovation are strongly correlated. They prove that ignoring innovation might lead to an overestimation of the CSR effect on financial performance. Hull and Rothenberg (2008) extend the study by McWilliams and Siegel (2000) based on the assumption that firms can differentiate themselves from others via innovation or CSR. Their results suggest that CSR positively influences FP, furthermore that innovation is an important explanatory variable in this context, and that innovation and CSR are substitutes in terms of firm differentiation. Another study based on McWilliams and Siegel (2000) is conducted by Lioui and Sharma (2012), who examine how environmental CSR affects return-on-assets (ROA) and Tobin's Q directly and indirectly via R&D. They explain the negative direct effect of CSR on FP with the costs of CSR measures and attribute the positive indirect effect to more efficient R&D (Lioui and Sharma 2012). The study by Cavaco and Crifo (2014) is the only example known to us using a complementarity approach in the CSR context. They analyze three dimensions of CSR: responsible behavior towards employees, customers and suppliers, and the environment. They find that firms caring about employees and at the same time about customers and suppliers can gain profits. Furthermore, firms should decide on either environmentally friendly behavior or responsible behavior towards customers and suppliers and not engage in both of them at the same time.

Beside these studies which mostly use scores as an overall indicator for a firm's social and environmental activities, research with a focus on environmental performance is related to our analysis. This literature is strongly connected to the Porter hypothesis, stating that environmental regulation implies innovation and in turn generates a competitive advantage (see Porter and van der Linde 1995). Rexhäuser and Rammer (2014) test the Porter hypothesis for German companies and find that it does not hold in general and depends on the type of environmental innovation. A

[4]The term Corporate Social Responsibility (CSR) is not commonly defined. In literature and practice, several definitions exist, but two common aspects of CSR can be found in these definitions: CSR activities relate to social and environmental issues and go beyond legal requirements. For our analyses, we apply this definition.

study closely related to the CSR context by Klassen and McLaughlin (1996) empirically examines the effect of environmental awards on financial performance. In their event study, they find significantly positive effects. Another study by Konar and Cohen (2001) shows that negative environmental performance (like emitted toxic chemicals and lawsuits) affects financial performance negatively. Russo and Fouts (1997) use environmental ratings as environmental performance indicators and observe a positive impact on financial performance. Research approaches using panel data have been developed in the last years to respond to the criticism of not considering unobserved firm heterogeneity in cross-sectional studies. Among them is a study on US firms by King and Lenox (2001), showing that firms' attributes may drive the effect on financial performance. Elsayed and Paton (2005) find a neutral effect of environmental performance on financial performance in a dynamic panel data analysis for UK firms, which is confirmed by Telle (2006) for Norwegian firms. Additionally, empirical studies on environmental management systems (EMS) are connected to research on environmental performance and its effect on a firm's performance. Studies on the determinants of environmental management systems reveal that consumer preferences have a strong impact on firms' engagement in EMS certification (Nakamura et al. 2001; Potoski and Prakash 2005; Nishitani 2009; Nishitani 2010).[5] In contrast, Harrington et al. (2008) find that firms' internal factors are the driving forces for implementing environmental management.

However, research especially on the ISO 9000 and ISO 14000 as certified standards (see Heras-Saizarbitoria and Boiral 2013 for a literature overview) is more related to the signaling concept. Heras-Saizarbitoria and Boiral (2013) categorize signaling models as non-technical, theoretical approaches. Terlaak and King (2006) apply certified standards to the signaling model and find that the ISO 9000 certification leads to a competitive advantage, especially in large and advertising-intensive industries with high information costs. However, there are studies that do not find support for the ISO 14000 certification having a positive effect on financial performance (e.g. Heras-Saizarbitoria et al. 2011 for Spanish firms) or even a negative effect on the market value (Cañón-de-Francia and Garcés-Ayerbe 2009). Nevertheless, Toms (2002) shows for UK firms that the disclosure of environmental activities in annual reports creates reputation, which in turn may lead to a competitive advantage.

These studies analyze whether a single pro-environmental activity leads to positive effects on financial performance. We, in contrast, focus on the effect of joint activities. We investigate whether a CSR measure signaling pro-environmental action is complementary to environmental innovations in such a way that the joint presence of the two would generate higher profits, or if a substitutive character can be verified.

[5]In contrast, Harrington et al. (2008) find that firms' internal factors are the driving forces for implementing environmental management.

8.1.2 Hypothesis and Basic Complementary Model

We analyze the joint effect of green innovation and CSR on FP. We expect that CSR increases the marginal financial returns by disclosing a firm's environmental innovation compared to a control group. The underlying assumption is that firms in which both CSR and environmental innovations are present can differentiate themselves from their competitors by verifying their environmentally friendly behavior through CSR and their ongoing dedication to the future through today's R&D spending. Thus, they may convince stakeholders such as consumers, investors, and trading partners with their reputation.

We relate this assumption to the signaling hypothesis and the resource-based view (RBV) of the firm. The resource-based view (RBV) of the firm goes back to Penrose (1959). It takes the perspective that a firm is a bundle of resources (Wernerfelt 1984) with the main assumptions of resource heterogeneity and resource immobility (see Bowen 2007). Therefore, the RBV takes the standpoint that competitive advantage is created within the firm depending on the characteristics and usage of its resources. Thus, these specific firm capabilities can generate a competitive advantage in the long-run. According to Branco and Rodrigues (2006), reputation is an intangible resource which could be affected by a firm's CSR activity. As such, CSR creates reputation and in turn increases financial performance (see e.g. McGuire et al. 1988; Waddock and Graves 1997), which is described as lead-lag relationship by Preston and O'Bannon (1997).[6] Also the returns to R&D efforts are hard to observe (see e.g. Aboody and Lev 2000; Chauvin and Hirschey 1994) and can be seen as intangible assets. We assume that CSR serves as a signal to overcome this information asymmetry by at least partially disclosing a firm's environmental activities and thus, to state its environmental commitment. This in turn helps a firm to create a positive reputation and trustworthiness and might also support the company in vertical differentiation from their competitors (see e.g. Fombrun 1996; Fombrun and Shanley 1990).

Theoretically we can base our assumptions on the complementarity approach. Complementarity (in the sense of Edgeworth) of firm strategies means that engaging more in one activity increases the marginal returns of engaging more in the other. In our case, certified CSR may signal environmental responsibility and thus increase the returns to green R&D by making stakeholders aware of this. In principle, there are two ways to estimate such a relationship: the adoption approach and the productivity approach, which is central to the present paper.[7] The productivity

[6]In the environmental context, Hart (1995) or Russo and Fouts (1997) are examples of empirical research on the RBV.

[7]Loosely speaking, the adoption approach relies on the correlation of two firm strategies in order to account for complementarity. Note that this approach is only valid in case of continuous strategic measures (Miravete and Pernías 2010) and thus not applicable in this paper. The adoption approach can be traced back to the work of Arora and Gambardella (1990). They show that a positive covariance among a pair of activity variables indicates complementarity if the activity variables are conditioned on any other firm-specific characteristics. For an overview of empirical studies, please see Brynjolfsson and Milgrom (2013).

approach is not restricted to continuous variables and accounts for the performance effects of the potentially complementary variables with respect to an objective function—in our case financial performance.

Based on Milgrom and Roberts (1990), who show that the concept of complementarity is directly related to the supermodularity of the objective function, we can impose an order on each pairwise combination of the variables. We begin with the smallest element in the order, {0,0}, which, in this case, means that neither CSR nor environmental innovation is carried out. Elements ranked higher in this order, denoted as {1,0} and {0,1}, represent either environmental innovation only or exclusive CSR engagement. Finally, the highest element in this order, {1,1}, stands for the joint use of both firm strategies. The condition implies that adopting both strategies jointly leads to a higher performance than adopting both of them in isolation, simply because one increases the marginal returns of the other. Formally, the condition for supermodularity and complementarity reads as follows:

$$f(1,0) + f(0,1) \leq f(1,1) + f(0,0), \qquad (8.1)$$

where *f*(.) represents the objective function (see Milgrom and Roberts 1990). Based on the complementary approach we formulate the following hypothesis:

> Firms engaged in both—environmental innovations as well as CSR—can gain ceteris paribus better financial performance.

As such, we analyze the joint presence of environmental innovation and CSR and compare their effect on financial performance to firms that have neither introduced environmental innovations nor CSR or only one of them.

8.2 Database and Choice of Variables

8.2.1 Database

We base our research on worldwide company panel data from the Thompson and Reuters ASSET4 (A4) database,[8] which allows us to get a better insight into the process organization of a company than CSR score data alone (as for example the Kinder, Lydenberg, Domini database, which is often used in the CSR context). The database mainly includes large companies based in the US. The unbalanced panel contains a collection of environmental, social, governance and financial data on more than 3000 global companies listed in major indices (e.g. S&P 500, MSCI World Index, Stoxx 300, Nasdaq, ASX 200). Publicly available information about a company (e.g. reports) is gathered yearly (beginning in 2002) by specially trained analysts with an increasing number of screened companies. Our sample extends over the years 2005 to 2009 as an unbalanced panel to ensure the availability of a two-year time lag for each variable (see

[8]The database can be accessed via the provider Thompson and Reuters.

Fig. 8.1, Appendix). The restricted sample for our estimations consists of 6737 observations including 1945 firms. Besides various factors for CSR characteristics, the dataset provides a rather limited amount of information on central firm-specific factors likely affecting firms' stock market value. Nevertheless, the panel data structure allows us to formally test whether CSR complements firms' environmental innovations in terms of financial performance or has a substitutive character.

8.2.2 Choice of Variables

8.2.2.1 Dependent Variable Financial Performance

The results of empirical analyses on the relation between a specific CSR activity and a specific FP strongly depend on how CSR and FP are measured: concerning CSR, if it is business-related or not and regarding FP, if it is an accounting-based or a market-based measurement (see Margolis et al. 2007). We analyze the effect of the joint presence of environmental innovation and CSR on the financial performance of a firm. This approach calls for market-based measurement of financial performance, replicating the long-term and future-oriented perspective. Therefore, we use the price-to-sales ratio (P/S_{it}) as a market-based way of measuring financial performance (see Pava and Krausz 1996, see Orlitzky et al. 2003, Margolis et al. 2007). It reflects the value placed on sales by past performance, other companies or the market. The profit margin affects the price-to-sales ratio and therefore, is an appropriate indicator for market power. In our sample the *P/S* variable shows a right-skewed distribution, so we use the logarithm of firm *i*'s *P/S*.

8.2.2.2 Environmental Innovation and CSR

Our complementary approach is based on the consideration that CSR signals a firm's engagement in environmentally friendly production to stakeholders. These stakeholders might be skeptical about firms' self-reported CSR activities. Certification from third parties of a company's CSR activity can serve as a signal which creates credibility and may thus support stakeholders in their decisions; moreover the respective firm can gain benefits from spreading this information. This is in line with findings by Terlaak and King (2006), who apply the signaling theory to certified management standards to overcome information asymmetries resulting in a competitive advantage.

In the database, the CSR indicator related to our approach is the dichotomous variable *Global Report Initiative* (CSR_{it}).[9] This variable reveals if companies publish their CSR report according to the Global Reporting Initiative (GRI) guidelines and it

[9]The A4 database offers other variables on CSR but the GRI fits our research question best and the database provides enough data points for the analysis.

serves as a proxy for firms' CSR performance. These guidelines are developed to standardize sustainability reporting and create transparency and comparability of companies around the world. Reports according to the GRI guidelines can be treated as a method of certification because GRI reporting comprises detailed information (over 100 indicators on firms' sustainability in the economic, environmental, social, and governance areas) and a third party ensures that the data is in accordance with the guidelines. The indicator may only suggest if the firm reports according to the GRI guidelines and not if the firm is really engaged in CSR activities. Nevertheless, we assume that only firms engaged in CSR activities actually report on them and in turn decide if they report in such detail as demanded by the GRI.

In our approach, we focus on the voluntary action of a firm in the environmental context. Therefore, the appropriate innovation variable available in the database is the dichotomous variable named environmental R&D ($Green_{it}$).[10] It provides information on whether the firm invested in R&D related to environmentally friendly products and processes to reduce emissions and resource consumption. This allows us to categorize environmentally and non-environmentally innovative firms. Nonetheless, the dichotomous variable is only an approximate value for innovation, as especially for large companies it is not clear which efforts they have made and how this affects the environmental activity of the respective firm.[11] Current environmental R&D investments are an indicator for the future market value of a firm. However, this kind of innovation is not easily observable for customers, investors or other business partners. Hence, it is plausible to assume that CSR might help to overcome this information asymmetry in the case of environmental R&D investment.

8.2.2.3 Explanatory and Control Variables

In the A4 database, additional variables illustrating market power, like market share or the Herfindahl index, are not available. This is especially true for the large firms listed on the stock exchange and included in the A4, which are operating globally and thus have no clear cut geographic market definition. Due to this diffuse market definition, using such market power indicators is questionable, as Aghion et al. (2005) point out. Alternatives like the Lerner index (see e.g. Aghion et al. 2005) or the price-to-cost ratio (see e.g. Gorodnichenko et al. 2010) are also not included in the A4 database. Nevertheless, the panel data structure allows controlling for unobservable but time consistent factors of market power. Nickell (1996) explains that the changes of the unobservable factors correlate with the changes of the observable variables. In our case the inclusion of the lagged price-to-sales ratio

---

[10]For a deeper discussion on R&D as innovation input, please see e.g. Kleinknecht et al. (2002) or Smith (2005).

[11]This is the only indicator in the A4 database for environmental R&D investments which is usable for the analyses. The A4 data on environmental R&D investment costs provides not enough data for the analysis.

($lag(P/S)_{it}$) controls for these unobservable factors and at the same time it considers that past financial performance may explain current financial performance.

Another explanatory variable for financial performance, especially from an investor's perspective and if measured by a market-based variable, is a risk parameter. The risk coefficient beta as parameter for stocks' volatility, which measures the risk of an investment, reflects the riskiness of the returns of a firm. Risk, which might affect future financial performance, has not been adequately taken into account in most previous empirical studies (see criticism by Orlitzky 2005, Margolis et al. 2007, Cochran and Wood 1984). CSR can serve as a risk management instrument to reduce a firm's risk (Husted 2005). Orlitzky and Benjamin (2001) provide evidence for this link between a firm's CSR and its financial risk in their meta-analysis. Therefore, we include the variable beta ($Beta_{it}$) measuring the market risk.

Moreover, business cycles, influencing stock market values, might cause stock market prices to differ across countries and time. Thus, we need a time trend control that varies across countries. This effect is assumed to have a very immediate impact, so no time lag is included. In addition, it is reasonable to consider the business cycle as exogenous. Therefore, information on real GDP (growth) by country and year is linked to the A4 database based on firms' country affiliation and included as a control variable ($GDP\text{-}Growth_{it}$).

Patents are indicators for a temporarily limited monopoly and an approximate measure for the stock of intangible assets. Companies holding patents have a technological advantage, which can be the reason for price differences resulting in better performance. The benefit of using US patents is their consistent measurement method and the relevancy for firms holding US patents. We use the logarithm of the number of patents ($ln(Patents)_{it}$) held by a company in a specific year as provided by the A4 database for our calculations because we assume that the stock of patents affects *P/S* in the same time period and is not time lagged.

Additionally, the age of a company might influence its financial performance either positively through learning effects, or negatively because of its inability to adjust to new challenges. Hopenhayn (1992) shows under which circumstances older firms can gain higher profits. Age is thus an important factor when measuring financial performance and we include the age of the company as an explanatory variable ($ln(Age)_{it}$). Furthermore, we control for the size of the company measured by sales in logarithmic form ($ln(Sales)_{it}$). Labor productivity is included as the logarithm of the number of employees by sales ($ln(Labor/Sales)_{it}$).

8.3 Results

8.3.1 Descriptive Statistics

As in the database, the restricted sample of 6737 observations mainly comprises the following countries: the United States (33%), Japan (18%), and the United Kingdom (14%). The European countries represent about 39% of the observations.

Concerning the distribution of observations by continents, Fig. 8.2 (Appendix) shows that most relate to Europe, the US, and Asia. Especially for Europe and Asia, there are more observations of firms engaged in both strategies. Furthermore, the sample covers the 12 industry sectors according to the Standard Industrial Classification (SIC) (Table 8.4, Appendix). The finance, insurance and real estate sector are highly represented in our sample (19%), followed by the transport sector (13%). Nevertheless, the various manufacturing industries account for almost half of the observations in the sample (43%). As expected, firms are innovators or are engaged in both firm strategies, particularly in the manufacturing sectors.

Table 8.5 (Appendix) provides an overview of the chosen variables with a short explanation, together with the mean, the standard deviation, and the minimum and maximum values. We focus on the two variables of interest—environmental innovators and CSR engagement—for the further descriptive analyses. There are 1039 observations of environmental innovators and 1501 observations on CSR engagement over the years 2005–2009 (Table 8.6, Appendix). As the number of observations varies in the unbalanced panel, also the number of firms that reported environmental R&D and CSR varies over the period. The share of green innovators adds up to over 5% in each year and the share of firms carrying out CSR accounts for more than 8% in each year.

Although the key variables of interest (i.e. whether CSR and environmental R&D are implemented) are binary indicators, they vary considerably over time within the firms. 33.26% of the 1945 firms in our sample implemented CSR for at least one year, whereas environmental R&D has been reported at least once for 24.27% of the firms. Approximately 9.51% of the firms reported to have CSR in place in all observed years. The respective number of environmental R&D is much lower, namely 6.84%. More interestingly, 22.51% of all firms introduced CSR in a certain year and stuck to CSR in all the following years. Approximately 67.70% of all firms implemented CSR for at least one year. 15.42% of all firms implemented environmental R&D in a certain year and stuck to it in all the following years. This means 63.56% of all the firms carried out environmental R&D at least in one year. 2% of all 1945 firms changed their engagement over time with respect to R&D and 1.23% with respect to CSR activities. Figure 8.3 (Appendix) shows the number of observations within the period of 2005–2009 for the four exclusive types of engagement. Throughout these years, most companies were not engaged in either one of the strategies. However, in all years there are observations for all four categories.

A more detailed descriptive analysis of environmental R&D and CSR indicates a correlation between both strategies (Table 8.1). In the sample, the joint realization of environmental innovation and CSR occurs more frequently than the implementation of environmental innovation alone. Table 8.1 also shows the frequency under the assumption that both firm strategies are independent in parentheses. Interestingly, if both strategy variables were independent, we would expect that only 232 firms had introduced both strategies jointly. However, the firms that actually implemented both strategies amount to more than twice the number we would expect in case of independency. Together with the very high coefficient of association (Kendall's tau-b), Table 8.1 offers strong evidence for a high correlation between

Table 8.1 Adoption decision and relative frequencies

CSR				*CSR*			
Green	0	1	Total	*Green*	0	1	Total (%)
0	4754 (4429)	944 (1270)	5698	0	83.43% (77.73%)	16.57% (22.29%)	84.58
1	482 (808)	557 (232)	1039	1	46.39% (77.67%)	53.61% (22.33%)	15.42
Total	5236	1501	6737	Total	77.72%	22.28%	100.00

Expected frequencies appear in parentheses. Pearson chi^2 (1) = 696.3276, Pr = 0.000. Kendall's tau-b = 0.3215, $P > z = 0.0000$

environmental innovation and CSR. This is in line with considerations by Terlaak (2007) that firms in R&D intensive industries can gain a competitive advantage thanks to certified standards.

Nevertheless, the correlation alone, of course, is insufficient to show the presence of complementarity. Whether this correlation survives multivariate statistics controlling for any other influencing factors and whether it really stems from complementarity is subject to the following empirical analysis.

8.3.2 *Estimation Strategy*

We assume the market capitalization (or market value) to depend on the sum of the firm's physical assets and intangible (knowledge) assets. In our data the information on the physical assets of firms is not available. This, in combination with the difficulties to measure intangible assets, motivates the use of lagged market capitalization information to account for assets. Scaling market capitalization by firms' total sales is a frequently used[12] size-independent measurement for firms' value created by each single dollar of sales—the price-to-sales ratio, henceforth, is P/S_{it}. The resulting regression equation reads as follows:

$$\ln(P/S_{it}) = \beta_0 + \beta_p \ln(P/S_{it-1}) + \beta_{10} Green_{it} + \beta_{01} CSR_{it} + \beta_{11} Both_{it} + \boldsymbol{C}_{it}\beta_c + \in_{it},$$

where $\boldsymbol{C}_{it}$ is a vector of controls described above, $\in_{it} = u_i + e_{it}$ with u_i denoting firm-specific fixed effects and e_{it} representing an idiosyncratic error component. $Green_{it}$ accounts for the choice in favour of green innovation alone (i.e. without introducing CSR). CSR_{it} denotes that only CSR is in place. The dummy $Both_{it}$ indicates that both strategies are present so that no implementation of either strategy serves as the

[12]Comparable studies use measures like Tobin's q to relate environmental regulation or environmental innovation to firms' market value and financial performance, such as Dowell et al. (2000) or Konar and Cohen (2001).

reference group and thus, β_{00} is necessarily zero. Note that the price-to-sales ratio (P/S_{it}) is measured as the year-end value. Furthermore, the literature assumes a rather short event window in which upcoming information on firms' CSR activities and green innovation affect the market value (see e.g. Cañón-de-Francia and Garcés-Ayerbe 2009). However, we assume a rather long event window of a whole year. This is simply due to data availability. In this sense, the key variables of interest, $Green_{it}$, CSR_{it}, and $Both_{it}$, enter the model in the same year as the dependent variable. The strategy variables ($Green_{it}$, CSR_{it}) cannot be considered strictly exogenous, as they are endogenous choices of firms, which may be dependent on firms' market value. If providing a credible signal for sustainability really complements the investment in green R&D and translates into higher firm values, clever managers are likely to be aware of this issue. Since it is probable that good management is correlated with higher market values, omitting a control for management may cause the strategy variables to be biased, as management remains an unexplained error component and thus, $cov(Green_{it}, \epsilon_{it,}) \neq 0$, $cov(CSR, \epsilon_{it,}) \neq 0$, and $cov(Both_{it}, \epsilon_{it,}) \neq 0$. Consequently, the empirical model needs to handle the endogeneity of the main variables of interest. An adequate solution does not consist in using one-year lagged values of the key variables of interest to rule out potential problems of endogeneity, as we assume a rather short-term event window in which upcoming information on CSR and green innovation can affect firms' market value. Therefore, we use lagged price-to-sales ratio information to control for physical and intangible assets.

However, incorporating a dynamic panel specification may cause potentially predetermined and thus not strictly exogenous regressors. Especially the lagged dependent variable is likely to be correlated with current errors via its correlation with past ones. Thus, it causes the classical linear regression model to be inconsistent, even if $\epsilon_{it,}$ is not autocorrelated. Furthermore, neither the lagged dependent variable nor the vector of controls ($\boldsymbol{C}_{it}$) allow controlling for all differences in the price-to-sales ratio across firms. These unexplained differences in the between-dimension of the panel data (i.e. across firms) may be correlated with at least some of the regressors, leading to bias of their coefficient estimates. Therefore, we apply the Arellano-Bond dynamic panel data difference GMM estimator, which uses all available lags as instruments.

8.3.3 Empirical Results

As a first step, we test whether the traditional variables explaining growth in price-per-sales are in line with previous research concerning direction and size (Table 8.2). Therefore, we estimate *Model 1* with the logged growth in price-per-sales as dependent variable and without the complementary variables of interest.[13] The

[13]We conducted preliminary tests on fixed effects versus random effects models. First, the F-test on the null hypothesis of no fixed effects is rejected. The Hausman-Test with the null hypothesis of no

results verify that our dynamic approach including the one-year time lag of price-per-sales ($ln(P/S)_{it-1}$) is appropriate, as they are highly significant and size as well as direction are comparable to previous studies. The high negative coefficient shows marginally decreasing growth rates in P/S. A one percent increase in the lagged price-to-sales ratio is significantly associated with a 0.83% smaller growth rate in the price-to-sales ratio between periods *t* and *t*–1. Or in other words, the higher the price-to-sales ratio of firm *i* already is, the lower the rates of growth. Furthermore, risk ($Beta_{it}$) affects P/S growth negatively as expected and is highly significant. GDP growth ($GDP\text{-}Growth_{it}$) is highly significant and shows that the effects of business cycles varying across countries and time do influence the stock market value (scaled by total sales) very strongly. Sales ($ln(Sales)_{it}$) affect P/S growth significantly negatively. The variables productivity ($ln(Labor/Sales)_{it}$), age ($ln(Age)_{it}$), and stock of patents ($ln(Patents)_{it}$) are not significant.

Next, we provide the results for the dependent variable in logarithmic form ($ln(P/S)_{it}$) in *Model 2*. Compared to the growth of P/S_{it} as a dependent variable, the lag of P/S_{it} exerts a positive effect. This shows that previous performance influences future performance positively and that the performance value is exactly the same coefficient as in growth *Model 1* plus 1, of course. The other coefficients necessarily equal each other in size and direction.

After these preparatory steps, we turn to our research focus if CSR and green innovations are complementarily affecting financial performance. To that end, we apply the previously explained estimation strategy and base the further estimation on *Model 2* by additionally include the variables of interest. In our main estimation approach *Model 3*, we concentrate on the variables representing environmental innovation ($Green_{it}$), CSR (CSR_{it}), and employing both strategies jointly ($BOTH_{it}$) by including them in the estimations (Table 8.2).

In general, the additional variables in *Model 3* do not change the basic model results of *Model 2*. All the variables show the same direction as well as significance level and are similar in size. We apply now our attention to the variables in our focus. *Model 3* shows that a firm strategy of either carrying out green innovation or CSR alone has no significant impact on the financial performance of the firm measured in P/S. However, using both strategies at the same time exerts a highly positive and significant effect on a firm's P/S. Firms that had both strategies in place enjoyed an 8.46% higher price-to-sales ratio compared to the control group, i.e. firms that neither engaged in environmental R&D activities nor in CSR. Based on our assumptions, this suggests that reporting according to the GRI guidelines might help to signal pro-environmental action. Based on these results, we apply the complementarity approach with a one-sided t-test for complementarity against the null, $Green_{it} + CSR_{it} – BOTH_{it} \geq 0$, which supports complementarity. However, we can only reject the null hypothesis with a 90% probability. This would imply

correlation is rejected, too. Therefore, we use a fixed effects model with robust standard errors for the following dynamic panel estimations in a base model.

Table 8.2 Estimation results base models

	Model 1	Model 2	Model 3
	FE robust	FE robust	FE robust
Dependent variable	$gr(P/S)_{it}$	$ln(P/S)_{it}$	$ln(P/S)_{it}$
$ln(P/S)_{it-1}$	−0.830***	0.170***	0.166***
	(0.0236)	(0.0236)	(0.0235)
$Beta_{it}$	−0.233***	−0.233***	−0.232***
	(0.0326)	(0.0326)	(0.0326)
$GDP\text{-}Growth_{it}$	3.460***	3.460***	3.492***
	(0.626)	(0.626)	(0.623)
$ln(Sales)_{it}$	−0.474***	−0.474***	−0.476***
	(0.0441)	(0.0441)	(0.0441)
$ln(Labor/Sales)_{it}$	−0.0567	−0.0567	−0.0545
	(0.0430)	(0.0430)	(0.0428)
$ln(Patents)_{it}$	0.00244	0.00244	0.00237
	(0.00517)	(0.00517)	(0.00514)
$ln(Age)_{it}$	−0.0781	−0.0781	−0.0715
	(0.105)	(0.105)	(0.104)
$Green_{it}$			0.00688
			(0.0261)
CSR_{it}			0.0285
			(0.0204)
$BOTH_{it}$			0.0846***
			(0.0280)
Constant	10.54***	10.54***	10.56***
	(0.967)	(0.967)	(0.966)
Observations	6737	6737	6737
R^2 within	0.620	0.532	0.533
Rho	0.819	0.819	0.818
Test for complementarity: H_0 (full test): $Green_{it} + CSR_{it} - BOTH_{it} \geq 0$			
Test statistic			2.10
p-value			0.0736

Note: The model includes four jointly significant year dummies. Robust standard errors in parentheses. *** $p < 0.01$, ** $p < 0.05$, * $p < 0.1$

complementarity of green innovations and the GRI as a CSR activity, but the result might not hold for all industries or firms. We address this concern in our robustness check on manufacturing firms in the next subchapter.

The inclusion of the lagged dependent variable in the *Models 1–3* might violate the strict exogeneity assumption as explained above. Furthermore, the management decision on carrying out green innovations or engaging in CSR might also be endogenous. As the database does not provide adequate instruments, we estimate models with a dynamic GMM approach suggested by Arellano and Bond (1991). These types of models use estimations in differences, which allows applying lags as instruments. The model is based on a two-step GMM procedure to yield more efficient, i.e. heteroscedasticity robust, estimates. As such, *Models 4* include as an instrumentation vector the controls C_{it} in differences (Table 8.3). The potentially endogenous variable $ln(P/S)_{it-1}$ is instrumented by the second and any further time lags. The variables $Green_{it}$, CSR_{it}, $BOTH_{it}$ are instrumented by the first and further time lags as well as the moving average in *Model 4*.[14]

The instrument of $ln(P/S)_{it-1}$ is subject to endogeneity. The instruments for our variables of concern show in *Model 4* that they are exogenous at the 5% ($p = 0.051$) or the 10% level ($p = 0.344$). This would imply that they are appropriate instruments. With respect to the previous fixed effects estimations and the *Model 3* results in combination with the tests on exogeneity of the instruments, *Model 4* seems the best available estimation strategy for our data. The results for the estimations in *Model 4* mostly confirm the previous results in direction and values. *Model 4* shows again a positive effect of employing both strategies (innovation and CSR) together and this time the coefficient estimate is significant at the 10% level. Thus, with the moving averages as instruments the joint innovation-CSR strategy becomes significant and the size of the coefficient estimate is also close to the previous models. Nevertheless, the instruments in our models are restricted and might violate the strict exogeneity assumption. Concerning complementarity, we test whether condition (8.1) holds. We can reject the null hypothesis with 98% or more for *Model 4*. This implies complementarity of green R&D and CSR of environmental innovation and CSR in form of reporting according to the GRI guidelines as our basic *Model 3* also suggests.

8.3.4 Robustness Checks

In a first robustness check, the sample is restricted to the manufacturing industries (see shaded area in Table 8.4, Appendix). This limits our sample to 2878 observation and 820 firms observed in the years 2005 to 2009. Our estimation models confirm the previous effects of the traditional explanatory variables also for this restricted sample (Table 8.8, Appendix). Although the instruments seem to work better in the limited sample, the effects of the variables of interest show no clear direction when comparing the different models. Furthermore, the complementarity test rather implies no complementarity. For our hypothesis, this would mean that the firm's GRI reporting is a rather poor signal for CSR in the manufacturing sector. However, the descriptive analysis shows that in some of the manufacturing industries both strategies are implemented at the same time. Additional data especially on sectors with green R&D activity would be needed to verify the results and reveal in which sectors CSR reporting might serve as a signal for clients.

[14] In the appendix, we additionally provide *Model 5* where the variables $Green_{it}$, CSR_{it}, $BOTH_{it}$ are instrumented by the first and any further time lags and *Model 6* with the moving average.

Table 8.3 Estimation Results GMM Model 4

	Model 4
	GMM
Dependent variable:	$\ln(P/S)_{it}$
$ln(P/S)_{it-1}$	0.366***
	(0.0435)
$Beta_{it}$	−0.133***
	(0.0356)
$GDP\text{-}Growth_{it}$	1.808**
	(0.774)
$ln(Sales)_{it}$	−0.581***
	(0.0457)
$ln(Labour/Sales)_{it}$	−0.134***
	(0.0465)
$ln(Patents)_{it}$	−0.00294
	(0.00544)
$ln(Age)_{it}$	0.0541
	(0.0996)
$Green_{it}$	−0.0335*
	(0.0189)
CSR_{it}	0.00903
	(0.0180)
$BOTH_{it}$	0.0470*
	(0.0270)
Observations	6,737
Instruments	*2-year and any further lags of* $ln(P/S)_{it}$, *1-year and any further lags and moving average of* $Green_{it}$, CSR_{it}, $BOTH_{it}$
Arellano-Bond test for AR(1) in first differences	0.000
Arellano-Bond test for AR(2) in first differences	0.775
Sargan-Hansen test: *2-year and any further lags of* $ln(P/S)_{it}$	
Excluding group Difference	0.102 0.000
Sargan-Hansen test: *1-year and any further lags of* $Green_{it}$, CSR_{it}, *and* $BOTH_{it}$	
Excluding group Difference	0.000 0.990
Sargan-Hansen test: *moving average of* $Green_{it}$, CSR_{it}, *and* $BOTH_{it}$	
Excluding group Difference	0.000 0.714
Test for complementarity: H_0 (full test): $Green_{it} + CSR_{it} - BOTH_{it} \geq 0$	
Test statistic *p*-value	5.28 0.0108

Note: The model includes four jointly significant year dummies. Robust standard errors in parentheses. *** $p < 0.01$, ** $p < 0.05$, * $p < 0.1$

Another issue consists in employing the GRI as CSR variable, as it is just an approximate measurement of a firm's CSR activity. The sample includes a huge number of firms in the service sector. This leads to the questions of whether it is appropriate to test for green innovation activity and its possible disclosure via CSR, and if CSR variables other than GRI reporting might serve as better signals for environmental R&D. A further limitation might be that the dichotomous indicator only tells us if the firm reports according to the GRI guidelines, which does not necessarily reveal much about the specific activities. As the indicators provided by the database that suit our research focus are mainly dichotomous, we cannot overcome most of these concerns. In the following robustness check we assess whether an alternative CSR indicator might also serve as a signal for environmental R&D or if the results depend on the choice of the CSR variable, as Margolis et al. (2007) and Orlitzky et al. (2003) stress.

For this purpose we use the variable *External Sustainability Audit* (CSR_{it}). This variable reveals if the company assigns its CSR/Sustainability report to an external auditor. As such, the variable could signal the firm's social and environmental CSR activities. The traditional explanatory variables confirm the results of the previous models with the GRI reporting as CSR variable. However, the *Models 11–14* show different effects of the variables of interest concerning direction but no significant ones (Table 8.9, Appendix). The instruments in *Model 14* seem to fit best in comparison to *Models 12* and *13*. The complementarity test shows no complementarity at the conventional significance levels. This implies that the CSR variable *External Sustainability Audit* is a poor signal for green innovation in our sample.

8.4 Discussion and Concluding Remarks

The signaling literature suggests that signals serve as information on unobservable attributes (Spence 2002) to overcome information asymmetries. We apply the signaling theory to the environmental engagement of a firm. As such, we use the firm's green R&D activities, which are hard to observe for stakeholders. Therefore, the firm needs a signal to communicate its environmental activities to differentiate itself from its competitors and gain an advantage. CSR, which has become increasingly important in the last years, can verify a firm's pro-social and pro-environmental engagement and serve as a signal. This is a signal in the sense of Wernerfelt's (1988) reputational economies of scale, which creates reputation not only for one product but for the firm as a whole. As such, CSR is a source of capabilities in the resource-based view of the firm: CSR creates reputation and, in turn, leads to higher financial performance. We analyze if CSR as a signal complements the environmental R&D activity of a firm and whether a joint strategy leads to higher financial performance.

Using data on global companies from the A4 database, we examine if environmental R&D and reporting according to the GRI guidelines are complementary, and consequently we research, if the joint strategy leads to better financial performance.

Our different analyses rather support the hypothesis that a joint strategy leads to higher financial performance, although the effects are rather small. In other words, green innovators can verify their activity through GRI reporting and attract clients. Further research may determine how this works with specific clients, such as consumers or investors.

However, our results do not allow a conclusion on CSR in general, which our additional analyses with a different CSR variable reveal. We cannot conclude that CSR per se is beneficial for green innovators. As Orlitzky and Benjamin (2001) and Margolis et al. (2007) have already pointed out, the effect of CSR on the financial performance of a firm depends on the measurement of CSR. Our results support this viewpoint and reveal that not every kind of CSR is suitable to transport unobservable signals of firms' environmental R&D engagement. Furthermore, the descriptive statistics might explain that the relation of green R&D and CSR depends on firm location and industry. On the one hand, this supports the resource-based view of firms that creating reputation which in turn leads to higher financial performance depends on the uniqueness of capabilities and their specific usage. Therefore, R&D as well as CSR are not advantageous in general. On the other hand, the behavioral view comes into play suggesting that personal values are needed for a social firm strategy.This becomes especially apparent when we interpret our results against the background of the descriptive statistics, which state that in Europe and Asia more firms carry out both strategies jointly. Further research would be necessary to analyze if such values are more expected or more accepted by firms' stakeholders in these countries than in other ones.

Further limitations of our analysis accrue from data constraints. In particular, the different firm strategies might be subject to endogeneity as they could contribute to the same personal management values. Instrumentation via the lagged variables and the moving averages is limited. The results need to be verified with additional data, which may be possible in the future as more and more data on GRI reporting will be available. Moreover, as the ASSET4 database mainly provides dichotomous variables, which might not change much over time, further research with more detailed data would provide better insights. Another drawback is the composition of the sample with a huge number of firms in the service sector, which might report their CSR activity, but are not engaged in innovation.

Nevertheless, our study provides the first results on complementarity of green R&D and CSR related to the signaling theory. As such, it tries to overcome the drawback of previous cross-sectional analysis, which consists in not accounting for unobservable factors by using panel data. We can verify that the signaling effect of CSR strongly depends on the type of CSR.

Acknowledgements This work is part of the project Impact Measurement and Performance Analysis of CSR (Corporate Social Responsibility), funded by the EU (7th Framework Program), Brussels, BE. We are indebted to the participants of the seminars and workshops held in Évora, Frankfurt (Oder), Mannheim, and Toulouse for their comments.

Appendix

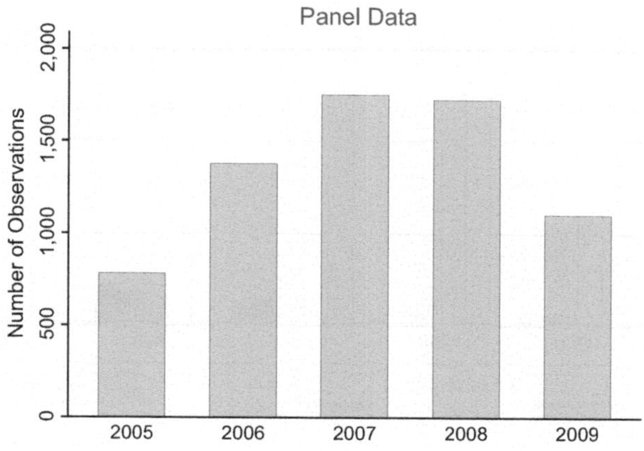

Fig. 8.1 Unbalanced panel data sample (6737 observations)

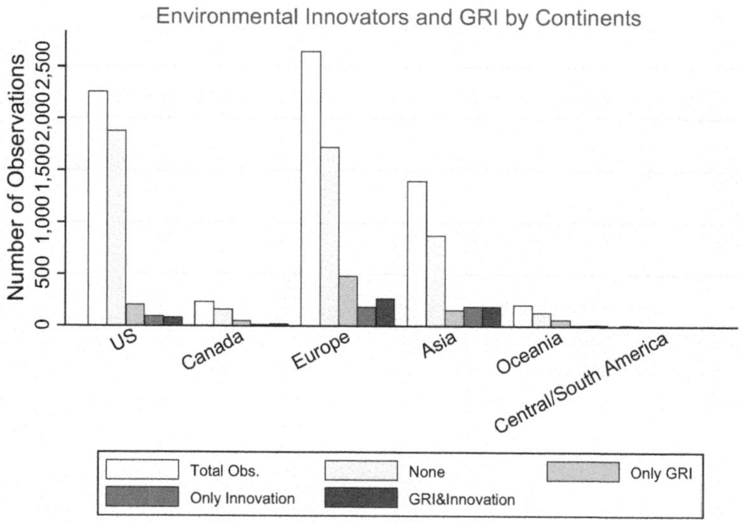

Fig. 8.2 Overview panel data and innovation and CSR categories by continents

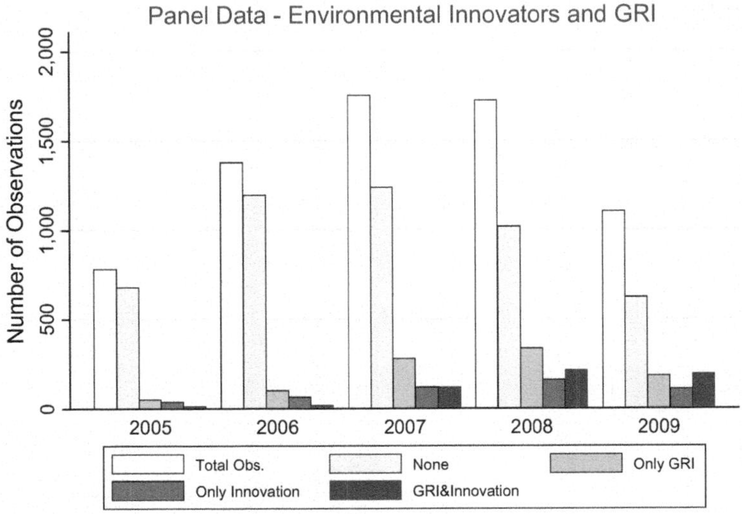

Fig. 8.3 Overview panel data and innovation and CSR categories

Table 8.4 Overview industry sectors (6737 observations)

Industry sectors SIC	No. of observations	Percent (%)
Mining	364	5.40
Construction	237	3.52
Manufacture food	297	4.41
Manufacture wood, paper, print	279	4.14
Manufacture chemicals	525	7.79
Manufacture metal, machinery, transport Eq.	698	10.36
Manufacture computers, electronic Eq.	480	7.12
Manufacture others	599	8.89
Transport, communication, electric	843	12.51
Wholesale and retail trade	583	8.65
Finance, insurance, real Estate	1281	19.01
Services	551	8.18
Total	6737	100.00

Table 8.5 Overview variables and descriptive statistics (6737 observations)

Variable	Definition	Mean (SD)	Min/Max
$ln(P/S)_{it}$	Logarithm ofprice-to-sales ratio	0.1759 (0.9624)	−2.2964/ 2.9498
$ln(Patents)_{it}$	Logarithm of US patents held by company	0.5109 (1.6150)	0/9.9739
$Beta_{it}$	Risk parameter beta	1.1060 (0.6208)	−2.4691/ 6.6454
$GDP\text{-}Growth_{it}$	Real GDP growth by country and year	0.0089 (0.0277)	−0.0854/ 0.1270
$ln(Age)_{it}$	Age of a company in logarithmic form	3.8521 (0.9357)	1.2528/ 6.2851
$ln(Sales)_{it}$	Logarithm of sales for company size	22.4038 (1.3987)	18.0315/ 26.7973
$Ln(Labor/Sales)_{it}$	Labor productivity of company as logarithm of employees by sales	−12.9804 (0.9110)	−18.1888/ −9.4769
$Green_{it}$	Dichotomous innovation variable environmental R&D	0.1542 (0.3612)	0/1
CSR_{it}	Dichotomous CSR variable on CSR reporting according to Global Reporting Initiative guidelines	0.2228 (0.4162)	0/1

Table 8.6 Overview environmental R&D and CSR variables

	Overall	Environmental R&D		CSR	
Year	No. of obs.	No. of observed firms with R&D	Share of observed firms with R&D (%)	No. of observed firms with CSR	Share of observed firms with CSR (%)
2005	781	51	6.53	66	8.45
2006	1377	82	5.95	119	8.64
2007	1751	234	13.36	394	22.50
2008	1724	372	21.58	547	31.73
2009	1104	300	27.17	375	33.97
Total	6737	1039	15.42	1501	22.28

GMM Estimators

Table 8.7 below provides the results for the basic GMM models, where the variables $Green_{it}$, CSR_{it}, $BOTH_{it}$ are instrumented by the first and any further time lags (*Model 5*) or the moving average (*Model 6*).

Compared to *Model 3*, the GMM estimation in *Model 5* shows smaller coefficients, except for the lagged P/S, and in some cases lower significance levels for the traditional variables influencing P/S. Nevertheless, the directions are comparable to the previous results. The coefficient estimates of the variables of interest, green innovation and CSR, again do not statistically differ from zero (however, their sign is

Table 8.7 Estimation Results GMM Models 5 and 6

	Model 5	*Model 6*
	GMM	GMM
Dependent variable	$\ln(P/S)_{it}$	$\ln(P/S)_{it}$
$ln(P/S)_{it-1}$	0.391***	0.409***
	(0.0491)	(0.0474)
$Beta_{it}$	−0.146***	−0.141***
	(0.0346)	(0.0348)
$GDP\text{-}Growth_{it}$	1.640**	1.606**
	(0.755)	(0.784)
$ln(Sales)_{it}$	−0.576***	−0.594***
	(0.0478)	(0.0473)
$ln(Labor/Sales)_{it}$	−0.111**	−0.130***
	(0.0449)	(0.0487)
$ln(Patents)_{it}$	0.000697	−0.000447
	(0.00547)	(0.00548)
$ln(Age)_{it}$	0.0837	0.0784
	(0.0932)	(0.0956)
$Green_{it}$	−0.00337	−0.0826**
	(0.0409)	(0.0359)
CSR_{it}	−0.0590	0.0118
	(0.0444)	(0.0246)
$BOTH_{it}$	0.114*	0.0166
	(0.0613)	(0.0345)
Observations	6737	6737
Instruments	*2-year and any further lags of* $ln(P/S)_{it}$, *1-year and any further lags of* $Green_{it}$, CSR_{it}, $BOTH_{it}$	*2-year and any further lags of* $ln(P/S)_{it}$, *moving average of* $Green_{it}$, CSR_{it}, $BOTH_{it}$
Arellano-Bond test for AR (1) in first differences	0.000	0.000
Arellano-Bond test for AR (2) in first differences	0.713	0.636
Sargan-Hansen test: *2-year and any further lags of* $ln(P/S)_{it}$		
Excluding group	0.041	0.000
Difference	0.000	0.000
Sargan-Hansen test: *1-year and any further lags of* $Green_{it}$, CSR_{it}, *and* $BOTH_{it}$		
Excluding group	0.000	
Difference	0.025	
Sargan-Hansen test: *moving average of* $Green_{it}$, CSR_{it}, *and* $BOTH_{it}$		
Excluding group		0.000
Difference		0.001
Test for complementarity: H_0 (full test): $Green_{it} + CSR_{it} - BOTH_{it} \geq 0$		
Test statistic	10.10	5.17
p-value	0.0007	0.0115

Note: The models include four jointly significant year dummies. Robust standard errors in parentheses. *** $p < 0.01$, ** $p < 0.05$, * $p < 0.1$

Table 8.8 Estimation results GMM models for manufacturing industries

	Model 7	*Model 8*	*Model 9*	*Model 10*
Dependent variable	FE robust	GMM	GMM	GMM
	$\ln(P/S)_{it}$	$\ln(P/S)_{it}$	$\ln(P/S)_{it}$	$\ln(P/S)_{it}$
$\ln(P/S)_{it-1}$	0.111***	0.287***	0.350***	0.292***
	(0.0322)	(0.0786)	(0.0724)	(0.0683)
$Beta_{it}$	−0.197***	−0.111**	−0.107**	−0.102**
	(0.0461)	(0.0447)	(0.0450)	(0.0415)
$GDP\text{-}Growth_{it}$	3.005***	2.188*	2.346*	2.423*
	(0.977)	(1.302)	(1.356)	(1.311)
$\ln(Sales)_{it}$	−0.434***	−0.532***	−0.563***	−0.596***
	(0.0917)	(0.0876)	(0.0974)	(0.0903)
$\ln(Labor/Sales)_{it}$	−0.122	−0.142	−0.175	−0.190*
	(0.0775)	(0.0875)	(0.109)	(0.0969)
$\ln(Patents)_{it}$	0.00259	−1.99e-05	−0.00371	−0.00455
	(0.00602)	(0.00627)	(0.00665)	(0.00678)
$\ln(Age)_{it}$	−0.190	0.149	0.100	0.169
	(0.168)	(0.146)	(0.146)	(0.151)
$Green_{it}$	0.0140	−0.0160	−0.0805**	−0.0366
	(0.0320)	(0.0437)	(0.0406)	(0.0240)
CSR_{it}	0.0176	−0.134**	0.0127	0.00599
	(0.0285)	(0.0613)	(0.0432)	(0.0265)
$BOTH_{it}$	0.0488	0.000594	−0.0221	0.0103
	(0.0382)	(0.0925)	(0.0512)	(0.0347)
Constant	9.257***			
	(2.092)			
Observations	2878	2878	2878	2878
R^2	0.531			
Rho	0.841			
Instruments		*2-year and any further lags of* $\ln(P/S)_{it}$, *1-year and any further lags of* $Green_{it}$, CSR_{it}, $BOTH_{it}$	*2-year and any further lags of* $\ln(P/S)_{it}$, *moving average of* $Green_{it}$, CSR_{it}, $BOTH_{it}$	*2-year and any further lags of* $\ln(P/S)_{it}$, *1-year and any further lags and moving average of* $Green_{it}$, CSR_{it}, $BOTH_{it}$
Arellano-Bond test for AR(1) in first differences		0.000	0.000	0.000
Arellano-Bond test for AR(2) in first differences		0.0639	0.0141	0.0248

(continued)

Table 8.8 (continued)

	Model 7	*Model 8*	*Model 9*	*Model 10*
	FE robust	GMM	GMM	GMM
Dependent variable	$\ln(P/S)_{it}$	$\ln(P/S)_{it}$	$\ln(P/S)_{it}$	$\ln(P/S)_{it}$
Sargan-Hansen test: *2-year and any further lags of* $\ln(P/S)_{it}$				
Excluding group		0.526	0.011	0.541
Difference		0.000	0.000	0.000
Sargan-Hansen test: *1-year and any further lags of* $Green_{it}$, CSR_{it}, *and* $BOTH_{it}$				
Excluding group		0.000		0.000
Difference		0.921		0.998
Sargan-Hansen test: *moving average of* $Green_{it}$, CSR_{it}, *and* $BOTH_{it}$				
Excluding group			0.000	0.000
Difference			0.300	0.503
Test for complementarity: H_0 (full test): $Green_{it} + CSR_{it} - BOTH_{it} \geq 0$				
Test statistic	0.15	4.76	0.78	1.02
p-value	0.3472	0.0146	0.1890	0.1569

Note: The model includes four jointly significant year dummies. Robust standard errors in parentheses. *** $p < 0.01$, ** $p < 0.05$, * $p < 0.1$

now negative). Concerning engagement in both strategies ($BOTH_{it}$), *Model 5* confirms the results from the basic models, as there is a positive effect on P/S but it is only significant at the 10% level. The test developed by Arellano and Bond (1991) for auto-correlation in an auto-regressive process of the first order (AR1) shows significant serial correlation but no significant evidence of serial correlation in the first-differenced errors at order two ($p = 0.713$). This allows us to use lags of more than two-years as instruments. The Sargan-Hansen test for over-identification against the null—i.e. that the vector of instruments is orthogonal to the vector of the errors (or against the null that the instruments are exogenous)—shows that the instrumentation of the variable $ln(P/S)_{it-1}$ is not strictly exogenous ($p = 0.000$). Unfortunately, the database does not provide better instruments and also estimations and tests with longer time lags reveal the same endogeneity problem. Therefore, we have to interpret the results with care. Although formal endogeneity is observable, it might not strongly affect the market value during the next year in reality as P/S is a year-end value. The tests for the subset of instruments with the lags of $Green_{it}$, CSR_{it}, $BOTH_{it}$ confirm exogeneity to be slightly over the 1% level. This might indicate that the strict exogeneity assumption could be violated.

Therefore, we use the moving averages of the innovation and CSR variables as instruments in *Model 6*. For the moving average we calculate the average of the sum of the current year, one-year, and two-year time lag for each of the variables $Green_{it}$, CSR_{it}, $BOTH_{it}$. In this model the traditional variables are similar to *Model 5*, except for the coefficient estimate of the stock of patents ($ln(Patents)_{it}$), which is not significant in all models. The variables of a pure CSR strategy and a joint CSR

Table 8.9 Estimation results GMM models for CSR variable sustainability external audit

	Model 11	*Model 12*	*Model 13*	*Model 14*
Dependent variable	FE robust ln(P/S)it	GMM ln(P/S)it	GMM ln(P/S)it	GMM ln(P/S)it
lag-ln(P/S)$_{it}$	0.168***	0.390***	0.430***	0.391***
	(0.0235)	(0.0477)	(0.0469)	(0.0465)
Beta$_{it}$	−0.232***	−0.160***	−0.153***	−0.152***
	(0.0326)	(0.0353)	(0.0361)	(0.0365)
GDP-Growth$_{it}$	3.485***	1.390*	1.351*	1.576**
	(0.623)	(0.783)	(0.775)	(0.799)
ln(Sales)$_{it}$	−0.474***	−0.595***	−0.612***	−0.603***
	(0.0443)	(0.0477)	(0.0479)	(0.0461)
ln(Labor/Sales)$_{it}$	−0.0565	−0.106**	−0.126***	−0.119***
	(0.0431)	(0.0445)	(0.0470)	(0.0455)
ln(Patents)$_{it}$	0.00248	0.00235	−0.000962	0.00104
	(0.00516)	(0.00558)	(0.00568)	(0.00566)
ln(Age)$_{it}$	−0.0780	0.0866	0.109	0.0885
	(0.104)	(0.0949)	(0.0974)	(0.102)
Green$_{it}$	0.0264	−0.0450	−0.0560	−0.0154
	(0.0251)	(0.0564)	(0.0393)	(0.0197)
CSR$_{it}$	−0.00980	−0.0211	0.00375	0.00390
	(0.0276)	(0.0509)	(0.0351)	(0.0210)
BOTH$_{it}$	0.0443	0.0186	−0.0449	−0.00942
	(0.0312)	(0.0744)	(0.0423)	(0.0302)
Constant	10.54***			
	(0.969)			
Observations	6737	6737	6737	6737
R^2	0.533			
Rho	0.819			
Instruments		*2-year and any further lags of ln(P/S)$_{it}$, 1-year and any further lags of Green$_{it}$, CSR$_{it}$, BOTH$_{it}$*	*2-year and any further lags of ln (P/S)$_{it}$, moving average of Green$_{it}$, CSR$_{it}$, BOTH$_{it}$*	*2-year and any further lags of ln(P/S)$_{it}$, 1-year and any further lags and moving average of Green$_{it}$, CSR$_{it}$, BOTH$_{it}$*
Arellano-Bond test for AR(1) in first differences		0.000	0.000	0.000
Arellano-Bond test for AR(2) in first differences		0.750	0.540	0.679

(continued)

Table 8.9 (continued)

	Model 11	Model 12	Model 13	Model 14
Dependent variable	FE robust	GMM	GMM	GMM
	ln(P/S)it	ln(P/S)it	ln(P/S)it	ln(P/S)it
Sargan-Hansen test: *2-year and any further lags of* $ln(P/S)_{it}$				
Excluding group		0.010	0.000	0.003
Difference		0.000	0.000	0.000
Sargan-Hansen test: *1-year and any further lags of* $Green_{it}$, CSR_{it}, *and* $BOTH_{it}$				
Excluding group		0.000		0.000
Difference		0.022		0.681
Sargan-Hansen test: *moving average of* $Green_{it}$, CSR_{it}, *and* $BOTH_{it}$				
Excluding group			0.000	0.000
Difference			0.004	0.453
Test for complementarity: H_0 (full test): $Green_{it} + CSR_{it} - BOTH_{it} \geq 0$				
Test statistic	0.53	1.58	0.02	0.00
p-value	0.2339	0.1042	0.4407	0.4766

Note: The model includes four jointly significant year dummies. Robust standard errors in parentheses. *** $p < 0.01$, ** $p < 0.05$, * $p < 0.1$

and R&D strategy are not significant. However, now green R&D affects P/S significantly negatively (5% level) but only with a small coefficient. Again the Arellano-Bond test at order one shows significant serial correlation, but no significant evidence of serial correlation in the first-differenced errors at order two ($p = 0.636$). We have to reject the null hypothesis of exogeneity of the subsets of instruments at the conventional 10% or 5% levels for both subsets.

Anderson-Hsiao Estimator

Table 8.10 below provides the results for the basic dynamic model setup, in which the endogeneity of the lagged dependent variables, but not the endogeneity of the key variables of interest, has been accounted for.

The most obvious insight from this table is that the coefficient estimate of the lagged dependent variable is far away from plausible values and also from the very basic OLS estimates provided in Table 8.5. The reason probably is a considerable instrumental variable bias due to a weak instrument problem. Recall that the results from Table 8.10 rely on a mode setup where all variables enter the model in differences. Although the correlation of the price-to-sales ratio in period t and $t-1$ is relatively high (0.883), the correlation of the first differences and lagged first differences is very small (−0.2354) making it a bad instrument. Also the first stage regressions support this view. The coefficient estimate of the excluded instrument in the structural equation is relatively low; let alone the fact that its level of significance

Table 8.10 Estimation results Anderson-Hsiao estimator

	(App. 1)	(App. 2)
	AH robust	First stage
Dependent variable	$\Delta\ln(P/S_{it})$	$\Delta\ln(P/S_{it-1})$
$\Delta ln(P/S)_{it-1}$	−2.754**	
	(1.093)	
$\Delta ln(P/S)_{it-2}$		0.0478**
		(0.0198)
$\Delta Beta_{it}$	−0.397***	−0.0268
	(0.0756)	(0.0198)
$\Delta GDP\text{-}Growth_{it}$	30.02***	8.003***
	(8.807)	(0.292)
$\Delta ln(Sales)_{it}$	0.349	0.357***
	(0.413)	(0.0338)
$\Delta ln(Labor/Sales)_{it}$	0.0353	0.0357
	(0.0828)	(0.0282)
$\Delta ln(Patents)_{it}$	−0.00836	−0.00137
	(0.0107)	(0.00383)
$\Delta ln(Age)_{it}$	−0.0645	−0.0667
	(0.308)	(0.114)
$\Delta Green_{it}$	0.0558	0.0456**
	(0.0719)	(0.0184)
ΔCSR_{it}	0.106	0.0570***
	(0.0724)	(0.0140)
$\Delta BOTH_{it}$	0.207**	0.0809***
	(0.104)	(0.0221)
Constant	0.0412	0.0123
	(0.0277)	(0.00786)
Observations	5633	5633
R-squared		0.219
rho		

Note: Robust standard errors in parentheses. *** $p < 0.01$, ** $p < 0.05$, * $p < 0.1$

is rather small, which supports the concern of a weak instrument problem. In particular, the F-statistic of the excluded instrument in the first stage regression ($F = 5.81$) is far away from areas considered to support non-weakness of instruments. Staiger and Stock (1997) propose a rule of thumb of a value of ten for the first stage F-statistic of a single excluded instrument to provide evidence for non-weakness. The central insight from this simple experiment is straightforward. Even in this basic setup, which only addresses the endogeneity of one variable, namely the lagged dependent variable, the Anderson-Hsiao estimator performs rather poorly given our data as lagged differences of the price-to-sales ratio, which is only loosely correlated with current values. Therefore, further lags as instruments might help mitigate this problem as in the Arellano-Bond GMM case. In this sense, the Arellano-Bond estimator seems to be a better choice allowing more consistent estimates, at least in part.

References

Aboody, D., & Lev, B. (2000). Information asymmetry, R&D, and insider gains. *The Journal of Finance, 55*(6), 2747–2766.

Aghion, P., et al. (2005). Competition and innovation: An inverted-U relationship. *The Quarterly Journal of Economics, 102*(2), 701–728.

Akerlof, G. A. (1970). The market for "lemons": Quality uncertainty and the market mechanism. *The Quarterly Journal of Economics, 84*(3), 488–500.

Arellano, M., & Bond, S. (1991). Some tests of specification for panel data: Monte Carlo evidence and an application to employment equations. *The Review of Economic Studies, 58*(2), 277–297.

Arora, A., & Gambardella, A. (1990). Complementarity and external linkages: The strategies of the large firms in biotechnology. *The Journal of Industrial Economics, 38*(4), 361–379.

Arora, S., & Cason, T. N. (1995). An experiment in voluntary environmental regulation participation in EPAs 33/50 program. *Journal of Environmental Economics and Management, 28*(3), 271–286.

Arora, S., & Gangopadhyay, S. (1995). Toward a theoretical model of voluntary overcompliance. *Journal of Economic Behavior & Organization, 28*(3), 289–309.

Bénabou, R., & Tirole, J. (2010). Individual and corporate social responsibility. *Economica, 77* (305), 1–19.

Bhattacharya, S. (1979). Imperfect information, dividend policy and "the bird in the hand" fallacy. *The Bell Journal of Economics, 10*(1), 259–270.

Bowen, F. (2007). Corporate social strategy: Competing views from two theories of the firm. *Journal of Business Ethics, 75*(1), 97–113.

Branco, M. C., & Rodrigues, L. L. (2006). Corporate social responsibility and resource-based perspectives. *Journal of Business Ethics, 69*(2), 111–132.

Brynjolfsson, E., & Milgrom, P. (2013). Complementarity in organizations. In R. Gibbons & J. Roberts (Eds.), *The handbook of organizational economics* (pp. 11–54). Princeton, NJ: Princeton University Press.

Cañón-de-Francia, J., & Garcés-Ayerbe, C. (2009). ISO 14001 environmental certification: A sign valued by the market? *Environmental and Resource Economics, 44*(2), 245–262.

Cavaco, S., & Crifo, P. (2014). CSR and financial performance: Complementarity between environmental, social and business behaviours. *Applied Economics, 46*(27), 3323–3338.

Chauvin, K. W., & Hirschey, M. (1994). Goodwill, profitability, and the market value of the firm. *Journal of Accounting and Public Policy, 13*(2), 159–180.

Cochran, P. L., & Wood, R. A. (1984). Corporate social responsibility and financial performance. *The Academy of Management Journal, 27*(1), 42–56.

Dowell, G., Hart, S. L., & Yeung, B. (2000). Do corporate global environmental standards create or destroy market value? *Management Science, 46*(8), 1059–1074.

Elsayed, K., & Paton, D. (2005). The impact of environmental performance on firm performance: Static and dynamic panel data evidence. *Structural Change and Economic Dynamics, 16*(3), 395–412.

Fombrun, C. J. (1996). *Reputation: Realizing value from the corporate image*. Boston, MA: Harvard Business School Press.

Fombrun, C. J., & Shanley, M. (1990). What's in a name? Reputation building and corporate strategy. *Academy of Management Journal, 33*(2), 233–258.

Gorodnichenko, Y., Svejnar, J., & Terrell, K. (2010). Globalization and innovation in emerging markets. *American Economic Journal: Macroeconomics, 2*(2), 194–226.

Griffin, J. J., & Mahon, J. F. (1997). The corporate social performance and corporate financial performance debate: Twenty-five years of incomparable research. *Business & Society, 36*(1), 5–31.

Harrington, D. R., Khanna, M., & Deltas, G. (2008). Striving to be green: The adoption of total quality environmental management. *Applied Economics, 40*(23), 2995–3007. https://doi.org/10.1080/00036840600994005

Hart, S. (1995). A natural-resource-based view of the firm. *Academy of Management Review, 20*(4), 986–1014.

Hart, S. L., & Ahuja, G. (1996). Does it pay to be green? An empirical examination of the relationship between emission reduction and firm performance. *Business Strategy and the Environment, 5*(1), 30–37.

Heras-Saizarbitoria, I., & Boiral, O. (2013). ISO 9001 and ISO 14001: Towards a research agenda on management system standards. *International Journal of Management Reviews, 15*(1), 47–65.

Heras-Saizarbitoria, I., Molina-Azorín, J. F., & Dick, G. P. M. (2011). ISO 14001 certification and financial performance: Selection-effect versus treatment-effect. *Journal of Cleaner Production, 19*(1), 1–12.

Hopenhayn, H. A. (1992). Entry, exit, and firm dynamics in long run equilibrium. *Econometrica, 60* (5), 1127–1150.

Hull, C. E., & Rothenberg, S. (2008). Firm performance: the interactions of corporate social performance with innovation and industry differentiation. *Strategic Management Journal, 29* (7), 781–789.

Husted, B. W. (2005). Risk management, real options, corporate social responsibility. *Journal of Business Ethics, 60*(2), 175–183.

King, A. A., & Lenox, M. J. (2001). Does it really pay to be green? *Journal of Industrial Ecology, 5* (1), 105–116.

Kitzmueller, M., & Shimshack, J. (2012). Economic perspectives on corporate social responsibility. *Journal of Economic Literature, 50*(1), 51–84.

Klassen, R. D., & McLaughlin, C. P. (1996). The impact of environmental management on firm performance. *Management Science, 42*(8), 1199–1214.

Kleinknecht, A., van Montfort, K., & Brouwer, E. (2002). The non-trivial choice between innovation indicators. *Economics of Innovation and New Technology, 11*(2), 109–121.

Konar, S., & Cohen, M. A. (2001). Does the market value environmental performance? *Review of Economics and Statistics, 83*(2), 281–289.

Kotchen, M. J. (2006). Green markets and private provision of public goods. *Journal of Political Economy, 114*(4), 816–834.

Lioui, A., & Sharma, Z. (2012). Environmental corporate social responsibility and financial performance: disentangling direct and indirect effects. *Ecological Economics, 78*, 100–111.

Lyon, T. P., & Maxwell, J. W. (2008). Corporate social responsibility and the environment: A theoretical perspective. *Review of Environmental Economics and Policy, 2*(2), 240–260.

Margolis, J. D., Elfenbein, H. A., & Walsh, J. (2007). Does it pay to be good? A meta-analysis and redirection of research on the relationship between corporate social and financial performance. *Ann Arbor, 1001*, 48109–41234.

McGuire, J. B., Sundgren, A., & Schneeweis, T. (1988). Corporate social responsibility and firm financial performance. *Academy of Management Journal, 31*(4), 854–872.

McWilliams, A., & Siegel, D. (2000). Corporate social responsibility and financial performance: Correlation or misspecification? *Strategic Management Journal, 21*(5), 603–609.

Milgrom, P., & Roberts, J. (1990). The economics of modern manufacturing: Technology, strategy, and organization. *The American Economic Review, 80*(3), 511–528.

Miravete, E. J., & Pernías, J. C. (2010). Testing for complementarity when strategies are dichotomous. *Economics Letters, 106*(1), 28–31.

Nakamura, M., Takahashi, T., & Vertinsky, I. (2001). Why Japanese firms choose to certify: A study of managerial responses to environmental issues. *Journal of Environmental Economics and Management, 42*(1), 23–52.

Nickell, S. J. (1996). Competition and corporate performance. *Journal of Political Economy, 104* (4), 724–746.

Nishitani, K. (2009). An empirical study of the initial adoption of ISO 14001 in Japanese manufacturing firms. *Ecological Economics, 68*(3), 669–679.

Nishitani, K. (2010). Demand for ISO 14001 adoption in the global supply chain: An empirical analysis focusing on environmentally conscious markets. *Resource and Energy Economics, 32* (3), 395–407.
Orlitzky, M. (2005). Social responsibility and financial performance: Trade-off or virtuous circle? *University of Auckland Business Review, 7*(1), 37–43.
Orlitzky, M., & Benjamin, J. D. (2001). Corporate social performance and firm risk: A meta-analytic review. *Business & Society, 40*(4), 369–396.
Orlitzky, M., Schmidt, F. L., & Rynes, S. L. (2003). Corporate social and financial performance: A meta-analysis. *Organization Studies, 24*(3), 403–441.
Pava, M. L., & Krausz, J. (1996). The association between corporate social-responsibility and financial performance: The paradox of social cost. *Journal of Business Ethics, 15*(3), 321–357.
Penrose, E. (1959). *The theory of the growth of the firm*. Oxford: Oxford University Press.
Poddi, L., & Vergalli, S. (2009). *Does corporate social responsibility affect the performance of firms?* (Fondazione Enri Enrico Mattei. Fondazione Eni Enrico Mattei 52 (9)), Brescia: Institutions and Markets.
Porter, M. E., & van der Linde, C. (1995). Toward a new conception of the environment-competitiveness relationship. *The Journal of Economic Perspectives, 9*(4), 97–118.
Potoski, M., & Prakash, A. (2005). Green clubs and voluntary governance: ISO 14001 and firms' regulatory compliance. *American Journal of Political Science, 49*(2), 235–248.
Preston, L. E., & O'Bannon, D. P. (1997). The corporate social-financial performance relationship: A typology and analysis. *Business & Society, 36*(4), 419–429.
Rexhäuser, S., & Rammer, C. (2014). Environmental innovations and firm profitability: Unmasking the porter hypothesis. *Environmental and Resource Economics, 57*(1), 145–167.
Riley, J. G. (2001). Silver signals: twenty-five years of screening and signaling. *Journal of Economic Literature, 39*(2), 432–478.
Ross, S. A. (1977). The determination of financial structure: The incentive-signalling approach. *The Bell Journal of Economics, 8*(1), 23–40.
Russo, M. V., & Fouts, P. A. (1997). A resource-based perspective on corporate environmental performance and profitability. *Academy of Management Journal, 40*(3), 534–559.
Smith, K. (2005). Measuring innovation. In J. Fagerberg, D. C. Mowery, & R. R. Nelson (Eds.), *The Oxford handbook of innovation* (1st ed., pp. 148–177). New York: Columbia University.
Spence, M. (1973). Job market signaling. *The Quarterly Journal of Economics, 87*(3), 355–374.
Spence, M. (2002). Signaling in retrospect and the informational structure of markets. *The American Economic Review, 92*(3), 434–459.
Staiger, D., & Stock, J. F. (1997). Instrumental variables regression with weak instruments. *Econometrica, 65*(3), 557–586.
Telle, K. (2006). It pays to be green – A premature conclusion? *Environmental and Resource Economics, 35*(3), 195–220.
Terlaak, A. (2007). Satisficing signaling: corporate social strategy and certified management standards. *Academy of Management Proceedings, August*(1), 1–6.
Terlaak, A., & King, A. A. (2006). The effect of certification with the ISO 9000 quality management standard: A signaling approach. *Journal of Economic Behavior & Organization, 60*(4), 579–602.
Toms, J. S. (2002). Firm resources, quality signals and the determinants of corporate environmental reputation: Some UK evidence. *The British Accounting Review, 34*(3), 257–282.
Waddock, S. A., & Graves, S. B. (1997). The corporate social performance- financial performance link. *Strategic Management Journal, 18*(4), 303–319.
Wernerfelt, B. (1984). A resource-based view of the firm. *Strategic Management Journal, 5*(2), 171–180.
Wernerfelt, B. (1988). Umbrella branding as a signal of new product quality: An example of signalling by posting a bond. *The RAND Journal of Economics, 19*(3), 458–466.

Chapter 9
Environmental Innovation and Corporate Sustainability: A 15-Year Comparison Based on Survey Data

Hüseyin Doluca, Benedikt Holzner, and Marcus Wagner

9.1 Introduction

The political sustainable development agenda initiated by the World Commission on Environment and Development (World Commission on Environment and Development 1987) has long reached the corporate sector. Consequently, non-financial goals such as environmental and social aspects have been integrated into corporate management, resulting in environmentally-related product or process innovations. Although there exist several databases and related analyses (Wagner 2007, 2008; Horbach 2008; Horbach et al. 2012; Schaltegger et al. 2013) the development of environmental innovation and corporate sustainability has not yet been analyzed over a long period in high detail. That is why we aim at addressing the research question of how corporate sustainability and environmental innovation activities developed over the past 15 years.

The answer to this question is of interest because environmental protection increasingly gains not only the attention of businesses and politics it also receives significant societal and media attention. Based on a unique dataset containing partly longitudinal survey data from 2001 and 2016, we present the status quo and the development of corporate sustainability efforts among manufacturing firms in Germany and the United Kingdom (UK). We further provide disaggregated results for seven different manufacturing industries while differentiating firm size (small-, medium- and large-sized firms). We find an overall increase of environmental activities and environmental management system (EMS) certification. However, some activities differ across industries and countries. Managerial activities and EMS implementation have overall greater popularity in Germany, while its level of adoption depends on the specific

H. Doluca (✉) · B. Holzner · M. Wagner
Chair of Management, Innovation and International Business, University of Augsburg, Augsburg, Germany
e-mail: hueseyin.doluca@wiwi.uni-augsburg.de; benedikt.holzner@wiwi.uni-augsburg.de; marcus.wagner@wiwi.uni-augsburg.de

J. Horbach, C. Reif (eds.), *New Developments in Eco-Innovation Research*, Sustainability and Innovation, https://doi.org/10.1007/978-3-319-93019-0_9

activity. Our results inform practitioners as well as researchers and politicians by providing insights about the development of environmental innovation and corporate sustainability activities across different manufacturing industries in these two countries.

The remainder of our analysis is organized as follows. We start with a brief summary of related research and an explanation of the research approach. Following this, results are presented in three parts: first, we compare corporate ecological sustainability between Germany and the UK as well as over time. Second, the status quo of corporate social sustainability is shown for both countries. Third, we examine corporate ecological sustainability for Germany more in-depth with regard to industry affiliation and firm size. At the end of the chapter, some general conclusions and a discussion of our results are provided.

9.2 Literature Review

Environmental innovation has been defined as the application or introduction of new products and processes contributing to the reduction of environmental burdens or to ecologically specified sustainability targets (Rennings 2000). Corporate sustainability behavior is more comprehensive and covers all corporate activities related to ecological, economic and social issues aiming at realizing a global and long-term sustainable development path. Existing literature suggests that environmental innovation and corporate sustainability behavior differ depending on different factors such as firm size, the main industry in which a firm is active and the type of environmental innovation. More specifically, since radical technological innovations are less likely pursued by larger firms (Almeida and Kogut 1997), environmental product and process innovations should be analyzed with respect to firm size, since process innovations tend to be more incremental. Analyzing product and process innovation separately is additionally necessary because the implementation of environmental management systems is found to be positively associated with environmental process innovation, whereas no empirical association is found with environmental product innovation (Wagner 2007). Only specific activities such as information of consumers and eco-labeling are shown to positively impact product innovations (Wagner 2008), indicating that a detailed activity-based analysis has to be performed.

Extant literature has not much addressed corporate sustainability as concerns of small and medium-sized enterprises (SME), since it either analyzed only large firms or the analysis focused on environmental aspects. Hence, existing analyses either do not provide a fully differentiated view with regard to firm size effects or do not cover all relevant sustainability aspects. For example, Schaltegger et al. (2013) focus on large companies only, yet empirical research indicates that SME differ in their approach to corporate sustainability (Wagner and Schrauth 2014; Wagner and Schaltegger 2003). By analyzing across all different firm sizes, we thus provide a more differentiated analysis and contribute novel insights to the literature. With our study we analyze a longer period than ever before and additionally provide a cross-country comparison of European countries. This enables us to point out national strengths and weaknesses and to assess the development of corporate sustainability by benchmarking it internationally. Furthermore, we can comment on the status quo of corporate sustainability and environmental innovation with our dataset and thus provides insights on recent trends. In this analysis, we describe the current situation of corporate sustainability and environmental innovation and compare the results with earlier studies.

9.3 Data and Method

Building on the European Business Environment Barometer (EBEB) of 2001 survey, to gain more recent insights, we collected data for 2016 in the context of the European Sustainability Management Barometer (ESMB) survey. The ESMB survey was conducted among manufacturing companies and thus continues the work of the EBEB. In the current round of 2016, the ESMB surveyed firms in the UK, Germany, Austria, Switzerland, Italy and Greece. In this report, we focus on a comparison of Germany and the UK, since for these countries sufficient responses were received in both years to make an exploratory statistical analysis feasible. As many questions in the 2016 survey are identical to those from 2001, we can assess the development of sustainability management over a 15-year period, which is unique in the context of large-scale studies on corporate sustainability management. We distributed the questionnaires to a random sample of manufacturing firms. The pooled dataset contains 783 observations 562 of which are from Germany. Based on this sample, we carry out an exploratory data analysis in order to establish trends and international differences as concerns corporate sustainability and environmental innovation as well as to identify the status quo in the industrialized countries Germany and UK. We methodologically build on frequency counts and Box-Whisker plots to assess first and second distributional moments in the data.

The composition of the dataset—especially regarding firm size (and to a lesser degree also with regard to industry structure) is not completely identical in the two countries as well as over time, which should be considered when making comparisons. We define the company size by the number of employees. To do so, we follow the European Union recommendation to classify companies into the following aggregate categories: small (below 50 employees), medium (50–249 employees) and large (at least 249 employees). While in 2001 the participants in Germany and the UK had a similar size distribution, this changed for 2016, when 64% of the German companies have at least 250 employees, whereas in the UK only around one-fourth of the responding firms reached this size. However, these differences in size distribution reflect however to a large degree the macro-industrial structure in each country and thus are mostly unavoidable in a comparative survey context such as ours (Fig. 9.1).

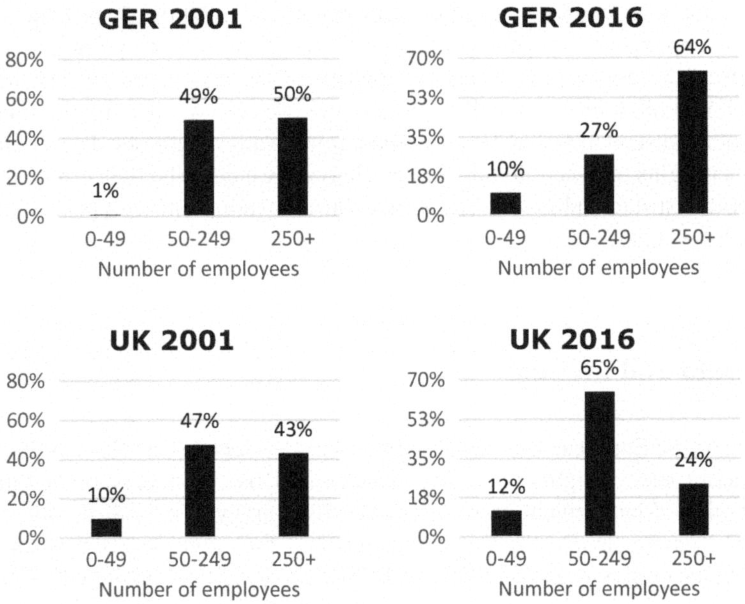

Fig. 9.1 Company size by year and country

9.4 Analysis

We first describe the results of technological and managerial ecological sustainability and environmental management activities as concerns the differences between Germany and the UK, and the development between 2001 and 2016. Second, the internal and external social sustainability activities are compared between the two countries. Finally, the German results are analyzed in more depth as concerns company size and industry affiliation.

9.4.1 Ecological Sustainability

In this part, we examine the results of 2001 and 2016 for Germany as well as for the UK. Various operational and managerial activities and environmental management system (EMS) adoption are focused upon.

9.4.1.1 Operational Environmental Activities

In the survey, 19 operational activities to improve environmental performance were listed and the participants had to state whether or not they implemented them in the

Table 9.1 Operational activities by country and year

Operational activity	GER 2001 (%)	GER 2016 (%)	UK 2001 (%)	UK 2016 (%)
Reduce waste	65	78	34	57
Substitution of hazardous input	46	76	35	62
Reduce material per unit	48	69	34	64
Reduce water use	52	69	32	68
Reduce air emission	53	66	39	52
Cleaner production technology	53	63	38	57
Packaging recycling	66	60	53	81
Reduce noise emission	46	58	34	48
Reduce transport energy	31	53	27	52
Material recycling	46	51	66	95
"Green" new product design	42	50	26	38
Product recycling	38	47	36	76
Reduce packaging per unit	46	38	30	62
Substitution of non-renewable materials	16	36	19	36
Reduce water emission	29	34	29	38
Biodiversity conservation		34		38
Emissions offsetting		30		38
Biodiversity restoration		29		38
Use of foreign waste streams	9	17	16	50

prior 3 years. As three activities premiered in the 2016 survey, only for the remaining 16 activities we show the 15-year comparison. For those operational activities Table 9.1 shows the adoption rates[1] of the different technological activities in descending order of the 2016 shares in Germany. Overall, the responding firms from the UK have made greater progress over the past 15 years. In 2001, the average responding firm in Germany adopted 41% of technological activities while in the UK it were only 34%. Fifteen years later, in 2016, the direction of this difference was reversed such that the average British firm adopted more technological activities (55%) than its German counterpart (50%). Differences also exist between German and British firms with respect to both, the most and less often adopted activities. In Germany, the top two activities are the reduction of waste and the substitution of hazardous products, while the latter activity increased the most (by 30%) within the last 15 years. In terms of the largest increase, substitution of hazardous products is subsequently followed by the activities reduction of transport energy and substituting non-renewable materials. It is conspicuous that in Germany recycling activities show little change compared with the remaining activities. The adoption rate of packaging recycling even decreased over the 15-year period.

[1]The adoption level is calculated by dividing the number of firms having undertaken the respective activity by the total number of firms.

The three activities being evaluated for the first time in 2016, namely emissions offsetting, biodiversity conservation, and biodiversity restoration, differ markedly in Germany from the activities already covered in 2001, with adoption rates ranging between 29 and 34%. In 2016, only the usage of foreign waste streams has been less often adopted.

In contrast to the German respondents, British firms place a stronger focus on recycling. The top three activities are material, packaging and product recycling with almost every company reutilizing materials (95%). It is notable that no activity adoption level decreased in the 15-year period but 9 out of 16 increased by more than 25%. Similar to Germany, in the UK activities concerning biodiversity restoration, biodiversity conservation and emissions offsetting are adopted by the lowest number of firms and have identical adoption levels of 38%, which are slightly larger than in Germany.

The activities around the implementation of cleaner technologies in the production process and "Green" design of new products reflects the environmental innovation performance of a company, corresponding to process and product innovation, respectively. Although the activities' adoption rates did increase in the 15-year period to a greater extent in the UK than they did in Germany, on average more German firms adopted innovation activities. In 2016, 50% of the German and 38% of the British firms had undertaken a "Green" product innovation activity within the past 3 years. A cleaner technology was applied by 63% of German and 57% of UK firms. In 2001, for almost every activity the share of adopting companies is higher in Germany. The only exceptions are material recycling, the use of foreign waste streams and the substitution of hazardous products. However, over the considered time period, British companies achieved higher adoption rates for almost every activity. Hence, in 2016 a differentiated picture is observed with regard to the leadership of the respective activities between the two countries. Firms in the UK seem to focus on recycling rather than focusing on more efficient production, the latter predominantly being done by German firms (which lead in substituting hazardous input as well as in reducing output in terms of waste).

9.4.1.2 Managerial Environmental Activities

We also surveyed 20 managerial environmental activities in the same manner as described for the operational activities. Table 9.2 shows the managerial activities covered, sorted in descending order by the 2016 results for German respondents. Overall, the managerial activities have higher adoption rates. Two-thirds of the activities have been adopted by at least two-thirds of the respondents in Germany and half of them of its UK counterparts. Responsibilities are most often adopted in Germany (90%), followed by environmental goals being part of a continuous improvement process and having measurable environmental goals (both 84%). In the UK, procedures to handle legal requirements (95%) and written environmental policies have the highest adoption rates (90%).

Table 9.2 Managerial activities by country and year

Managerial activity	GER 2001 (%)	GER 2016 (%)	UK 2001 (%)	UK 2016 (%)
Clear responsibilities	74	90	53	86
Improvement process for environm. goals	54	84	44	81
Measurable environm. goals	52	84	41	81
Procedure to handle legal requirements	57	84	70	95
Written environm. policy	53	83	69	90
Programs for environm. goals	47	83	38	81
Environm. performance indicators	38	82	26	62
Environm. program audit	43	81	38	67
Review EMS efficiency	40	77		38
Separate environm./HSE report	42	75	25	43
Environm. staff trainings	54	71	36	67
Initial environm. review	57	70	62	67
Supplier selection by environm. performance	51	66	39	62
Environm. data in annual report	38	66	30	76
Demand suppliers to take environm. actions	45	64	32	43
Consumer information about environm. effects	33	48	32	43
Eco-labeling	16	44	14	19
Life cycle assessment for products	18	39	13	48
Benchmarking with other companies	16	27	22	19
Market research on 'Green' products	15	23	18	24

Environm. = Environmental
HSE = Health and Safety Executive

Manufacturing companies in both countries rarely benchmark their own environmental performance with other companies. Furthermore, market research for specifically environmental-friendly ("Green") products is also rare (19 and 27%, respectively). Eco-labels show a growing popularity in Germany (44%). In the UK, they are less widespread (19%). Reviews of EMS efficiency, environmental performance indicators and placing a demand on suppliers to take environmental activities are also topics, that get substantially more attention from German companies. They also publish a separate environmental report more often (+32%). In comparison, British companies put this information in the annual report more often (+10%), which makes the difference less pronounced. Except for this and procedures for identification and evaluation of relevant legal requirements, managerial activities are generally adopted to a greater degree by German companies, as compared to firms in the UK.

9.4.1.3 Environmental Management Systems

In this section, we show the trend of implementing an EMS for Germany and the UK. EMS cover the implementation, organization, and advancement of operational environmental management. Besides, a certified EMS can signal the level of environmental performance to outside parties, which is used to reliably assess the benefits that result from corporate environmental activities. Furthermore, we show the relation between environmental activities other than EMS and the implementation of an EMS.

The diffusion of EMSs has grown over the past 15 years (Fig. 9.2). German manufacturers tend to implement them rather more frequently than UK ones. In 2016, 79% of the responding firms had implemented an EMS in Germany, compared to 52% in the UK. Even in 2001, the German share was bigger: with 45% of the firms having implemented an EMS, it was 17% higher. Non-certified systems are very rare in both countries.

EMS theoretically should support the implementation of environmental activities. Therefore, it is expected that companies with an EMS perform more activities. We can further assume companies have a stronger motivation for environmental protection if they are willing to implement an EMS. We present the relation between the presence of an EMS and the number of managerial and operational activities for German and British companies in this section. We do not differentiate between the different types of certification. The Box-and-Whisker plot (Fig. 9.3) confirms our expectations and shows further information about the distribution of the number of implemented activities: companies with an EMS are found to have implemented more operational environmental activities. In 2001 and in 2016, the number of activities was clearly higher, even though there were some time- and country-specific peculiarities.

For German companies with an EMS, the median remained almost unchanged.[2] The above discussed overall increase is due to the increase of both, the lower and upper quartiles. In 2016, the number of implemented activities was on average higher in the UK, where the median reached a value of 11 for companies with EMS and 9 for companies without one. Surprisingly, the firms without an EMS raised both the median and the quartile values in both countries. This suggests a somewhat limited role of certification, since evidently other factors such as regulations or increased public awareness must have driven the average number of technological activities up in firms without a certified EMS, a finding that is consistent with earlier research (Hertin et al. 2008; Tyteca et al. 2002).

For managerial activities, the existence of an EMS makes a big difference (Fig. 9.4). In all surveys, the median of participants with an EMS is at least 14. For those without an EMS, only the UK 2016 survey reached a value above four. It is

[2]Note that also we have checked for outliers, for robustness reasons we prefer using the median instead of using the mean.

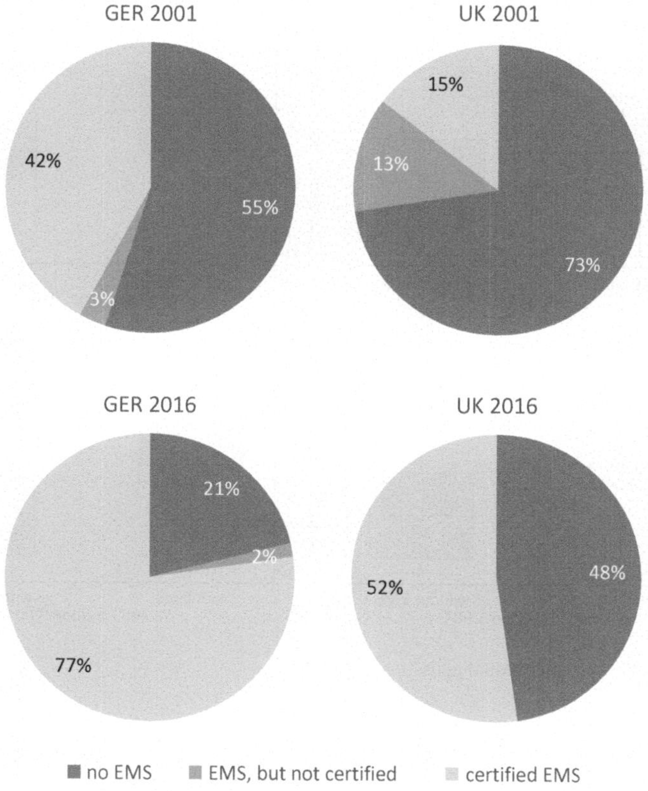

Fig. 9.2 Adoption of EMS

worth mentioning, that some of the activities enquired about are required to certify an EMS. Therefore, their adoption rate is 100% in the companies with an EMS.

9.4.2 *Internal and External Social Sustainability*

Alongside the ecological activities, companies are also concerned about social issues. This section examines what internal and external efforts companies make in support of social sustainability. We describe the current dissemination levels of 17 internal and 21 external activities in total. The results are presented for Germany and the UK in 2016.

The results show the internal activities in descending order for the share in Germany. The most frequent activities in Germany and the UK are the "offer for health protection" and "general education and training programs for employees", followed by "employee suggestion scheme". While in Germany nearly every company is implementing those activities (91–93%), in the UK about two-third of the

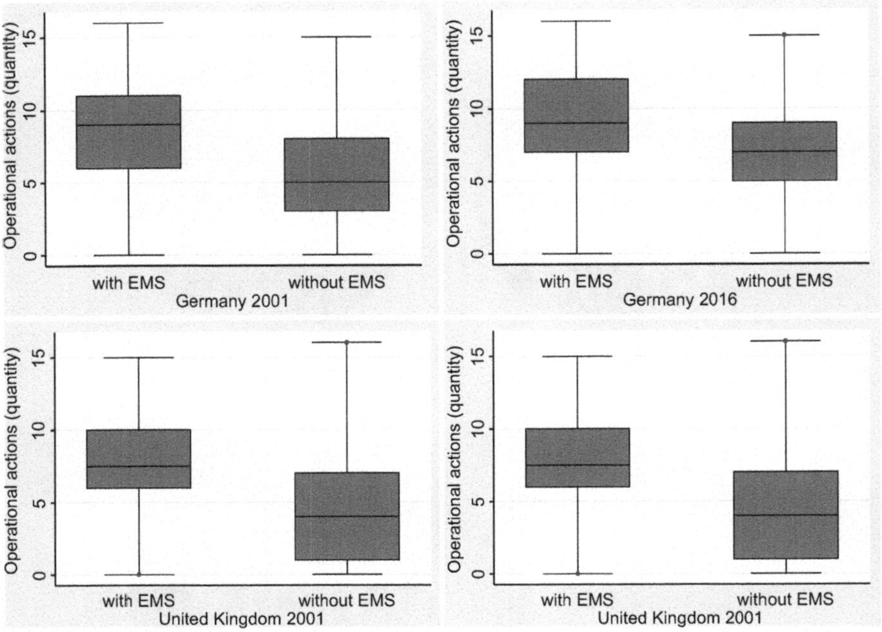

Fig. 9.3 Boxplots of operational activities vs. EMS

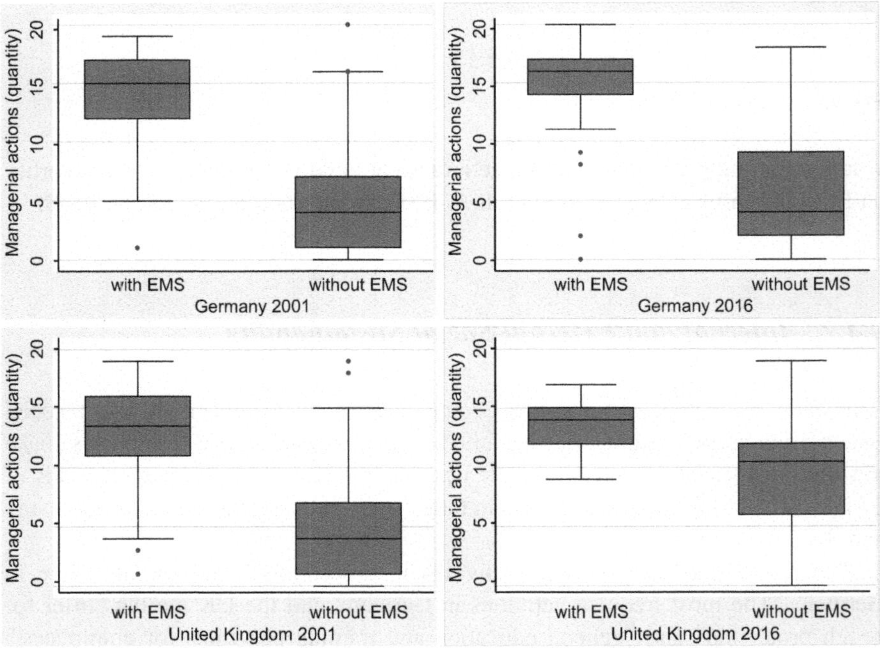

Fig. 9.4 Boxplots of managerial activities vs. EMS

participants do so. These activities bring direct benefits such as less sick days or a better qualification of employees for the companies. However, British firms lead in terms of support for the childcare of employees and support of gender diversity. Other activities for the equal treatment of all employees like "ethnic diversity plans" (77% vs. 56%) or the "fair distribution of wages" (68% vs. 53%) have an average adoption level (across all activities): in Germany and the UK, respectively 51 and 33% of the participants apply social standards like the "Recommendation 146" of the International Labour Organisation (ILO). A balanced scorecard incorporating sustainability aspects was implemented by 28 and 19% of the responding firms, respectively. The UN Global Compact is an initiative for socially responsible business policies. With its ten principles, it is meant to promote a sustainable economy worldwide. Nineteen percent of the German manufacturers and 13% of the British ones joined this agreement. Among the firms in Germany and the UK, 26 and 19% respectively use a quality management system based on the European Foundation for Quality Management (EFQM) model that includes societal and employee welfare goals. Only 10% of the German and none of the UK companies implemented the ISO 26000 standard for social responsibility (Table 9.3).

Most of the firms offer apprenticeship positions under their external social sustainability activities. Fair trading relationships are especially supported in Germany (87%). Sport and cultural sponsoring is also more popular in Germany with shares being around twice the level of those in the UK. Support for the local community is important in both countries (71 and 65%, respectively). A company's regional integration has a positive effect when it comes to recruiting or retaining employees, but also with regard to support for sustainable regional development.

Table 9.3 Internal social sustainability activities

	GER (%)	UK (%)
Health protection	93	76
General education program	92	76
Employee suggestion scheme	91	76
High level social benefits	85	41
Individual work time models	83	71
Ethnic diversity plans	77	56
Qualification activities for job returners	76	38
Flexible work place design	70	65
Fair distribution of wages	68	53
Gender diversity support	68	75
Social standards (e.g. ILO 146)	51	33
Time for education on issues relevant for society at large	49	38
Support with child care by the company	47	59
Sustainability balanced scorecard	28	19
EFQM-based management system	26	19
UN Global Compact membership	19	13
ISO 26000 implementation	10	0

Table 9.4 External social sustainability activities

	GER (%)	UK (%)
Apprenticeship positions	93	71
Fair trading relationships	87	41
Support/sponsoring of sport events	76	35
Community support	71	65
Support for regions of company's locations	67	47
Justice-marked commodities	64	24
Cultural sponsoring	56	29
Social issues reporting or sustainability reporting	54	44
Support of education initiatives	50	19
Stakeholder dialogue initiatives	43	47
Social performance indicators	43	40
Promotion of human rights	42	41
Corporate volunteering	41	47
Corporate citizenship activities	34	35
OECD guidelines for multinational enterprises	30	7
Social justice programs abroad	30	31
Aid to homeless	19	12
Social marketing	16	27
Social accounting	16	20
Social Accountability (SA) 8000 standard	15	6
"Fair Trade" declaration of products	13	14

This might explain the support for the region in which the firms' operations are located (67 and 47%). Sixty four percent support justice-marked commodities (i.e. commodities for which just trading relations with customers, suppliers and other business partners exist), while only 13% declare their own products to be "Fair Trade" certified. Apart from this, only the adoption of the Social Accountability 8000 standard, which deals with social accounting in general and social marketing, is less than or equal to 15% in both countries (Table 9.4).

9.4.3 Detailed Analysis by Industry and Size for Germany

In this third part, the results for Germany will be analyzed in more depth with respect to effects of the participating companies' size and industry. Further, we will examine which companies cooperate during product planning and development as concerns environmental aspects. Since some questions were not asked in the UK in 2001, we can no longer consider British firms in this part of the analysis.

9.4.3.1 Categorizations via Industry and Company Size

In order to provide improved comparability, we defined 7 industry classes out of 21 options offered to the participants for identifying their main industry at a detailed level. These aggregate industry classes are "Consumer industry", "Wood, paper, publishing and printing products", "Chemical industry", "Glass, ceramic and metal products", "Engineering and vehicle construction", and "Electric and electronic devices". Additionally, there is the class of "Other manufacturing industries", which contains firms assigning themselves to this class. In order to remain parsimonious, due to low number of participants in the utilities, transport and recycling sectors they have been assigned as other manufacturing industries.

The distribution of the participants across the so-defined aggregate industry classes is as follows. In 2001 "Glass, ceramic and metal products" and the "Consumer industry" had the biggest shares each with 18%. In the 2016 survey, most companies were part of the "Chemical industry", representing 16%. At the expense of all remaining industry classes, the other manufacturing companies are larger representated in 2016 (30% against 16%). Apart from this, the structure is similar in both surveys, enabling a meaningful comparison over time. As concerns the distribution of the total population of German manufacturing firms, in 2014, the largest industry in terms of the number of companies was "Glass, ceramic and metal products", followed by "Consumer goods" and "Engineering and vehicle construction" (Statistisches Bundesamt 2016). These three industries represent the second, third, and fourth largest industries in our responses. Therefore, the results reported in the following are broadly representative for the German manufacturing sector as a whole as concerns industry distribution.

We further aggregated the companies by size, based on the categories reported in the introduction for the number of employees. Overall, the size of respondents ranges from 6 to 610,000 employees. Whilst in 2001 half of the participants had between 50 and 250 employees, in 2016 firms with more than 250 employees are having the largest share (47%). Small companies with less than 50 employees were represented only in smaller numbers in 2001. In the 2016 survey, they account for 10%. This change in size composition in our responses may affect the results. The distribution in each size category based on the aggregated categories is displayed in Table 9.5.

9.4.3.2 Operational Environmental Activities

We now take a closer look at the operational activities taken to diminish or prevent negative environmental impacts. As can be seen from Fig. 9.5, independent of firm size more activities have been implemented over time. Furthermore, in both periods, the quantity of operational activities rises with the size of the company. In 2016, the median of the medium-sized firms is approaching that of large firms. Some large firms implemented all 16 activities for the first time in 2016. The influence of size is

Table 9.5 Firm size distribution by aggregate industry category

Year	2001			2016		
Industry	01–49 (%)	50–249 (%)	≥250 (%)	01–49 (%)	50–249 (%)	≥250 (%)
Consumer industry	80	19	17	18	11	11
Wood, paper, publishing, and printing	0	14	6	14	14	6
Chemical industry	0	13	13	23	14	16
Glass, ceramic and metal products	20	21	15	0	19	12
Engineering and vehicle construction	0	12	17	0	0	17
Electric and electronic devices	0	11	10	14	13	8
Other manufacturing industries	0	10	20	32	29	31

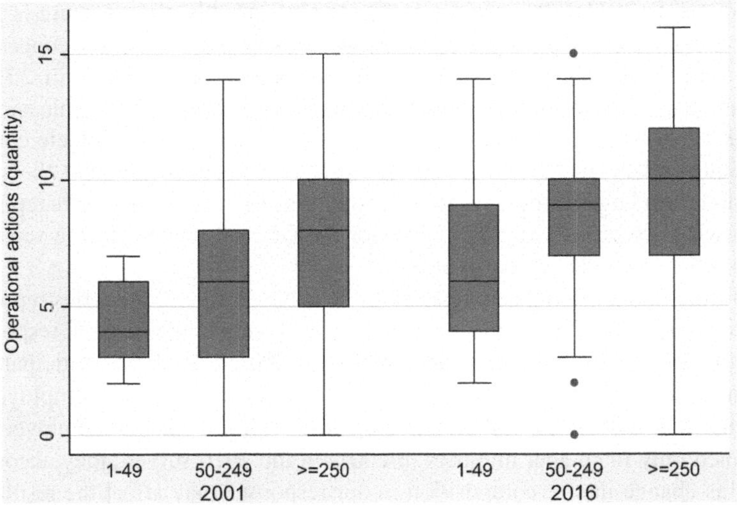

Fig. 9.5 Boxplot of operational activities by aggregate size category in Germany

particularly visible when examining the small firms. Small firms turn out to implement a distinctively smaller amount of activities. This is likely the case because, typically, they have less resources and thus activity pursuance is more challenging. However, even for the small firms, the maximum number of activities implemented increased from 7 to 15, in doing so catching up remarkably with the maximum number of activities of the medium-sized firms in 2016. This underscores the considerably increased relevance of sustainable operations in every size category in 2016.

As can be seen from Fig. 9.5 these size-specific results are essentially determined by the adoption of individual activities: in most cases, larger companies are more likely to implement operational activities. Nevertheless, there are exceptions, especially small enterprises have the biggest share in product recycling and in reducing the packaging per unit of product and have generally caught up very strongly over the years. Overall, medium-sized companies have as well caught up with large enterprises over the past 15 years, as indicated by the shrinking gap of many activity levels. However, as concerns the crucial activities for environmental innovation of implementation of cleaner production technologies and "Green" new product designs large firms have kept their lead from 2001 to 2016. Thus, whilst encouragingly over all aggregate size categories interest in environmental innovation has considerably increased between 2001 and 2016, large companies keep their leading edge. As concerns the newly introduced items on ecosystem services, small and large enterprises interestingly engage more often in the restoration or conservation of biodiversity than medium-sized ones. Opposed to this, emissions offsetting as a comparatively new tool, is to date much more often implemented by larger companies (Table 9.6).

To consider industry-specific differences in operational environmental activities, we compare the seven aggregated industry categories defined above. In all of them, the number of implemented operational activities in 2016 is higher than 15 years ago. The "Chemical industry" and the "Engineering and vehicle-constructing sector" are leading in both periods. They only swapped their position over time. Except for the "Other manufacturing industries", all remaining aggregate industries share the same median of six operational activities in 2001. In 2016, they still share a median, but now it is at nine activities in which firms are engaged. Together with the second quartile now being engaged at not less than five activities, the progress to a higher operational level is unambiguously visible across all industries (Fig. 9.6).

To reveal differences between the aggregate industries, their specific adoption level for every activity is shown in Table 9.7. "Engineering and vehicle construction" has most often the highest level of adoption, especially as concerns the reduction of various emissions. However, every aggregate industry is leading in terms of adoption for at least one activity, confirming that environmental exposure is industry-specific. The "Consumer industry", for example, is most concerned about recycling issues. The reduction of transport energy is an activity of increasing importance across all aggregate industries. Other activities, like packaging-related ones, show less consistent patterns. While gaining larger shares in the "Wood, paper, publishing and printing products" industry, they lose shares in many other aggregate industries. Overall, companies in the "Wood, paper, publishing and printing products" industry, followed by firms in the "Consumer industry" have made the biggest progress over the last 15 years resulting in average adoption rates of 53.9 and 51.7%, respectively. Firms in the "Engineering and vehicle construction" industry (60.5%) and the "Chemical industry" (56.0%) have even higher values and thus the best environmental performance in 2016. In contrast in the "Glass, ceramic and metal products" industry, the average adoption rate per activity is 46.8% and is thus the lowest one across all aggregated sectors.

Table 9.6 Adoption of operational activities by aggregate size category in Germany

	Small enterprises			Medium enterprises			Large enterprises		
	2001 (%)	2016 (%)	Δ (%)	2001 (%)	2016 (%)	Δ (%)	2001 (%)	2016 (%)	Δ (%)
Reduce water consumption	40	37	−3	42	62	20	59	78	19
Reduce material per unit	40	50	10	43	70	27	48	75	**27**
Material recycling	40	41	1	34	44	10	54	53	−1
Use of foreign waste streams	20	14	−6	8	12	4	7	15	8
Substitution of non-renewable materials	0	23	23	10	40	30	20	40	20
Substitution hazardous input	20	41	21	29	76	47	59	83	24
Reduce air emission	0	36	**36**	40	56	15	66	76	10
Reduce water emission	40	18	−22	17	27	10	38	40	2
Reduce noise emission	20	36	16	40	62	22	50	59	9
Reduce waste	20	82	**62**	51	75	24	74	77	3
Product recycling	20	59	**39**	26	48	21	47	47	0
Packaging recycling	60	68	8	61	65	4	66	59	−7
Reduce packaging per product unit	40	50	10	40	33	−7	48	42	−6
Reduce transport energy	40	41	1	25	43	18	35	61	**26**
Cleaner technology	0	46	**46**	38	54	16	55	73	18
"Green" new product design	40	41	1	32	46	14	48	60	12
Biodiversity restoration		32			21			34	
Biodiversity conservation		36			24			41	
Emissions offsetting		14			19			36	

Differences bigger 25% in bold

9.4.3.3 Managerial Environmental Activities

The following section analyses in more depth the managerial activities in terms of size and industry differences. As Fig. 9.7 shows, the number of implemented managerial activities increased in all three size categories over the 15-year period from 2001 to 2016. Especially the mid-sized companies increased their median values from 4 to 15 activities. While the number of implemented activities differed strongly between the company sizes in 2001, the median in 2016 is almost equal across all three size categories and the differences manifest mostly in the quartile values. Still, large companies implement more activities and only outliers in this size category implement less than 11 of the managerial activities as Fig. 9.7 shows. This also implies that competitive differentiation based on managerial activities becomes increasingly difficult for large firms.

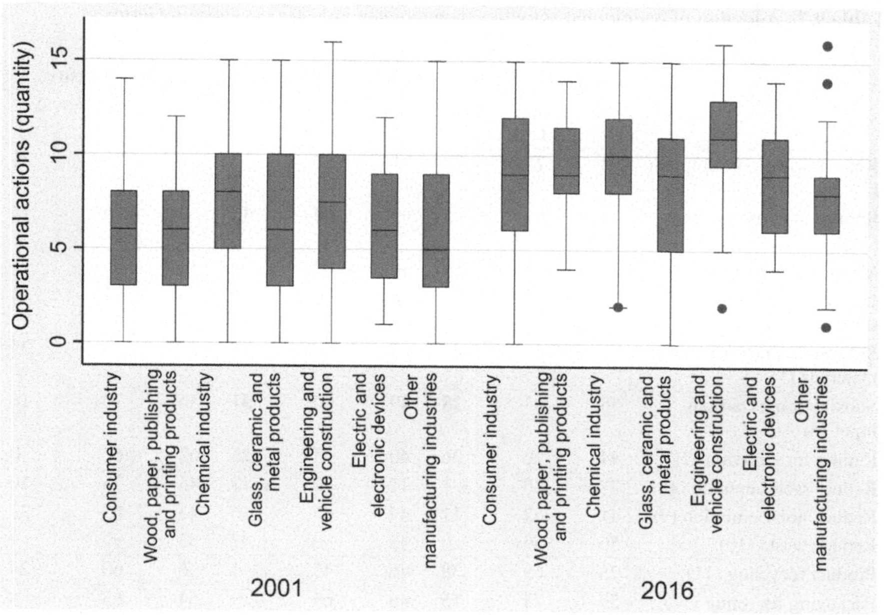

Fig. 9.6 Boxplots of operational activities by aggregate industry category in Germany

This development can also be seen in Table 9.8. The adoption share of almost every activity increased for medium-sized and large enterprises. Environmental performance indicators, reviews of the EMS efficiency or an audit of the environmental program have been activities which are rarely adopted in manufacturing companies with 50–249 employees in 2001 (23–25% adoption rates). Over time, these activities have become more common in this size category as evidenced by higher adoption shares in 2016 (71–78% adoption rates). Measurable environmental goals and a separate report for environmental, health and safety topics evolved similarly. In large firms, eight activities are essentially standards in practice with about nine out of ten firms implementing them, partly because they are mandatory elements required for EMS certification. Eco-label usage increased massively in companies with more than 50 up to 250 employees where it now has an adoption share of 59%, which is the highest across all three size categories. As in 2001, the adoption share of medium-sized companies for this activity is a little higher than the one of large companies (46% vs. 43%). One explanation for this may be that eco-labelled products are often more regional products and therefore more often produced by smaller (and local) companies. It is possibly also harder for large companies to establish the environmental quality level required for an eco-label uniformly over a much larger volume of inputs, as it is the case in food production and paper manufacturing.

Finally, Fig. 9.8 shows that the differences across aggregate industry categories decreased in a way that in 2016 the median of almost every industry is the same and at a level of 15 implemented managerial activities. The "Wood, paper, publishing

Table 9.7 Adoption of operational activities by aggregate industry category in Germany

	Consumer industry			Wood/paper/ publishing/printing			Chemical industry		
	2001 (%)	2016 (%)	Δ (%)	2001 (%)	2016 (%)	Δ (%)	2001 (%)	2016 (%)	Δ (%)
Reduce water consumption (1)	52	74	22	51	60	9	55	76	21
Reduce material per unit (2)	40	63	23	43	90	**47**	62	76	14
Material recycling (3)	33	44	11	40	60	20	48	65	17
Use of foreign waste streams (4)	0	7	7	8	20	11	10	14	4
Substitution non-renewable materials (5)	8	52	**44**	17	50	**33**	21	46	25
Substitution hazardous input (6)	35	63	**28**	49	80	**31**	56	70	16
Reduce air emission (7)	44	70	**26**	40	65	**25**	62	62	0
Reduce water emission (8)	27	26	−1	17	30	13	45	35	−10
Reduce noise emission (9)	41	52	11	49	65	16	43	65	22
Reduce waste (10)	56	59	0	43	90	**47**	83	92	9
Product recycling (11)	25	63	**38**	46	45	−1	36	60	24
Packaging recycling (12)	59	74	15	46	65	19	64	65	1
Reduce packaging per unit (13)	49	44	−5	26	55	**29**	50	51	1
Reduce transport energy (14)	46	63	17	17	35	18	26	60	**34**
Cleaner technology (15)	40	70	16	43	70	27	55	76	22
"Green" new product design (16)	32	56	**24**	31	55	**39**	50	65	**15**
Biodiversity restoration (17)		37			20			27	
Biodiversity conservation (18)		37			30			38	
Emissions offsetting (19)		26			40			24	

	Glass/ceramic/metal products			Engineering/vehicle construction			Electric/electronic devices			Other manufacturers		
	2001 (%)	2016 (%)	Δ (%)	2001 (%)	2016 (%)	Δ (%)	2001 (%)	2016 (%)	Δ (%)	2001 (%)	2016 (%)	Δ (%)
(1)	59	63	4	50	88	**38**	39	52	13	40	70	**30**
(2)	43	77	**34**	54	80	**25**	53	70	17	38	61	24
(3)	43	40	−3	52	54	2	44	57	13	45	41	−4
(4)	18	20	2	12	17	5	3	0	−3	9	16	6
(5)	14	20	6	17	25	8	8	48	**40**	23	36	13
(6)	30	87	**57**	56	92	**36**	50	78	**28**	43	77	**34**
(7)	66	73	7	39	83	**45**	44	57	12	57	63	6
(8)	31	30	−1	25	54	**29**	14	26	12	32	36	4
(9)	51	70	19	54	71	17	14	39	**25**	49	48	−0
(10)	59	60	1	64	92	**28**	66	91	25	57	69	12
(11)	36	43	7	52	58	6	44	61	17	25	33	8
(12)	69	57	−12	77	71	−6	75	70	−5	51	49	−2
(13)	38	30	−8	56	50	−6	55	44	−11	32	30	−2
(14)	21	50	**29**	32	63	**31**	17	48	**31**	34	54	20

(continued)

Table 9.7 (continued)

	Glass/ceramic/metal products			Engineering/vehicle construction			Electric/electronic devices			Other manufacturers		
	2001 (%)	2016 (%)	Δ (%)	2001 (%)	2016 (%)	Δ (%)	2001 (%)	2016 (%)	Δ (%)	2001 (%)	2016 (%)	Δ (%)
(15)	54	63	−1	48	80	**32**	33	78	**45**	43	50	−7
(16)	33	53	**31**	60	75	15	61	65	4	23	37	**14**
(17)		13			29			26			40	
(18)		10			41			26			49	
(19)		30			29			22			33	

Differences greater than 25% in bold

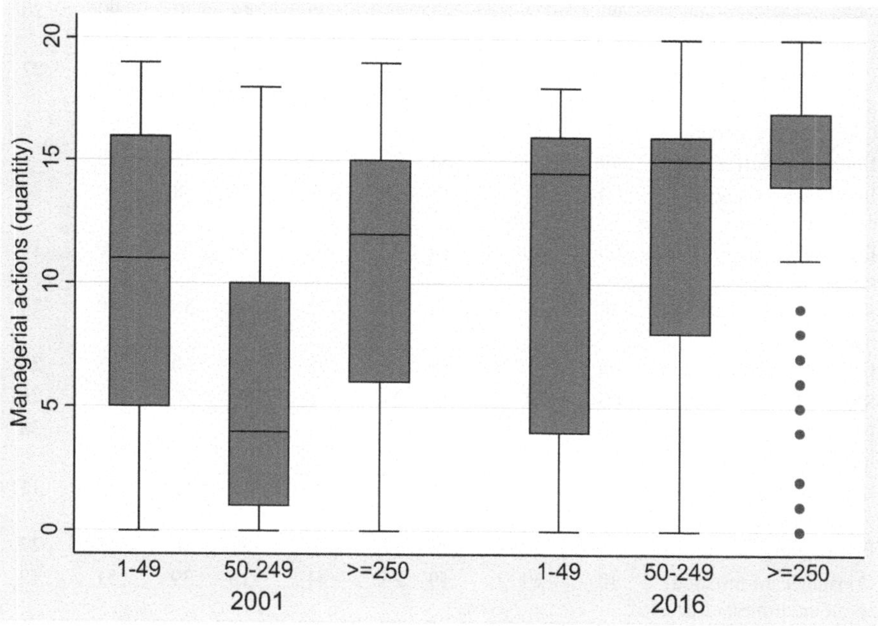

Fig. 9.7 Boxplot of managerial activities by aggregate size category in Germany

and printing product" industry's median is with a median value of 16 even a bit higher. Fifteen years ago, the average value for this sector was four and thus the lowest across all aggregate industries, which witnesses a remarkable improvement. The "Electric and electronic devices" industry and the "Consumer industry" have a larger quartile spread than the other five aggregate industries, which suggests that in the former the variability with regard to environmental management is considerably bigger. This suggests that some firms in these two industry categories still lag more behind, in particular since 15 years ago, they already had the second and third lowest median across all seven aggregate industries. Still, for the manufacturing sector in Germany overall, we see a remarkable shift towards increased adoption of

Table 9.8 Adoption of managerial activities by aggregate size category in Germany

	Small enterprises			Medium enterprises			Large enterprises		
	2001 (%)	2016 (%)	Δ (%)	2001 (%)	2016 (%)	Δ (%)	2001 (%)	2016 (%)	Δ (%)
Supplier selection by environm. performance	60	55	−5	42	62	20	59	70	11
Demand suppliers to take environm. actions	40	32	−8	31	54	23	58	72	15
Written environm. policy	60	68	8	36	78	**42**	70	92	22
Procedure to handle legal requirements	80	64	−16	39	78	**39**	74	93	19
Initial environm. review	40	68	**28**	43	64	21	73	73	0
Measurable environm. goals	80	73	−7	33	79	**46**	70	90	20
Programs for environm. goals	60	68	8	31	75	**44**	63	90	**27**
Clear responsibilities	80	77	−3	63	91	**28**	87	93	6
Environm. staff trainings	40	73	**33**	35	62	**27**	73	77	4
Improvement process for environm. goals	60	77	17	41	78	**37**	68	90	22
Environm. data in annual report	60	50	−10	24	64	**40**	52	70	18
Separate environm./HSE report	60	68	8	28	73	**45**	56	80	24
Environm. program audit	60	68	8	25	78	**53**	60	88	**28**
Review EMS efficiency	60	64	4	24	76	**53**	56	82	**26**
Environm. performance indicators	60	73	13	24	71	**47**	51	89	**38**
Benchmarking with other companies	20	9	−11	11	24	13	19	30	11
Eco-labeling	0	59	**59**	17	46	**29**	16	43	**27**
Consumer information about environm. effects	40	59	19	25	41	16	39	51	12
Market research 'Green' products	20	9	−11	11	16	5	18	28	10
Product life cycle assessment	40	23	−17	8	27	19	26	46	20

Differences greater 25 % in bold
Environm. = Environmental
HSE = Health and Safety Executive

managerial activities supporting environmental protection. However, it is somewhat less evident that this has resulted in a complementary increase in the adoption rate of operational and technological activities to protect the environment. We cannot ascertain from our survey that this increase in activities has also improved actual environmental performance in terms of lower emissions and resource consumption. This is particularly difficult to establish since in the last two decades globalization

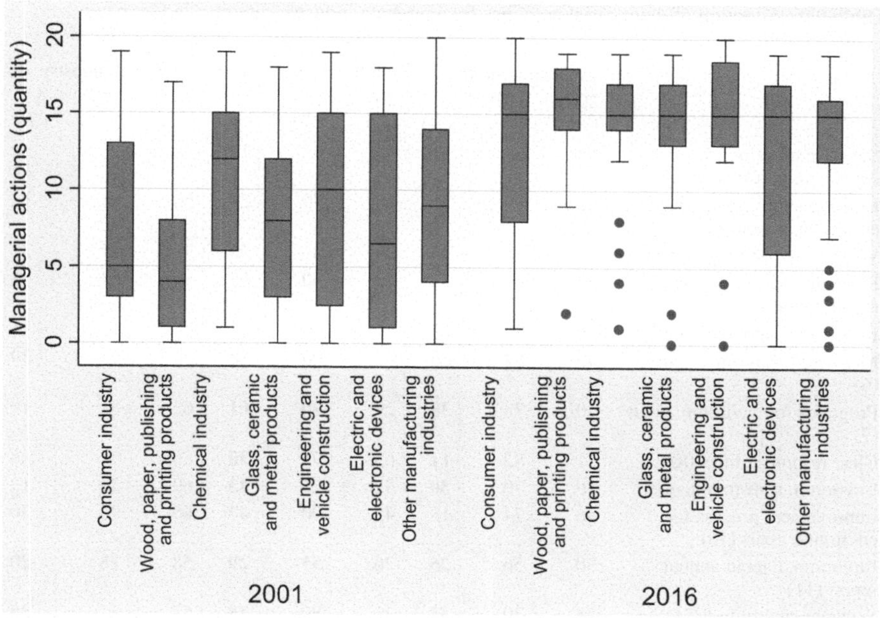

Fig. 9.8 Boxplots of managerial activities by aggregate industry category in Germany

processes have continued to fragmentize value chains by means of outsourcing and offshoring. This makes a reliable assessment of actual environmental performance very difficult since this would require accounting for a shift of polluting activities which increasingly move beyond the direct firm boundaries, which is highly challenging due to constraints in data availability.

Table 9.9 shows the adoption shares by aggregate industry for individual managerial activities in detail, which supports the results derived from the Box-Whisker plots in Fig. 9.7. All activities' adoption shares have (often significantly) increased in all industries. However, there are still industry-specific differences. The companies in the "Wood, paper, publishing and printing products" industry raised their shares the most, which corresponds to the earlier observation in this respect. More specifically, for 15 out of the 20 activities surveyed, the firms increased their adoption share by more than 25%. Firms in the "Chemical industry" and "Glass, ceramic and metal products" industry have most often the largest adoption share. Manufacturers of "Electric and electronic devices" show the least improvement. In half of the activities surveyed, they have a lower adoption share than in any of the other aggregate industries. The definition and introduction of measurable environmental goals, as well as the publication of environmental reports, are examples of activities in which the industry lags behind. Some activities are still not broadly implemented by manufacturing firms in Germany, as for example conducting market research for "Green" products is an activity with relatively low adoption rates across all industries. In only three aggregate industries, more than a quarter of the participants stated

Table 9.9 Adoption of managerial activities by aggregate industry category in Germany

	Consumer industry			Wood/paper/ publishing/printing			Chemical industry		
	2001 (%)	2016 (%)	Δ (%)	2001 (%)	2016 (%)	Δ (%)	2001 (%)	2016 (%)	Δ (%)
Supplier selection by environm. performance (1)	52	63	11	52	65	13	57	65	8
Demand suppliers to take environm. actions (2)	48	59	11	46	70	24	50	68	18
Written environm. policy (3)	44	78	**34**	29	85	**56**	74	95	21
Procedure to handle legal requirements (4)	50	89	**39**	28	90	**62**	81	83	3
Initial environm. review (5)	46	70	25	47	60	13	76	65	−11
Measurable environm. goals (6)	45	82	**36**	32	90	**58**	62	92	**30**
Programs for environm. goals (7)	40	78	**38**	29	90	**61**	62	81	19
Clear responsibilities (8)	71	82	11	65	95	**30**	86	92	6
Environm. staff trainings (9)	40	70	**30**	32	75	**43**	69	81	12
Improvement process for environm. goals (10)	56	74	18	43	90	**47**	65	95	**30**
Environm. data in annual report (11)	30	56	**26**	26	55	**29**	58	78	20
Separate environm./HSE report (12)	38	70	**32**	32	80	**48**	57	87	**30**
Environm. program audit (13)	36	74	**38**	23	90	**67**	61	89	**28**
Review EMS efficiency (14)	37	63	**26**	16	85	**69**	57	81	24
Environm. performance indicators (15)	38	86	**47**	23	85	**62**	56	89	**33**
Benchmarking with other companies (16)	15	22	7.0	21	45	24	22	27	5
Eco-labeling (17)	32	67	**35**	17	80	**63**	17	38	21
Consumer inform. about environm. effects (18)	37	56	19	26	55	**29**	46	46	0
Market research 'Green' products (19)	23	33	10	13	40	**27**	22	24	2
Product life cycle assessment (20)	10	22	12	22	40	18	31	60	**29**

	Glass/ceramic/metal products			Engineering/vehicle construction			Electric/electronic devices			Other manufacturers		
	2001 (%)	2016 (%)	Δ (%)	2001 (%)	2016 (%)	Δ (%)	2001 (%)	2016 (%)	Δ (%)	2001 (%)	2016 (%)	Δ (%)
(1)	50	63	13	55	79	24	60	78	18	40	63	23
(2)	39	80	**41**	53	75	22	46	61	15	39	51	13
(3)	54	90	**36**	58	88	**30**	53	70	17	65	87	22
(4)	61	90	**29**	58	92	**34**	63	83	20	67	83	16
(5)	66	70	4	64	58	−6	50	70	20	65	80	15
(6)	57	87	**30**	62	88	**26**	50	70	20	54	87	**33**
(7)	51	87	**36**	60	88	**28**	44	70	**25**	46	89	**42**
(8)	72	90	18	82	92	10	72	87	15	81	94	13
(9)	46	63	17	70	79	9	58	65	7	67	73	6
(10)	59	90	**31**	71	88	16	58	65	7	58	87	**29**
(11)	34	70	**36**	51	63	12	33	52	19	52	70	18
(12)	33	87	**54**	46	67	21	46	52	7	52	80	**28**

(continued)

Table 9.9 (continued)

	Glass/ceramic/metal products			Engineering/vehicle construction			Electric/electronic devices			Other manufacturers		
	2001 (%)	2016 (%)	Δ (%)	2001 (%)	2016 (%)	Δ (%)	2001 (%)	2016 (%)	Δ (%)	2001 (%)	2016 (%)	Δ (%)
(13)	41	90	**49**	50	88	**38**	44	61	16	53	84	**31**
(14)	36	90	**54**	46	88	**42**	46	61	15	47	80	**33**
(15)	33	90	**57**	48	88	**40**	34	61	**27**	40	80	**40**
(16)	16	20	4	16	50	**34**	3	22	19	19	20	1
(17)	12	30	18	10	33	23	9	30	21	16	47	**32**
(18)	34	40	6	29	42	13	26	52	**26**	37	53	16
(19)	10	20	10	8	33	**25**	12	22	10	17	11	−6
(20)	21	27	6	21	50	**29**	11	48	**36**	16	31	16

Differences greater 25% in bold
Environm. = Environmental
HSE = Health and Safety Executive

that they pursue this type of research. As stated before, the overall trend is that companies adopt more environmentally-related managerial activities and have used the past 15 years for the implementation of a growing number of different activities. There are still industry-specific differences though, and in future the focus needs to be on implementing more of those activities that only received minimum attention so far, because these are sometimes also qualitative game-changers in terms of contributors to sustainable development as well as in terms of enabling competitive differentiation.

9.5 Summary and Discussion

In our analysis, we found an overall increase of environmental activities and of EMS implementation levels in both countries, Germany and the UK. Our 15-year comparison shows an increasing effort of manufacturing firms regarding environmental concerns. Nevertheless, some environmental activities are actually less widely diffused than they were 2001. Managerial activities and EMSs are more popular in Germany, while operational activities have to be analyzed individually to observe in which country they are implemented more often. Although EMS implementation increased over time, its positive impact on the probability of operational activities decreased while the impact on managerial ones remained unchanged. In terms of the distinguishing sub-classes of different manufacturing industries, the differences in implementing managerial ecological activities disappeared mostly within the last 15 years. Nevertheless, operational activities are still predominantly performed by engineering and vehicle constructing firms.

Moreover, we find that ecological sustainability increases with company size, most likely because of higher availability of resources. In addition to this, the differences between the different firm sizes regarding managerial activities have decreased over time. The influence of both size and industry is low for those activities. It is noticeable that the presence of an EMS influences the environmental performance in Germany more than in the UK. Regarding social activities, the picture is different: social activities are very widespread across firms. Especially in Germany, some social activities have become standards. However, in both countries there is still potential to improve the companies' social performance. When it comes to topics which are more distant to the core business, adoption rates decrease, likely because of missing incentives.

Contrary to Schaltegger et al. (2013), overall we found higher adoption rates in Germany than in the UK. Schaltegger and colleagues stated the corporate sustainability performance of UK firms to be above whilst that of German firms was rated to be below the international average. The difference in our findings can possibly be explained by our broader sample, which contains firms of all size categories. Given the comparable firm structure, our results are similar to those of Wagner and Schrauth (2014) who cover data for only 10 years, thus suggesting that our findings indeed represent long-term trends. Nevertheless, our dataset covers 5 more years and thus our analysis certainly captures the status quo better and provides a more comprehensive and current overview of the development as well as status quo of corporate sustainability and environmental innovation in German manufacturing firms.

Our findings can serve as impulse for future research in terms of calling for further examination of a number of phenomena. To start with, the lack of interest in environmental benchmarking with other companies could imply that the environmental activities are not implemented to differentiate from competitors. This could support the argument that fulfilling regulations or consumer expectations could be the main motivation for environmental activities. Moreover, country-specific differences can at least partly be attributed to national regulations and practices. For example the British firms' focus on recycling or their effort for the childcare of employees can at least partly be attributed to waste regulations or the existence of a governmental childcare system. Although European legislation is the same for Germany and the UK as EU member states, the precise implementation is often left to the individual member state. Hence, it is national legislation as well as a path dependent and country specific corporate culture that can cause the differences.

Our findings, therefore, provide useful insights to practitioners as well as for researchers and politicians. Knowing the sustainability behavior in a specific industry helps practitioners determining own strengths and weaknesses to remain competitive. Policymakers interested in a better understanding of the variation between individual firms with regard to environmentally related and socially beneficial innovation activities can use our findings to substantiate and ameliorate policy initiatives. Specifically, they may use the detailed information about firms' behavior to set incentives in favor of those activities being less adopted but relevant for achieving sustainable development. Especially our analysis equips them with detailed information regarding the specific conditions for a different size or industry categories.

Finally, our findings also have implications for researchers. Analyses on corporate sustainability and environmental innovation should be interpreted with regard to time as well as to the spatial scope of the data. Moreover, our findings suggest that generalizing results from an analysis being based on one industrialized country to other industrialized countries is not always possible and thus such transfers need to be considered with heightened care.

References

Almeida, P., & Kogut, B. (1997). The exploration of technological diversity and geographic localization in innovation: Start-up firms in the semiconductor industry. *Small Business Economics, 9*(1), 21–31.

Hertin, J., Berkhout, F., Wagner, M., & Tyteca, D. (2008). Are EMS environmentally effective? The link between environmental management systems and environmental performance in European companies. *Journal of Environmental Planning and Management, 51*(2), 255–280.

Horbach, J. (2008). Determinants of environmental innovation – New evidence from German panel data sources. *Research Policy, 37*(1), 163–173.

Horbach, J., Rammer, C., & Rennings, K. (2012). Determinants of eco-innovations by type of environmental impact – The role of regulatory push/pull, technology push and market pull. *Ecological Economics, 78*, 112–122.

Rennings, K. (2000). Redefining innovation—Eco-innovation research and the contribution from ecological economics. *Ecological Economics, 32*(2), 319–332.

Schaltegger, S., Harms, D., Hörisch, J., Windolph, S. E., Burritt, R., Carter, A., Truran, S., Crutzen, N., Ben Rhouma, A., Csutora, M., Tabi, A., Kokubu, K., Kitada, H., Haider, M. B., Kim, J.-D., Lee, K.-H., Moneva, J. M., Ortas, E., Álvarez-Etxeberria, I., Daub, C.-H., Schmidt, J., Herzig, C., & Morelli, J. (2013). *International corporate sustainability barometer: A comparative study of 11 countries*. Lüneburg: Centre for Sustainability Management.

Statistisches Bundesamt. (2016). *Beschäftigung und Umsatz der Betriebe des Verarbeitenden Gewerbes*. https://www.destatis.de/DE/Publikationen/Thematisch/IndustrieVerarbeitendesGewerbe/ThemaIndustrie.html. Accessed 29 Jan 2018.

Tyteca, D., Carlens, J., Berkhout, F., Hertin, J., Wehrmeyer, W., & Wagner, M. (2002). Corporate environmental performance evaluation: Evidence from the MEPI project. *Business Strategy and the Environment, 11*(1), 1–13.

Wagner, M. (2007). On the relationship between environmental management, environmental innovation and patenting: Evidence from German manufacturing firms. *Research Policy, 36*(10), 1587–1602.

Wagner, M. (2008). Empirical influence of environmental management on innovation: Evidence from Europe. *Ecological Economics, 66*(2–3), 392–402.

Wagner, M., & Schaltegger, S. (2003). How does sustainability performance relate to business competitiveness. *Greener Management International, 44*, 5–16.

Wagner, M., & Schrauth, F. (2014). *Umwelt- und Nachhaltigkeitsmanagement in deutschen Unternehmen: ein Zehnjahresvergleich unter besonderer Berücksichtigung von kleinen und mittelständischen Unternehmen und Branchenunterschieden*. Augsburg: Chair for Management, Innovation und International Business.

World Commission on Environment and Development. (1987). *Our common future*. Oxford: Oxford University Press.

Chapter 10
Effects of Innovation and Domestic Market Factors on OECD Countries' Exports of Wind Power Technologies

Joachim Schleich and Rainer Walz

10.1 Introduction

In several OECD and emerging countries, the increasing deployment of renewable energy sources (RES) for electricity generation is a key strategy for tackling global climate change, preserving resources, and securing energy supply. For example, the European Union (EU) has set a binding target of 20% for the share of RES in final energy consumption in 2020, and an indicative target of 27% for 2030. Wind power is typically considered to exhibit the largest future potential among RES power technologies (IEA 2014), with high growth potentials and ambitious targets not just in OECD countries. For example, in 2015, India wanted to increase its RES electricity capacity fivefold by 2022. More than a third of this capacity is expected to be wind power (target = 60 GW). China's 13th Five Year Plan on Energy Development of 2017 foresees an increase in wind capacity by more than 50% until 2020 compared to 2015 levels (target is 200 GW). Finally, Brazil plans to triple its wind power capacity to meet its target for 2024 of 24 GW). Thus, international competitiveness remains crucial for exporters of wind power technologies in order to capture substantial shares in this growing market.

The academic literature indicates that innovation and domestic market factors play an important role for countries' success in technology exports. Notably, the product life cycle theory (Vernon 1966; Krugman 1979; Fagerberg 1987) concludes

J. Schleich (✉)
Grenoble Ecole de Management, Univ Grenoble Alpes ComUE, Grenoble, France

Fraunhofer Institute of Systems and Innovation Research, Karlsruhe, Germany
e-mail: Joachim.Schleich@isi.fraunhofer.de

R. Walz
Competence Center Sustainability & Infrastructure Systems, Fraunhofer Institute for Systems and Innovation Research ISI, Karlsruhe, Germany
e-mail: rainer.walz@isi.fraunhofer.de

J. Horbach, C. Reif (eds.), *New Developments in Eco-Innovation Research*, Sustainability and Innovation, https://doi.org/10.1007/978-3-319-93019-0_10

that—because developing countries catch up over time—developed countries need to innovate continuously to enjoy high levels of exports and revenues. More recently, the lead market concept, which has been developed and refined in the academic literature of business economics (Beise 2004; Meyer-Krahmer 2004; Beise and Cleff 2004; Beise and Rennings 2005; Walz 2006; Rennings and Smidt 2010; Walz and Köhler 2014), stresses demand conditions like domestic market size or the characteristics of domestic consumers, the general export orientation of countries, and innovation friendliness. The lead market concept foresees an active role for public policy in creating favorable framework conditions. In addition to demand and endowment factors, the systems of innovation (SI) literature (Lundvall 1985, 1992, 2007; Fagerberg 1992; Freeman 1987, 1995; Nelson 1993; Malerba 2002, 2005) highlights the role of user-producer processes, institutions and regulation, as well as the technological capabilities for competitiveness in international markets. The empirical literature on non-energy technologies has long since established the importance of innovation, expenditures for research and development (R&D) (including government support for R&D), and patent activity for countries' export performance (e.g. Hirsch and Bijaoui 1985; Fagerberg 1987; Soete 1987; Greenhalgh 1990; Wakelin 1998; Roper and Love 2002; Levinson 2009; Wang et al. 2010). However, only a few studies have explored the role of domestic market factors and innovation for RES exports. In particular, Sawhney and Kahn (2012) find that exporting countries' domestic RES generation is positively related to renewable energy technology exports to the USA. For a large set of developed and developing countries, Diederichs (2016) finds that more intensive environmental regulation and a larger domestic knowledge stock drive exports of renewable energy technology. In addition, the effects of policies on exports may fully materialize only after several years. Countries with a long history of renewables support tend to export more renewable technology goods than countries which only began supporting RES a few years ago.

For renewable electricity technologies, the domestic market factors include domestic support policies, in particular. Such support may be justified by environmental benefits, knowledge and technology spillovers or imperfect competition in electricity generation (e.g. Rennings 2000; Walz 2007). Without policy intervention, these market failures lead to lower provision of RES than socially desired.[1] Historically, the competitiveness of RES depended on various types of support mechanisms, which varied across countries and time. Most prominently, feed-in tariffs (FITs) provide fixed payments to electricity generators for each kWh of electricity generated from RES. Other support measures include investment subsidies or tax exemptions, production tax credits (PTCs), quota obligations for the share of RES electricity generated or distributed, and tradable green certificate (TGC) schemes.

Groba (2014) finds a positive effect of FIT (but not of other support mechanisms) on PV exports from OECD countries. In contrast, Diederichs (2016) finds renewable

[1]Among others, Horbach et al. (2012) note that, since environmental policies are also demand-side innovation policies, environmental and innovation policies should be explored together.

energy quotas and tax credits have a stronger effect on exports than feed-in-tariffs, which seem to play an important role especially in the earlier phases of renewable energy support policies.

In this paper, we empirically explore the effects of innovation and domestic market factors on a country's exports of wind power technologies. Our panel econometric analysis relies on data for twelve OECD countries over the time span from 1991 to 2011. The factors considered distinguish between input and output measures of innovation, i.e. between technology-specific R&D and patenting activity in wind power technologies. Domestic market factors include the size of the domestic wind power market, and RES support policies. We also allow for differences between various types of policy measures, i.e. between feed-in tariffs and other support measures. Last, but not least, we also consider the effects of the policy process on exports of wind power, i.e. the existence of policy targets and stability. While the scant empirical literature has looked at the effects of innovation and domestic market factors—but not in a multi-country framework—to our knowledge, so far, no study has explored the effects of policy characteristics on countries' RES export performance.

The extant conceptual and empirical literature has explored the role of policy process characteristics for innovation in RES technologies. In particular, the policy analysis literature indicates that a stable regulatory framework is favorable to innovation in RES (e.g. Jänicke and Lindemann 2010; Bergek et al. 2008). Schleich et al. (2017) find that patenting activities in wind power technologies in OECD countries are positively correlated with the presence of production or capacity targets for wind power or renewable energy sources and a stable policy environment. Rogge and Schleich (2018) conclude that the consistency and credibility of the German policy mix are positively associated with the level of RES manufacturers' innovation expenditures.

The remainder of our study is organized as follows. Section 10.2 presents the methodology, including a description of the data, the variables, and the econometric approach. The results are shown and discussed in Sect. 10.3. Section 10.4 summarizes the main findings and policy implications.

10.2 Methodology

We employ panel econometrics to estimate the impact of innovation and policy measures on exports of wind power technologies, relying on a time series of cross-sectional data for 12 OECD countries: Austria (AT), Denmark (DK), France (FR), Germany (DE), Italy (IT), Japan (JP), The Netherlands (NE), Spain (SP), Sweden (SE), Switzerland (CH), the United Kingdom (UK) and the United States (US). The country choice was motivated by their relevance for exports of wind power technologies as well as data availability. Our sample includes countries with relatively low exports such as Switzerland or Austria as well as countries with relatively high exports such as Denmark and Germany. In total, the countries included in our sample

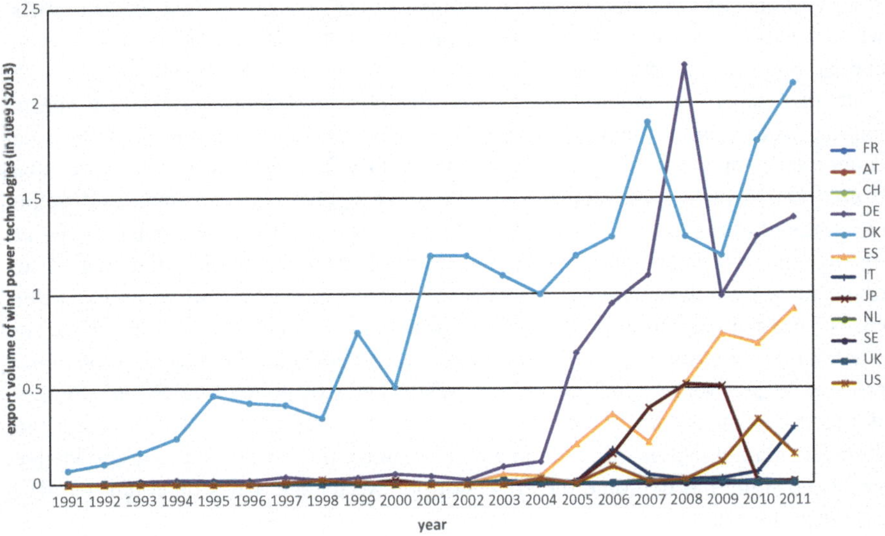

Fig. 10.1 Annual export volume of wind power technology for 12 OECD countries

account for 79–99% of global exports of wind power technologies in any given year of the analyzed period 1991–2011.

10.2.1 Dependent Variable

The export volume of wind power technologies (export) is the dependent variable (see Fig. 10.1). The data were retrieved from the UN-COMTRADE database. The Harmonized Systems classification disaggregates the classification number 8502 "Electric generating sets and rotary converters" into six-digit classifications, which differ with regard to primary energy. We used the six-digit classification 850231 "Wind power generating sets" to retrieve the data.

10.2.2 Explanatory Variables

The set of explanatory variables includes various proxies for countries' innovation activities and for domestic market factors.

Table 10.1 Definition of variables

	Definition	Data sources
Dependent variable		
Windexport	Export volume of wind power technologies (10e9 $2013)	UN-COMTRADE for HS classification number 850231 "Electric generating sets and rotary converters—Wind-powered"
Explanatory variables		
Windpatents	Number of international patents for wind technologies	Patent families with at least a Patent Cooperation Treaty (PCT) application or an EPO application; EPO and WIPO data retrieved with Questel www.questel.com using the International Patent Classification (IPC)
R&D	Public R&D for wind power including onshore and offshore technologies and wind energy systems and other technologies (Group 32) (million $2013)	IEA RDD online data service: http://www.iea.org/statistics/RDDonlinedataservice/
FIT	Dummy, value of one if a FIT or FIP is implemented	IEA/JRC Global Renewable Measures Database, data for instrument were taken primarily from European Renewable Energies Federation and the literature
NOFIT	Dummy, value of one if a support measure other than a FIT or a FIP is implemented	IEA/JRC Global Renewable Measures Database, data for instrument were taken primarily from European Renewable Energies Federation and the literature
Windcap	Installed additional wind power capacity (GW = 1000 MW)	Global wind energy council global statistics
Target	Dummy, value of one if target for wind energy or capacity exists	Reports by IEA, IRENA, national agencies, academic and grey literature
Stability	Dummy, value of one if regulatory environment considered stable	Reports by IEA, IRENA, national agencies, academic and grey literature

10.2.2.1 Measures of Innovation

We include two measures of innovation activity. First, we use public R&D expenditures for wind power technologies (see Table 10.1). While extensively used in the empirical literature to capture innovation activities, R&D expenditures are a measure of input. Thus, not all R&D expenditures may actually lead to innovations.[2] Second, we use patents in wind power technology as an output measure. Patents have been widely used in the empirical literature, but are also regarded as an incomplete measure of innovation (Griliches 1990). For example, patents only capture codified knowledge, not tacit knowledge. Likewise, not all of the generated innovations are

[2]Due to a lack of data, private R&D expenditures for wind power technologies could not be used.

actually patented. In particular, inventors may use other methods such as secrecy or lead time to protect their technological innovations (e.g. Hall et al. 2014). Finally, patents may not adequately reflect the commercial value of an innovation (Harhoff et al. 1999). Instead, companies may use patents for strategic reasons, for example, to preclude competitors from patenting in related areas (Cohen et al. 2000). Wind power technologies form the patent sub-class F03D, which includes mainly masts, motors, and rotors.[3] Our patent data collection followed the transnational patent approach (Frietsch and Schmoch 2010)[4] and refers to applications and country assignment based on the country where the inventor lives (rather than the location of the company headquarter).

10.2.2.2 Domestic Market Factors

We consider two types of variables as proxies for domestic market factors. First, to reflect the impact of the domestic market size, we use the additionally installed capacity of (offshore) wind power in a country per year. Second, we account for the existence of RES support mechanisms, distinguishing between FITs and non-FITs. The types and design of the support mechanisms for renewable electricity differ across countries and over time. Several countries have switched mechanisms over time, primarily from FITs and TGCs to feed-in premium (FIP) systems.[5]

Demand-side instruments for RES include measures supporting deployment such as feed-in tariffs (FITs), which make fixed payments to electricity generators for each kWh of electricity supplied from RES. Other support mechanisms include investment subsidies or tax exemptions, production tax credits (PTCs), quota obligations for the share of RES electricity generated or distributed, and tradable green certificate (TGC) schemes. The thrust of the conceptual and empirical literature suggests that FITs are more favorable for RES employment and also for innovation than other support measures, because FITs mean more predictable remuneration for investors (see also Haas et al. 2004; Schmidt et al. 2012; Bergek and Berggren 2014; Polzin et al. 2015). For investors in wind power, investment security is particularly important, since capital costs account for around 80% of the total costs (e.g. Kleßmann et al. 2013).

[3]More specifically, FD03 comprises of: (1) wind motors with rotation axis substantially parallel to the flow of air entering the machine; (2) wind motors with rotation axis substantially at a right angle to the flow of air entering the machine; (3) other wind motors; (4) controlling wind motors; (5) adaptations of wind motors for special use; (6) combinations of wind motors with apparatus driven thereby; and (7) other details, component parts, or accessories of wind motors.

[4]See Schleich et al. (2017) for further details. For additional details, we also refer to the IEA 'Renewable Energy Policies and Measures Database' (http://www.iea.org/policiesandmeasures/renewableenergy/).

[5]FIP means that electricity producers receive a premium payment on top of the electricity wholesale price. To improve the compatibility of RES support systems with the electricity markets, in 2014 the EU adopted the "Environmental and Energy State Aid Guidelines for 2014–2020" (European Commission 2014). Accordingly, FIPs elicited via bidding systems will become the central RES support mechanism in all EU countries.

Table 10.2 Descriptive statistics of dependent and explanatory variables (1991–2011)

Variable	Unit	Obs.	Mean	SD	Min	Max
Windexport	10e9 $2013	252	0.14	0.37	0.00	2.20
Windpatents	Count	252	25.90	49.94	0.00	284
R&D	Million $2013	250	13.81	20.23	0.19	197.21
FIT	0/1-Dummy	252	0.47	0.50	0.00	1.00
NOFIT	0/1-Dummy	252	0.24	0.43	0.00	1.00
Windcap	GW	240	0.53	1.19	0.02	2.25
Target	0/1-Dummy	252	0.62	0.49	0.00	1.00
Stability	0/1-Dummy	252	0.51	0.50	0.00	1.00

We include a dummy variable, *FIT*, which takes on the value of one if a FIT was in place in a specific year.[6] Similarly, *NOFIT* is equal to one if other-than-FIT support mechanisms—typically TGCs—were implemented.

Following Schleich et al. (2017), we capture the effects of the domestic policy process via two dummy variables.[7] First, *target* equals one if a domestic target is in place for electricity generated from wind power or, more generally, from renewable energies. Second, *stability* equals one if there is a stable regulatory framework in place and a supportive regulatory framework exists, and if there are information and education programs in place.

Table 10.1 provides an overview of the variables and references to the data sources. The descriptive statistics of all variables (in levels) appear in Table 10.2.

10.2.2.3 Econometric Model and Estimators

We employ a panel econometrics model to analyze the factors driving innovation activity in wind power technologies:

$$\begin{aligned} windexports_{i,t} = {} & constant + \beta_1 windpatents_{i,t-2} + \beta_2 R\&D_{i,t-2} \\ & + \beta_3 FIT_{i,t-2} + \beta_4 FIT_{i,t-2} + \beta_5 windcap_{i,t-2} \\ & + \beta_6 target_{i,t-2} + \beta_7 stability_{i,t-2} + \alpha_i + \varepsilon_{i,t} \end{aligned} \tag{10.1}$$

where $i = 1,\ldots,12$ indexes the cross-sectional units (countries) and $t = 1991,\ldots,$ 2011 indexes time; α_i represents an unobserved country-specific effect, and $\varepsilon_{i,\,t}$ is an idiosyncratic error term. The explanatory variables enter with a lag of one or two periods, recognizing that companies need time to respond to policy and market factors which then translate into exports.[8] Lagging the explanatory variables also

[6] *FIT* also equals one if a FIP or a PTC was in place, since the incentives for investors are similar to those of FITs.

[7] Schleich et al. (2017) also provide details and examples on how these variables were constructed.

[8] Findings are quite robust to alternative specifications of the lags. Specifically, lagging all explanatory variables by 1 year (rather than 2 years) hardly changes the results.

establishes causality. The dependent variable and all the explanatory variables except the dummy variable enter Eq. (10.1) as natural logarithms. Hence, the coefficients may be interpreted as elasticities and quasi-elasticities for the dummy variables.

Compared to a purely cross-sectional analysis, a panel analysis allows for more general heterogeneity across countries. In particular, omitted country characteristics which affect a country's propensity to patent and which are correlated with other regressors do not result in inconsistent parameter estimates in panel data models as long as these unobserved country-specific effects [i.e. α_i in Eq. (10.1)] are roughly constant over the period in question. We employ random effects and fixed effects estimators. The latter use variation within countries only (i.e. deviation of variables from the country means). If the unobserved effects are not correlated with the observed explanatory variables, then the RE estimator (i.e. treating unobserved effects as random) yields more efficient parameter estimates than the FE estimator that treats these effects as country-specific. The random effects estimator uses the variation of variables within countries as well as variation among countries. However, if unobserved effects are correlated with observed explanatory variables, then the RE estimator yields inconsistent estimates. The FE estimator yields consistent estimates in both cases. We conduct a standard Hausman test to decide between a RE and a FE estimator.

10.3 Results

Table 10.3 displays the results from estimating Eq. (10.1). Results for the FE estimator appear in the column titled Model 1a and the column titled Model 1b shows the results for the RE estimator.[9] A first glance suggests that both estimators produce quite similar parameter estimates. A formal Hausman tests confirms this [$\chi^2(7) = 6.10$; p-value $= 0.53$). Thus, the RE estimator should be preferred. We, therefore, limit our interpretation of the findings to the results of the RE estimator, i.e. Model 1b. Accordingly, an increase in patenting activity in wind power technologies increases exports of wind power technologies. The point estimate implies that a 1% increase in the number of patents raises exports by 0.11% with a lag of 2 years, but the coefficient is just shy of being statistically significant at conventional levels. In comparison, the coefficient associated with *R&D* exhibits the expected sign and is statistically significant. An increase in R&D expenditures by 1% leads to an increase in patenting by 0.366%. Neither FITs nor non-FIT support schemes appear to affect exports. In contrast, the coefficient associated with installed wind capacity turns out to be statistically significant. An increase in *windcap* by 1% increases exports by about 0.4% with a lag of 2 years. On average, wind technology exports roughly double if

[9]The average VIF of the explanatory variables in Table 10.3 and all VIFs are below five. Based on the standard cut-off point of ten, collinearity does not appear to be an issue.

Table 10.3 Results for fixed-effects and random-effects estimators (*p*-values in parentheses)

	Fixed effects (Model 1a)	Random effects (Model 1b)	Fixed effects (Model 2a)	Random effects (Model 2b)
Windpatents (t−1)	0.091 (0.211)	0.110 (0.127)	0.091 (0.177)	0.120 (0.076)*
R&D (t−2)	0.291 (0.079)*	0.366 (0.022)**	0.167 (0.264)	0.323 (0.022)**
FIT (t−2)	−0.193 (0.724)	0.083 (0.876)		
NOFIT (t−2)	0.449 (0.376)	0.536 (0.285)		
Instruments (t−2)			0.528 (0.185)	0.807 (0.043)**
Windcap (t−2)	0.291 (0.016)**	0.343 (0.003)***		
Target (t−2)	1.178 (0.003)***	1.060 (0.007)***	1.035 (0.005)***	0.984 (0.008)***
Stability (t−2)	0.921 (0.020)**	0.755 (0.052)*	1.125 (0.002)***	0.979 (0.006)***
Constant	−6.491 (0.000)***	−6.704 (0.000)***	−7.509 (0.000)***	−7.889 (0.000)***
R^2 (overall)	0.40	0.433	0.301	0.401
Number of groups	12	12	12	12
Sample size	188	188	215	215

*Significant in two-tailed t-test at $p = 10\%$
**Significant in two-tailed t-test at $p = 5\%$
***Significant in two-tailed t-test at $p = 1\%$

targets for wind power or other renewable electricity sources are in place compared to a situation where such targets do not exist. Similarly, a stable (rather than an unstable) policy environment raises wind exports by about 75%. Thus, the size effects appear to be rather high for the factors reflecting the policy process.

We re-estimated Eq. (10.1) where—to save degrees of freedom—we combine the policy support variables *FIT* and *NOFIT* to form a new variable, *instruments*. We also drop *windcap*, since the domestic installation of wind capacity is likely to be driven by the policy support measures, thus also avoiding a potential "bad controls" problem[10] (Angrist and Pischke 2009). The results of this model appear in Table 10.3 in the column Model 2a for the FE estimator and in column Model 2b for the RE estimator. Having conducted a Hausman test [$\chi^2(7) = 1.56$; *p*-value $= 0.91$), we again prefer the RE estimator. We also note that the point estimates hardly differ between Model 1 and Model 2. However, unlike in Model 1, the coefficients associated with *windpatents* and with *instruments* are both statistically significant in Model 2.

[10]By "bad controls" we mean control variables that may themselves be outcome variables.

10.4 Conclusions

The results of our panel econometric analysis using a panel of twelve OECD countries over a period of more than two decades suggests that innovation activities and domestic market factors affect exports of wind power technologies. Akin to the existing empirical findings for other technologies, both input (i.e. R&D) and output (i.e. patents) measures of innovation were found to be positively correlated with exports of wind technology. These findings support the premises derived from the systems of innovation literature, in particular, since patents may also be interpreted as an indirect measure of learning effects and of the technological capabilities of a country.

Similar to the findings of Sawhney and Kahn (2012) for photovoltaics, we conclude that a larger domestic market is associated with higher wind power exports. This finding offers empirical corroboration of the insights of the lead market literature. While we found that exports are higher if domestic support measures are in place, we found no differences in wind power technology exports between *FIT* and *NOFIT* support measures. Arguably, the dummies employed to reflect the various support measures are rather crude. *FIT* and *NOFIT* only capture differences in the types of support mechanisms, but not in other design characteristics such as support levels or duration.

Last, but not least, our findings suggest that characteristics of the domestic policy process affect countries' exports of wind power technologies. Similar to the findings by Schleich et al. (2017) concerning differences in countries' patenting activities, the existence of RES deployment targets and a stable policy environment were found to explain differences in countries' export performance in wind power technologies. Notably, the size effects were rather large. Thus, our results support the notion that providing a long-term, stable regulatory framework is more relevant for exports of wind power technologies than the actual type of support scheme applied. Together with the results for the market size, our findings for target setting and stability of the regulatory framework emphasize the relevance of domestic market factors for exports of wind power technologies. The present debate about the future support for renewable electricity in the EU beyond 2020 also revolves around the need for national targets and support schemes for mature technologies like wind onshore (Held et al. 2015). In this context, we infer that EU Member States' commitments to credible wind power expansion trajectories as an integral part of their energy and climate plans would be conducive to EU exports of wind power technologies.

Acknowledgements The authors would like to thank Rouven Emmerich for his thorough research assistance, Frank Marscheider-Weidemann for his help in retrieving patent data, and Mario Ragwitz for sharing his insights on renewable energy policies. Special thanks go to Gillian Bowman-Köhler for proofreading our paper.

References

Angrist, J., & Pischke, J.-S. (2009). *Mostly harmless econometrics*. Princeton: Princeton University Press.

Beise, M. (2004). Lead markets: Country-specific drivers of the global diffusion of innovations. *Research Policy, 33*, 997–1018.

Beise, M., & Cleff, T. (2004). Assessing the lead market potential of countries for innovation projects. *Journal of International Management, 10*(4), 453–477.

Beise, M., & Rennings, K. (2005). Lead markets for environmental innovations: A framework for innovation and environmental economics. *Ecological Economics, 52*, 5–17.

Bergek, A., & Berggren, C. (2014). The impact of environmental policy instruments on innovation: A review of energy and automotive industry studies. *Ecological Economics, 106*, 112–123.

Bergek, A., Hekkert, M., & Jacobsson, S. (2008). Functions in innovation systems: A framework for analysing energy system dynamics and identifying system building activities by entrepreneurs and policy makers. In T. Foxon, J. Köhler, & C. Oughton (Eds.), *Innovations in low-carbon economy* (pp. 79–111). Cheltenham: Edward Elgar.

Cohen, W., Nelson, R., & Walsh, J. (2000). *Protecting their intellectual assets: Appropriability conditions and why U.S. manufacturing firms patent, or not*. NBER working paper no. 7552. National Bureau of Economic Research, Cambridge.

Diederichs, H. (2016). *Environmental policy and renewable energy equipment exports*. Wiesbaden: Springer Gabler.

European Commission. (2014). Guidelines on state aid for environmental protection and energy 2014–2020, 2014/C 200/01. http://eur-lex.europa.eu/legal-content/EN/TXT/HTML/?uri=CELEX:52014XC0628%2801%29&from=EN. Accessed 6 Jun 2016.

Fagerberg, J. (1987). A technology gap approach to why growth rates differ. *Research Policy, 16* (2–4), 87–99.

Fagerberg, J. (1992). The home market hypothesis re-examined: The impact of domestic-user producer interaction in exports. In B.-Å. Lundvall (Ed.), *National systems of innovation: Towards a theory of innovation and interactive learning*. London: Pinter.

Freeman, C. (1987). *Technology and economic performance: Lessons from Japan*. London: Pinter.

Freeman, C. (1995). The "national system of innovation" in historical perspective. *Cambridge Journal of Economics, 19*(1), 5–24.

Frietsch, R., & Schmoch, U. (2010). Transnational patents and international markets. *Scientometrics, 82*, 185–200.

Greenhalgh, C. (1990). Innovation and trade performance in the UK. *Economic Journal, 100*(400), 105–118.

Griliches, Z. (1990). Patent statistics as economic indicators: A survey. *Journal of Economic Literature, 28*, 1661–1707.

Groba, F. (2014). Determinants of trade with solar energy technology components – Evidence on the porter hypothesis? *Applied Economics, 46*, 503–526.

Haas, R., Eichhammer, W., Huber, C., Langniss, O., Lorenzoni, A., & Madlener, R. (2004). How to promote renewable energy systems successfully and effectively. *Energy Policy, 32*(6), 833–839.

Hall, B., Helmers, C., Rogers, M., & Vania, S. (2014). The choice between formal and informal intellectual property: A review. *Journal of Economic Literature, 52*(2), 375–423.

Harhoff, D., Narin, F., & Vopel, K. (1999). Citation frequency and the value of patented inventions. *Review of Economics and Statistics, 81*(3), 511–515.

Held, A., Ragwitz, M., Resch, G., Genoese, F., Liebmann, L., Pató, Z., & Szabo, L. (2015). *Implementing the EU 2030 climate and energy framework – A closer look at renewables and opportunities for an Energy Union*. Issue paper, towards 2030 project. Centre for European Policy Studies, Brussels.

Hirsch, S., & Bijaoui, I. (1985). R&D intensity and export performance: A micro view. *Weltwirtschaftliches Archiv, 121*(2), 238–251.

Horbach, J., Rammer, C., & Rennings, K. (2012). Determinants of eco-innovations by type of environmental impact – The role of regulatory push/pull, technology push and market pull. *Ecological Economics, 78*, 112–122.

International Energy Agency, IEA. (2014). *Energy technology perspectives 2014*. Paris: OECD.

Jänicke, M., & Lindemann, S. (2010). Governing environmental innovations. *Environmental Politics, 19*(1), 127–141.

Kleßmann, C., Held, A., Rathmann, M., de Jager, D., Gazzo, A., Resch, G., Busch, S., & Ragwitz, M. (2013). Policy options for reducing the costs of reaching the European renewables target. *Renewable Energy, 57*, 390–403.

Krugman, P. R. (1979). A model of innovation, technology transfer, and the world distribution of income. *Journal of Political Economy, 87*(2), 253–266.

Levinson, A. (2009). Technology, international trade, and pollution from US manufacturing. *American Economic Review, 99*(5), 2177–2192.

Lundvall, B. (1985). *Product innovation and user-producer interaction*. Industrial development research series 31. Aalborg University Press, Aalborg.

Lundvall, B. (Ed.). (1992). *National systems of innovation: Towards a theory of innovation and interactive learning*. London: Pinter.

Lundvall, B. (2007). National innovation systems – Analytical concepts and development tool. *Industry and Innovation, 14*, 95–119.

Malerba, F. (2002). Sectoral systems of innovation and production. *Research Policy, 31*, 247–264.

Malerba, F. (2005). Sectoral systems – How and why innovation differs across sectors. In J. Fagerberg, D. C. Mowery, & R. R. Nelson (Eds.), *The Oxford handbook of innovation* (pp. 380–406). Oxford: Oxford University Press.

Meyer-Krahmer, F. (2004). Vorreiter-Märkte und Innovation. In F.-W. Steinmeier & M. Machnig (Eds.), *Made in Deutschland 21* (pp. 95–110). Hamburg: Hoffmann und Campe.

Nelson, R. R. (Ed.). (1993). *National innovation systems: A comparative analysis*. New York: Oxford University Press.

Polzin, F., Migendt, M., Täube, F., & von Flotow, P. (2015). Public policy influence on renewable energy investments – A panel data study across OECD countries. *Energy Policy, 80*, 98–111.

Rennings, K. (2000). Redefining innovation – Eco-innovation research and the contribution from ecological economics. *Ecological Economics, 32*, 319–332.

Rennings, K., & Smidt, W. (2010). A lead market approach towards the emergence and diffusion of coal-fired power plant technology. *Economia Politica, 27*(2), 303–328.

Rogge, K., & Schleich, J. (2018). Do policy mix characteristics matter for low-carbon innovation? A survey-based exploration for renewable power generation technologies. *Research Policy* (forthcoming).

Roper, S., & Love, J. H. (2002). Innovation and export performance: Evidence from the UK and German manufacturing plants. *Research Policy, 31*(7), 1087–1102.

Sawhney, A., & Kahn, M. E. (2012). Understanding cross-national trends in high-tech renewable power equipment exports to the United States. *Energy Policy, 46*, 308–318.

Schleich, J., Walz, R., & Ragwitz, M. (2017). Effects of policies on patenting in wind-power technologies. *Energy Policy, 108*, 684–695.

Schmidt, T., Schneider, M., Rogge, K., Schuetz, M., & Hoffmann, V. (2012). The effects of climate policy on the rate and direction of innovation: A survey of the EU ETS and the electricity sector. *Environmental Innovation and Societal Transitions, 2*, 23–48.

Soete, L. G. (1987). The impact of technological innovation on international trade patterns: The evidence reconsidered. *Research Policy, 16*(2–4), 101–130.

Vernon, R. (1966). International investment and international trade in the product cycle. *Quarterly Review of Economics, 88*, 190–207.

Wakelin, K. (1998). The role of innovation in bilateral OECD trade performance. *Applied Economics, 30*(10), 1135–1346.

Walz, R. (2006). Impact of strategies to increase RES in Europe on employment and competitiveness. *Energy and Environment, 17*(6), 951–975.

Walz, R. (2007). The role of regulation for sustainable infrastructure innovations: The case of wind energy. *International Journal of Public Policy, 2*(1/2), 57–88.

Walz, R., & Köhler, J. (2014). Using lead market factors to assess the potential for a sustainability transition. *Environmental Innovation and Societal Transitions, 10*, 20–41.

Wang, C., Wei, Y., & Liu, X. (2010). Determinants of bilateral trade flows in OECD countries: Evidence from gravity panel data models. *World Economics, 33*(7), 894–915.

Chapter 11
Exploring the Role of Instrument Design and Instrument Interaction for Eco-Innovation: A Survey-Based Analysis of Renewable Energy Innovation in Germany

Karoline S. Rogge and Joachim Schleich

11.1 Introduction

Addressing the various environmental sustainability challenges associated with issues such as climate change, biodiversity loss, or resource constraints requires the redirection and acceleration of innovation towards sustainable solutions.[1] The prevalence of various market, structural and transformational system failures that hinder eco-innovation and wider sustainability transitions (Weber and Rohracher 2012) has led to an emerging literature acknowledging the need for policy intervention in the form of policy mixes (OECD 2015; Rogge and Reichardt 2016). Large discrepancies remain, however, between this acknowledgement and the mainstreaming of such thinking into innovation policy and research (OECD/IEA/NEA/ITF 2015; Rogge et al. 2017).

For such an endeavor, much can be learned from the literature on eco-innovation, which has long recognized the important role of policy in spurring such innovation (Rennings 2000; Jaffe et al. 2002; OECD 2011). Building on the notion of "double

[1]This chapter draws in parts on earlier work published by the authors as Rogge and Schleich (2018) [and the corresponding working paper Rogge and Schleich (2017)], but the focus of our analysis here differs: whereas this earlier work examines the impact of policy mix characteristics on green innovation, here we focus on the relevance of instrument design and instrument interaction.

K. S. Rogge (✉)
SPRU – Science Policy Research Unit, University of Sussex, Brighton, UK

Fraunhofer Institute for Systems and Innovation Research ISI, Karlsruhe, Germany
e-mail: k.rogge@sussex.ac.uk

J. Schleich
Fraunhofer Institute for Systems and Innovation Research ISI, Karlsruhe, Germany

Grenoble Ecole de Management Univ Grenoble Alpes ComUE, Grenoble, France
e-mail: Joachim.Schleich@isi.fraunhofer.de

J. Horbach, C. Reif (eds.), *New Developments in Eco-Innovation Research*,
Sustainability and Innovation, https://doi.org/10.1007/978-3-319-93019-0_11

externalities", past empirical research has provided important insights into the measurement and determinants of eco-innovation (del Río 2009; OECD 2009; Kemp and Pontoglio 2011; Bergek and Berggren 2014). One of the key policy insights of this literature is that eco-innovation depends more on the design than on the type of a policy instrument (Kemp 1997; Vollebergh 2007), with environmental stringency standing out as a particularly relevant design feature (Frondel et al. 2008; Ghisetti and Pontoni 2015).

Important insights have been gained into the role of instrument mixes for eco-innovation. Most prominently, studies conclude that combining demand-pull and technology-push instruments (Peters et al. 2012; Costantini et al. 2015; Schleich et al. 2017) is conducive to eco-innovation. In addition, research has stressed the importance of employing systemic instruments (Taylor 2008; Wieczorek and Hekkert 2012). However, most of the literature exploring instrument interactions and the role of design features for such interactions assumes an ex-ante perspective (Spyridaki and Flamos 2014; del Río and Cerdá 2017). Only a few studies to date have employed an ex-post perspective (Guerzoni and Raiteri 2015; Cantner et al. 2016).

A recent review of econometric survey-based analyses shows that regulation is one of the few generally statistically significant determinants of eco-innovation (del Río et al. 2016). Because of data limitations, however, econometric models typically capture the effect of a particular policy instrument by including a dummy variable (del Río et al. 2016). In contrast, some specialized eco-innovation surveys have provided more in-depth insights into the link between policy and eco-innovation by including environmental policy stringency as a policy variable, for instance (Johnstone 2007; Kammerer 2009). Others have considered climate, energy and innovation policy instruments simultaneously (Schmidt et al. 2012), but have not allowed for their interaction.

Despite recent progress in gathering more detailed policy data alongside other innovation measures, to the best of our knowledge, survey-based analyses have not yet simultaneously looked at instruments, their design and their interaction. Especially large-scale innovation surveys, such as the Community Innovation Survey (CIS) conducted within the European Union, tend to cover policy to a limited extent, and often have a narrow focus on public support for research and development (R&D), appropriation methods, or obstacles to innovation. Similarly, the Oslo Manual, which provides guidelines for innovation surveys, puts little emphasis on the measurement of policy instruments, their design and interaction as a determinant of innovation, despite stressing the important role of innovation survey data for guiding policy (OECD 2005).

A notable exception to this apparent neglect of policy in mainstream innovation surveys is the question block on eco-innovation, which was introduced as a supplement to the 2008 CIS wave, following suggestions made by the 'Measuring Eco-Innovation' (MEI) project (Kemp and Pearson 2007). Since then, these large-scale surveys have collected and analyzed information on eco-innovation and its drivers that explicitly includes (environmental) policy for the participating countries such as Germany, Spain, Italy and France. Using the CIS survey as a key data source allowed a better understanding of the factors related to eco-innovation in general, and the role of policy in particular (Rennings and Rammer 2011; Horbach et al. 2013; Borghesi et al. 2015). However, these studies have not focused on the role of instrument design and instrument interaction for eco-innovation.

In this paper, we take a first step towards addressing this current shortcoming based on the example of the decarbonization of the energy system, in which renewable energies play a key role (Jacobsson and Bergek 2004; Gallagher et al. 2012). Given that innovation in the energy sector is dominated by suppliers, we focus on the manufacturers of renewable power generation technologies (Pavitt 1984; Rogge and Hoffmann 2010). We limit the scope of our explorative study to the German *Energiewende* because of its pioneering role in renewable energy policy and innovation (Bruns et al. 2011; Pegels and Lütkenhorst 2014; Strunz 2014; Quitzow et al. 2016).

Building on a recent study examining the impact of policy mix characteristics on innovation (Rogge and Schleich 2018), this chapter aims to empirically explore the role of policy instrument design (e.g. regarding the level of support) and instrument interactions using data from a survey among manufacturers of renewable energy technologies. In particular, we are interested in answering the question of whether innovation in these technologies is related to respondents' perceptions about the EU emissions trading system (EU ETS), the (design of the) German Renewable Energy Sources Act (EEG) and the interaction between them. We have chosen the EU ETS and the EEG as the two core demand-pull instruments for our analysis. This is in line with the wider literature on instrument interaction, which has largely focused on interactions between climate policies and renewable energy policies (IEA 2011; Gawel et al. 2014; Spyridaki and Flamos 2014). Our empirical analysis relies on data collected using a redesigned CIS questionnaire, which explicitly captures the current policy mix and innovation in renewable power generation technologies. The resulting unique dataset collected in 2014 allows us to econometrically analyze the links between the instrument mix and eco-innovation.

The remainder of the chapter is structured as follows. In Sect. 11.2, we develop our analytical framework from the literature. Section 11.3 presents the research case of the German *Energiewende*, and Sect. 11.4 introduces our methodological approach. In Sect. 11.5, we present our results, and discuss these in Sect. 11.6 and offer concluding remarks.

11.2 Analytical Framework

In line with the eco-innovation literature, our interdisciplinary analytical framework as presented in Fig. 11.1 differentiates between firm-external and firm-internal determinants of eco-innovation (del Río 2009). This framework draws on environmental economics, innovation studies and policy analysis and builds on earlier work by the authors (Rogge and Schleich 2018). However, here we introduce a new focus on multiple instruments and their interactions, thereby contributing to policy mix

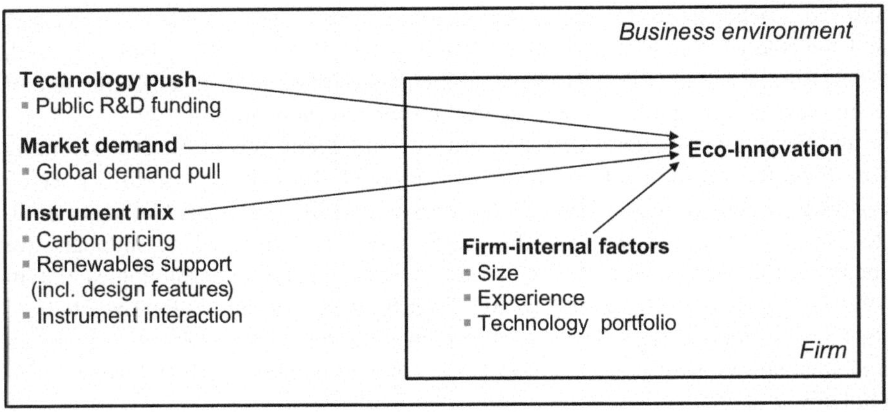

Fig. 11.1 Analytical framework for exploring the determinants of eco-innovation. Source: adapted from Rogge and Schleich (2018)

studies focusing on instrument interactions (del Río 2006; Spyridaki and Flamos 2014).

We include both technology-push and demand-pull factors as classical firm-external determinants of innovation (Mowery and Rosenberg 1979; Di Stefano et al. 2012). In addition, we focus on the influence of multiple instruments and their interaction, thereby following calls for a greater focus on instrument mixes (Cunningham et al. 2013). In doing so, we pay particular attention to instrument design features such as the level of support or its predictability (Hoffmann et al. 2008; Reichardt and Rogge 2016). It has been suggested that such instrument design features play a key role in driving eco-innovation and in determining instrument interactions (del Río et al. 2016; del Río and Cerdá 2017).

We focus on the role of firm size, a firm's technology portfolio, and its experience with green technologies as the firm-internal determinants of eco-innovation. In the following, we elaborate on these potential firm-external and firm-internal determinants of eco-innovation, and focus in particular on the role played by the instrument mix.

11.2.1 Firm-External Determinants of Eco-Innovation

Our analytical framework includes *technology-push* and *demand-pull* factors as firm-external determinants of eco-innovation (Mowery and Rosenberg 1979; Di Stefano et al. 2012). In view of the strong relevance of policy for eco-innovation, we focus on policy-driven technology push and demand pull (Horbach 2008). Over the past decade, strong evidence has been accumulated that both aspects matter for eco-innovation (Schmidt et al. 2012; Veugelers 2012; Costantini et al. 2015; Schleich et al. 2017). More specifically, recent evidence has highlighted that demand-pull effects result from a combination of demand at home and abroad, whereas technology-push effects seem to stem from public R&D support in the domestic market only (Peters et al. 2012; Dechezleprêtre and Glachant 2014).

While it is often possible to identify technology-push policies by looking at public R&D funding, it is more challenging to evaluate demand-pull policies (Edler et al. 2012). Therefore, it may be useful to focus on the core demand-pull instruments. In the case of low-carbon innovation, the core demand-pull instrument concerns *carbon pricing*, for example in the form of an emissions trading systems (Borghesi et al. 2015). The innovation impact of the EU emissions trading system (EU ETS) has been studied extensively, with most findings indicating that it has had a small but positive effect so far (Rogge 2016). In contrast, studies focusing on innovation in renewable energies include *renewables support* as the core demand-pull instrument. Del Río and Peñasco (2014) systematically reviewed the empirical literature on national renewable energies support policies, and conclude that feed-in tariffs are the most appropriate promotion instrument to spur innovation and early diffusion in renewable energy sources for electricity generation.

Recent conceptual and empirical work suggests that the innovation effects of policy instruments are not just driven by the type of the instrument, but also and in particular by its *design features* (Kemp 1997; Vollebergh 2007). Perhaps most prominently among these, the relevance of policy stringency has been well established for eco-innovation (Kemp and Pontoglio 2011; Ghisetti and Pontoni 2015). However, other design features have also received attention, such as the level of support, the predictability of an instrument, or its flexibility (Hoffmann et al. 2008; Hašcic et al. 2009; Reichardt and Rogge 2016). For renewable energies, a number of descriptive design features have been suggested for potential consideration, such as the duration of support, decline of support levels over time, quantitative limits for installed capacities (e.g. in GW per year), or technology-specific and geographical differentiation (del Río 2012; Hoppmann et al. 2013).

In addition, with the introduction of the EU ETS, it has been increasingly argued that the *interaction of instruments* needs to be taken into consideration when evaluating the impact of climate policy and energy policy (Sorrell and Sijm 2003; del Río 2006). Similarly, innovation studies have stressed the need to account for instrument interactions when evaluating innovation policy (Flanagan et al. 2011; Cunningham et al. 2013). So far, however, both lines of research have produced only a few empirical studies with ex-post assessments on the role of such instrument interaction for policy effectiveness (Guerzoni and Raiteri 2015; Cantner et al. 2016). Finally, it has been pointed out that when studying such interaction effects, the design of the interacting instruments should also be considered (del Río 2010; del Río and Cerdá 2017). This is consistent with earlier findings on the relevance of design features as one of the determinants of eco-innovation. However, previous empirical ex-post evaluations exploring the role of instrument interactions for eco-innovation have not yet accounted for design features.

Therefore, we are particularly interested in whether instrument design and instrument interaction matter for low-carbon innovation.

11.2.2 Firm-Internal Determinants of Eco-Innovation

Regarding the firm-internal determinants of innovation, we draw on insights from evolutionary economics and the resource-based view of the firm (Nelson and Winter 1982; Wernerfelt 1984; Barney 2001). These suggest that a firm's resources, capabilities and competencies matter for innovation, which is why we include three firm characteristics in our analytical framework (Teece et al. 1997; Helfat et al. 2007; del Río et al. 2015).

The first concerns *firm size*, which has typically been found to affect eco-innovation positively, with larger firms spending more on innovation (Kesidou and Demirel 2012; del Río et al. 2016).

Second, we include a firm's *experience* with producing renewable energy technologies in order to capture its accumulated resources as well as its technological and organizational capabilities and competencies in using the respective green technology as determinants of innovation (Kammerer 2009; Horbach et al. 2012). While others have included a firm's age to capture this effect (del Río et al. 2016), we argue that experience may be the better proxy for this phenomenon for firms which are diversifying their portfolio.

Finally, our framework considers a firm's *technology portfolio* to control for differences between renewable energy technologies and the relative importance of its green branch, as this may affect a firm's perceptions of and responses to policy stimuli (Schmidt et al. 2012; Huenteler et al. 2016).

11.3 Research Case

We chose to focus on innovation in renewable energy because decarbonizing the global energy system is expected to involve the massive deployment of renewable energies (IRENA 2013; IEA and IRENA 2017). Specifically, we use the case of Germany as a pioneering country for renewables support through feed-in tariffs introduced by the German Renewable Energy Sources Act (EEG) in 2000 and adapted over time in line with socio-technical and socio-political challenges (Grau 2014; Hoppmann et al. 2014; Lauber and Jacobsson 2016). As such, the EEG serves as the core renewables instrument within a rich instrument mix which is expected to help Germany meet its national target for the share of renewables in the electricity mix of 40–45% by 2025 and of at least 80% by 2050. In addition, Germany has set itself ambitious short and long-term targets for the reduction of greenhouse gas emissions of 40% by 2020 and 80% by 2050 compared to 1990 levels. The EU ETS is typically considered the core climate policy instrument here within a rich instrument mix (Matthes 2017). This instrument mix, among others, also includes public support for R&D to facilitate the decarbonization of the energy system. The German government's support here has climbed to above 800 million euros per year since

2014, with a good third of this dedicated to supporting renewable energies (BMWi and BMU 2010; BMWi 2015, 2016b).

Several studies have analyzed the German *Energiewende* in general and core demand-pull instruments in particular (Strunz 2014; Quitzow et al. 2016). However, only a few have looked at instrument interaction and at the role played by instrument design (Lehmann 2010; Gawel et al. 2014). To the best of our knowledge, no quantitative ex-post evaluation has explicitly addressed the role of instrument design for instrument interaction regarding innovation in renewable energies. Given that innovation in the power sector has traditionally been dominated by suppliers, we focus on the innovation activities of manufacturers of renewable power generation technologies in Germany, because of its strong and export-oriented manufacturing base (Pavitt 1984; Rogge and Hoffmann 2010). We rely on survey data as it allows us to capture firm-specific assessments of the instrument mix.

Dedicated company surveys addressing the links between policy and low-carbon innovation in the German energy sector are rare. Two relevant exceptions are the studies of Schmidt et al. (2012) and Doblinger et al. (2015), who surveyed, among others, German manufacturers of renewable power generation technologies in 2009 and 2012, respectively. For non-emitting technologies, i.e. primarily for renewable energy technologies, Schmidt et al. (2012) find that the firms' perceptions of long-term climate targets, technology policies and their expectations about the third phase of the EU ETS are related to their R&D decisions. Doblinger et al. (2015) conclude that stronger demand-pull policies reduce the realization of high-risk R&D projects in favor of smaller improvements; a finding that was reinforced by perceived higher levels of regulatory uncertainty. However, neither study addresses instrument interactions or uses a conventional innovation survey questionnaire.

The year 2013, i.e. the year before we conducted our survey, was characterized by considerable regulatory uncertainty. Following the Fukushima accident in 2011 and the resulting decision by the German government to phase out nuclear energy by 2022 (Hermwille 2016), and due to the decline in technology costs, particularly for PV modules (Hoppmann et al. 2014), the share of renewable energies in the German electricity mix grew strongly in 2012 (BMWi 2015). The subsequent increases in the EEG surcharge led to high-level debates about a retrospective downward adjustment of the guaranteed feed-in tariffs (set for 20 years). Such a retrospective adjustment had previously been unthinkable (Bröcker 2013). Although, ultimately, no such adjustment was made, its very debate is likely to have tarnished the predictability and associated investment security of the EEG, the core demand-pull instrument. Moreover, in light of the federal elections, which took place in the fall of 2013, the next regular reform of the EEG was postponed until the formation of a new government coalition. This resulted in substantial uncertainty about the ambition of the *Energiewende* in general and the future of the EEG in particular.

Eventually, the new coalition government subsumed all *Energiewende*-related activities under one roof at the new Federal Ministry of Economics and Energy (BMWi). In early 2014, the BMWi published the first pillars for the revision of the EEG (BMWi 2016a). However, the uncertainty about important design features of the new EEG 2.0 remained high until the Federal Cabinet adopted the amended

Renewable Energy Sources Act on April 8, 2014.[2] Planned design changes included, among others, the reduction of feed-in tariffs and the introduction of auctions to determine support levels as an alternative to feed-in tariffs. Further design changes concerned the introduction of technology-specific binding expansion corridors as well as the step-wise expansion of direct marketing and the reduction of privileges for self-consumed power.

Amidst these policy mix developments, the share of renewables in the electricity mix had reached 27.4% by the end of 2014, and Germany was on track to meet its 2025 target (BMWi 2014).

11.4 Methodology

Our empirical analysis relies on a novel dataset from a survey of German manufacturers of renewable power generation technologies. We briefly describe the data collection in Sect. 11.4.1, and refer to Rogge and Schleich (2018) for further details, especially on the construction of the company database and the implementation of the survey. Section 11.4.2 then presents the econometric model and the variables used.

11.4.1 Data

Our data relies on computer-assisted telephone interviews (CATI), which were carried out by the research institute SOKO between April 9, 2014 and July 22, 2014 with 390 CEOs or top-level managers responsible for company strategy, R&D or sales and with an overview of products, innovation and corporate policy.[3] On average, these phone interviews lasted for 30 min. All data were anonymized by SOKO for further processing.

To a large extent, our questionnaire draws on the Community Innovation Survey (CIS), an established tool for measuring corporate innovation activities in European Member States. Since the CIS includes only few items on policy, we added supplementary questions to the policy mix.[4] Among others, we asked for the respondents' perceptions regarding specific policy instruments including the EU ETS and the

[2]This uncertainty was fully resolved after approval was given by the Federal Parliament (Bundestag) on July 4, 2014.

[3]The 390 participants correspond to a response rate of 35.7% of all German manufacturers of renewable power generation technologies.

[4]In case companies had more than one renewable power generation technology in their portfolio, respondents were asked to answer questions concerning their main renewable power generation technology so as to be able to gather technology-specific information, in particular regarding the instrument mix and innovation expenditures.

Table 11.1 Operationalization of variables for instrument design features of the EEG

EEG design feature	Statement [Own translation from German into English] In early April, the Federal Cabinet adopted an Amendment to the Renewable Energy Sources Act. In your mind, to which extent did changes in the following elements negatively affect sales of your main renewable power generation technology in Germany[a]	Variable name
Feed-in tariffs	The lowering of the feed-in tariffs	Lower_FIT
Auctions	The introduction of auctions to elicit support levels	Auction
Binding expansion corridor	The introduction of technology-specific binding expansion corridors	Corridor
Mandatory direct marketing	The stepwise introduction of mandatory direct marketing	Market
Self-consumed power	The disadvantaging of self-consumed power	Self_cons

[a]Response categories ranged from one (do not expect negative effects at all) to six (expect very negative effects); and "don't know"

EEG. In particular, we also added questions relating to changes to key design features of the EEG as the core demand-pull instrument for renewable energy technologies. The specific questions on design features refer to the amended EEG of April 8, 2014 (see Table 11.1).[5]

Our sample of 390 responses includes approximately 70% responses from small and medium-sized enterprises (SMEs). More than half concern solar PV (ca. 37%); other large shares relate to biogas (ca. 22%) and onshore wind power (ca. 17%). The large majority of these produce components for renewable power generation technologies (ca. 71%), with the remainder either producing final products for generating power from renewable energy sources (ca. 24%) or the respective production plants (ca. 5%). In 2013, only 11.1% of the companies in our sample operated exclusively in the German market; on average, exports accounted for almost 40% of sales.

The majority of respondents were fairly active with regard to innovation. More than 80% of companies had carried out innovation activities in the 3 years prior to the survey (2011–2013). In addition, three out of four companies had introduced product innovations during this period and two-thirds had introduced process innovations for the selected renewable power generation technology. Between 2011 and 2013, about a quarter of the respondents had received public R&D funding (from Germany or the EU) to pursue innovation activities in the main renewable power generation technology.

[5]A more detailed account of the questionnaire is available in Rogge and Schleich (2018).

11.4.2 Econometric Model

11.4.2.1 Dependent Variable

The econometric analysis follows Rogge and Schleich (2018) in that we use innovation expenditures as the dependent variable in our multivariate analysis. Specifically, we used survey information on actual or estimated innovation expenditures for each company's main renewable power generation technology in 2014 and 2015.[6] A substantial portion of the companies reported innovation expenditure of zero in one or both years, i.e. 25.6% for 2014 and 31.3% for 2015. We therefore employ the "corner solution" Tobit model to specify the regression equation for innovation expenditures in a particular year (y). Relying on the "latent variable" approach, truncation (from below) is modeled as:

$$\begin{aligned} &y_i^* = \beta X_i + u_i; \\ &u_i \tilde{} N(0, \sigma^2) \\ &y_i = y_i^* \; if \, y_i^* \geq 0 \\ &y_i = 0 \; if \, y_i^* < 0 \end{aligned} \tag{11.1}$$

where y_i^* reflects the latent (i.e. desired) level of innovation expenditures of firm i in a given year. The vector of explanatory variables X_i allows us to test our hypotheses and capture other factors related to firms' innovation expenditures. Thus, positive values for innovation expenditures are observed if y^* exceeds the threshold level of zero; otherwise companies report zero expenditures.

Estimating innovation expenditures separately for 2014 and 2015 may lead to biased and inconsistent parameter estimations (Greene 2012). We therefore estimate a bivariate Tobit model, where the error terms capture possible correlations between innovation expenditures in 2014 and 2015. Simulated maximum likelihood methods as implemented in Stata 14 are used to estimate the model.

11.4.2.2 Explanatory Variables

We include five groups of explanatory variables to capture the effects of: (1) technology push and demand pull (TP&DP), (2) policy instruments, (3) EEG design features, (4) instrument interaction, and (5) control variables to reflect company- and technology-specific effects. Table 11.2 presents the descriptive statistics for the sample used in the econometric analysis.

[6]Respondents were asked about their expenditures for innovation activities (including intramural—in-house—and extramural R&D, acquisition of machinery, equipment and software, acquisition of other external knowledge, and other preparation).

Table 11.2 Descriptive statistics of dependent and explanatory variables ($N = 160$)

Variables	Unit	Mean	Standard deviation	Minimum	Maximum
Innovation expenditures 2014[a]	In 1000 euros	2115	95,600	0	75,000
Innovation expenditures 2015[a]	In 1000 euros	2308	97,471	0	75,000
TechPush	In 1000 euros	130.2	651.1	0	6000
DemandPull	Dummy	0.36	0.481	0	1
SupportETS	Dummy	0.61	0.490	0	1
SupportEEG	Dummy	0.72	0.451	0	1
Lower_FIT	Dummy	0.575	0.496	0	1
Auction	Dummy	0.669	0.472	0	1
Corridor	Dummy	0.738	0.441	0	1
Market	Dummy	0.519	0.501	0	1
Self_cons	Dummy	0.606	0.490	0	1
ETSxFIT	Dummy	0.338	0.474	0	1
Size (sales)[a]	In million euros	298.38	711.99	0.05	5500
Experience[a]	In years	15.33	11.76	0	64
Wind	Dummy	0.22	0.41	0	1
RE_share	In %	50.81	37.47	0.04	100

[a]The natural logarithm is used in the econometric estimation. Since the logarithm of zero is not defined, using the logarithm meant losing one observation (where magnitudes were zero). No observation in our final sample had zero experience. When public R&D (TechPush) or innovation expenditures were zero, we assigned the value of zero to the undefined logarithm. Taking the logarithm did not lead to negative values for the dependent variables, because all positive innovation expenditures in 2014 and 2015 exceeded 1000 euros (Note that innovation expenditures are measured in units of 1000 euros)

1. Technology push and demand pull

Regarding technology push, we use the amount of public R&D funding (in euros) each company had received between 2011 and 2013 from German or EU funding bodies for the main technology (*TechPush*). Most of the companies in our sample had identified Germany ($n = 360$) and Europe ($n = 333$) as their home market.

Regarding demand pull, we relied on a dummy variable (*DemandPull*), which takes the value of one if the respondent expected the sum of domestic sales and exports of the main technology in 2014 to be higher than in 2013 and zero otherwise. This variable can be interpreted as a proxy for the effect of global demand-pull instruments because of the strong dependence of the market demand for renewable power generation technologies on such instruments (Peters et al. 2012; Hoppmann et al. 2013; Dechezleprêtre and Glachant 2014).

2. Policy instruments

To specifically capture the effect of the core demand-pull instruments on innovation activities in renewable power technologies, we consider companies' perceptions towards the two core demand-pull instruments EU ETS and EEG.

First, the survey asked participants to evaluate to which extent the EU ETS supports the development of renewable energies. Response categories ranged from one (does not support it at all) to six (supports very strongly). To construct *SupportETS*, we first calculate the median value of the responses. Then, *SupportETS* takes on the value of one if the response category was at least as high as the median value and zero otherwise.

Second, the survey asked participants to evaluate to which extent the EEG supports the development of renewable energies. Response categories again ranged from one (does not support it at all) to six (supports very strongly). As before, to construct *SupportEEG*, we first calculate the median value of the responses. Then, *SupportEEG* takes on the value of one if the response category was at least as high as the median value and zero otherwise.

3. EEG design features

The set of EEG design features include the five variables described in Table 11.1, i.e. *Lower_FIT, Auction, Corridor, Market, and Self_cons*. These variables are all constructed in the same way. We first calculate the median value of the responses to the statement presented in Table 11.2. The variable is then coded as one if the response category was at least as high as the median value and zero otherwise.

4. Instrument interaction

To explore possible interaction effects between the EU ETS and the EEG, we focus on the reduced level of support as the key EEG design feature. We chose *LowerFIT* because this was the most important of the five EEG design features considered according to the descriptive statistics for the original items. Consequently, we constructed *ETSxFIT* by multiplying *SupportETS* and *Lower_FIT*.

5. Company- and technology-specific factors

We include four variables to control for firm-internal effects. First, *size* captures the total sales of each firm in 2013 in domestic and foreign markets (i.e. for diversified firms, this includes business fields other than the main renewable energy technology). The second variable, *experience*, is calculated as the number of years each firm had been offering products for the main renewable power generation technology (using 2014 as the reference year). Finally, we capture each firm's technology portfolio with two explanatory variables: *wind* takes the value of one if a firm's responses referred to either onshore or offshore wind and zero otherwise[7]; *RE_share* measures the share of

[7]Including dummies for other renewable energy technologies yielded coefficients which were far from being statistically significant. We therefore only incorporate *wind*.

employees working in the main renewable power generation technology in 2013 relative to all employees.

11.5 Results

Our econometric analysis involves estimating three alternative model specifications. The *base model* includes the company- and technology-specific factors together with the technology-push and demand-pull variables. The *instruments and design* model also includes the two core demand-pull instruments EU ETS and EEG and the EEG design features. The *interaction model* differs from the *instruments and design model* by the addition of the interaction variable ETSxFIT. Table 11.3 displays the estimation results and reports heteroskedasticity-robust *p*-values in parentheses below the parameter estimates.

We first note that all three models produce very similar results for the company- and technology-specific factors. We further note that, in all three models, the correlation between the two equations is high and positive and statistically significant, thus corroborating the use of the multivariate Tobit framework.[8] Also, collinearity does not appear to be a problem.[9]

11.5.1 Base Model

In general, the coefficients in the base model all exhibit the expected signs and are almost all statistically significant. They are also very similar to the respective findings in Rogge and Schleich (2018), although the samples differ slightly between the two studies because of differences in missing observations for the different sets of explanatory variables. To allow for better comparability across the models presented in this study, we limit all analyses to observations where participants responded to the items on the EEG design features.[10]

More specifically, the findings confirm the positive relationship of European technology-push and global demand-pull effects with innovation expenditures in 2014 and 2015. Calculating the marginal effect for *TechPush* in the R&D 2014 equation implies that, on average, a 1% increase in public subsidies for R&D received for a manufacturer's main renewable power generation technology between

---

[8]For example, for the *base* model $\rho = 0.902$. Based on a Likelihood-Ratio test, the null hypothesis ($\rho = 0$) can be rejected at $p < 0.01$ [$\chi^2(1) = 216.801$]. Similar conclusions can be drawn for the *instruments and design* and the *interaction* models.

[9]When considering all explanatory variables used in the *interaction* model, the average variance inflation factor (VIF) is 1.74 and all VIFs of the individual variables are below 4.5, and thus well below 10, which is the critical threshold value commonly used in the literature.

[10]Our findings are virtually the same if we relax this condition.

Table 11.3 Regression results

Category	Variable	Base model		Instruments and design		Interaction	
		2014	2015	2014	2015	2014	2015
TP & DP	TechPush	**0.194**** **(0.016)**	**0.212**** **(0.047)**	0.154 (0.110)	0.158 (0.218)	0.139 (0.146)	0.136 (0.278)
	DemandPull	**2.445***** **(0.005)**	**4.585***** **(0.000)**	**2.802***** **(0.003)**	**4.830***** **(0.000)**	**2.747***** **(0.004)**	**4.736***** **(0.000)**
Instruments	SupportETS			0.694 (0.521)	1.283 (0.389)	−1.325 (0.281)	−2.318 (0.164)
	SupportEEG			−0.790 (0.476)	−0.389 (0.805)	−0.866 (0.435)	−0.517 (0.740)
EEG design features	Lower_FIT			**−1.885**** **(0.033)**	**−3.008**** **(0.016)**	**−4.092***** **(0.007)**	**−7.065***** **(0.001)**
	Auction			1.939 (0.114)	**2.905*** **(0.087)**	1.891 (0.131)	2.816 (0.108)
	Corridor			0.894 (0.373)	0.526 (0.692)	1.303 (0.200)	1.218 (0.370)
	Market			**−2.847**** **(0.019)**	**−3.993**** **(0.016)**	**−2.984**** **(0.015)**	**−4.243**** **(0.011)**
	Self_cons			−0.674 (0.480)	−1.376 (0.324)	−0.650 (0.493)	−1.305 (0.342)
Interaction	ETSxFIT					**3.560*** **(0.053)**	**6.531**** **(0.011)**
Firm characteristics	Size	**1.038***** **(0.000)**	**1.142***** **(0.000)**	**1.012***** **(0.000)**	**1.130***** **(0.000)**	**1.014***** **(0.000)**	**1.125***** **(0.000)**
	Experience	**1.092*** **(0.075)**	1.056 (0.185)	**1.290*** **(0.053)**	**1.618*** **(0.066)**	**1.288**** **(0.047)**	**1.667*** **(0.051)**
	Wind	**2.736***** **(0.005)**	**2.807**** **(0.030)**	**2.187**** **(0.024)**	**2.318*** **(0.069)**	**2.470**** **(0.013)**	**2.862**** **(0.033)**
	RE_share	**3.110**** **(0.026)**	**3.465*** **(0.052)**	**2.973**** **(0.034)**	2.853 (0.126)	**2.688*** **(0.055)**	2.316 (0.213)

Constant	**−14.743*** (0.001)**	**−18.785*** (0.001)**	**−12.859*** (0.004)**	**−17.214*** (0.004)**	**−11.516*** (0.008)**	**−14.829** (0.010)**
Log (Pseudo)likelihood	−772.380 (76.28)		−709.492 (94.50)		−705.781 (96.72)	
Rho (Chi-Squared)	0.902 (216.801)		0.919 (217.99)		0.920 (216.115)	
Observations	165		156		156	

Note: robust p-values in parentheses; $^{***}p < 0.01$, $^{**}p < 0.05$, $^{*}p < 0.1$

2011 and 2013 is associated with an increase in firm-level innovation expenditures by 0.147% in 2014 for firms with positive innovation expenditures in 2014.[11]

For our firm characteristics, we find positive and significant correlations with innovation expenditures. As expected, our results show that larger firms (in terms of sales) are positively related to higher innovation expenditures in 2014 and 2015. For example, a 1% increase in sales is associated with an increase in innovation expenditures of about 1% in both years. More experienced firms (in terms of years of activity in the main renewable power generation technology) spend more on innovation, but the coefficient for 2014 is not statistically significant at conventional levels. Firms active in wind technologies are associated with statistically significantly higher innovation expenditures in 2014 and 2015 compared with firms that focus on other renewable electricity technologies, indicating strong differences across technologies. Finally, the coefficient associated with the share of employees working in the main renewable power generation technology takes on the expected positive sign, and turns out to be significant for both 2014 and 2015.

11.5.2 Instruments and Design Model

Regarding our variables for the two core demand-pull instruments, *SupportETS* and *SupportEEG*, we found no support for a correlation with innovation expenditures for either year.

The results for the *instruments and design* model suggest that three of the five EEG design features, i.e. *Lower_FIT*, *Market* and *Auction*, are statistically significantly related with innovation expenditures for both years. First, a perceived negative effect of lowering the feed-in tariffs on manufacturers' sales of renewable energy technologies is negatively related to innovation expenditures in 2014 and 2015. Second, perceiving the stepwise introduction of mandatory direct marketing to have negative effects on sales is negatively associated with innovation expenditures in both years. Finally, *auction* and thus the policy design change to determine the level of support through auctions rather than predetermining it through feed-in tariffs turned out to be statistically significant for innovation expenditures in 2015, while the coefficient for 2014 is just shy of being statistically significant.

In comparison, the coefficients associated with *Corridor*, and *Self_cons* were far from being statistically significant for both years.

Finally, we note that our technology-push variable no longer exhibits statistical significance at conventional levels, most likely because of the lower degrees of freedom in the *instrument and design* model compared to the *base model*.

[11]Consistent marginal effects were derived from running a single Tobit model for innovation expenditures in 2014.

11.5.3 Interaction Model

Our final model explores the potential effects of interaction between our two core demand-pull policy instruments. Specifically, we consider the interaction between the EU ETS as the most important EU climate policy instrument and the reduced feed-in tariffs as the main EEG design feature. Since the findings for the other explanatory variables in the *interaction* model are almost identical to those of the *instruments and design* model, we focus on the findings for this interaction term. The positive and statistically significant coefficient associated with *ETSxFIT* suggests that the positive effect of the EU ETS on innovation expenditures is stronger, the more negative the perceived effects of lowering the FITs are on sales. By the same token, strong negative perceived effects of lowering the FITs are stronger, the less positive the perceived effects of the EU ETS are on sales.

11.6 Discussion and Conclusions

Bearing in mind the explorative nature of our study, our econometric analysis provides evidence that changes in instrument design and instrument interaction matter for eco-innovation in the case of renewable power generation technologies in Germany. In particular, if companies believe a change in certain EEG design features (such as lowering the feed-in-tariff, introducing auctions, or mandatory direct marketing) will negatively affect their domestic sales, they are likely to spend less on low-carbon innovation. In addition, when considering the interaction between the EU ETS and the most relevant design feature of the EEG (the feed-in tariffs), we find that favorable perceptions about the impact of the EU ETS on innovation have a stronger positive effect on innovation expenditures, the more negative the perceived effects of lowering the FIT are on sales. In contrast, we find no correlation between innovation and these perceived effects of the EU ETS and EEG per se.

These results confirm earlier findings in the literature suggesting that instrument design rather than instrument type matter for eco-innovations (Kemp and Pontoglio 2011). In addition, our study confirms theoretical considerations concerning the importance of accounting for design features of policy instruments when investigating their interactions (del Río and Cerdá 2017). That is, we find that both policy instrument design and interaction matter for innovation in renewable energies, and thus should be considered in future studies on the links between policy and eco-innovation.

Turning to *technology push*, we find that public financial support for innovation projects is linked with higher private innovation expenditures in the future, which is generally in line with the literature (Johnstone et al. 2010; Costantini et al. 2015).

Regarding *demand-pull* effects, our study supports earlier findings that market growth—which in the case of renewable energies at the time of our survey was still mainly policy-induced—is positively associated with eco-innovation (Horbach 2008; Hoppmann et al. 2013; Schleich et al. 2017). In our case, technology providers who expect their sales of green technologies to increase compared with the previous year tend to spend more on low-carbon innovation. Of course, this growth expectation not only depends on policy-induced market growth, but also on the competiveness of firms. For example, in recent years, PV module manufacturers in Germany have been particularly challenged by Chinese competitors (Quitzow 2015). Ultimately, global and not only domestic market expectations matter.

In terms of our *control variables*, we find strong evidence that firm *size* (measured in total sales in 2013) positively affects low-carbon innovation expenditures. These results are in line with others reported in the eco-innovation literature (Kammerer 2009; Kesidou and Demirel 2012; del Río et al. 2016). In addition, we also find evidence that *experience* with the main renewable power generation technology (measured in years) positively correlates with innovation expenditures for renewable power generation technologies, suggesting that early movers spend more on green innovation. This also underlines the importance of technological and organizational capabilities found in the eco-innovation literature (Kammerer 2009; Demirel and Kesidou 2011; Horbach et al. 2012). Regarding the *technology portfolio*, our findings suggest possible differences across technologies (Huenteler et al. 2016), with those companies active in on- and offshore wind power committing to higher innovation expenditures than the rest. Furthermore, firms with a higher share of employees working in the main renewable power generation technology were found to spend more on innovation in renewable power generation technologies.

Overall, we argue that our explorative study provides empirical support for going beyond aggregated technology-push and demand-pull variables in studies examining the links between policy and eco-innovation, and extending policy coverage to include the design of core policy instruments, and instrument interaction. In particular, we find strong evidence for a positive relationship between innovation expenditures on renewable power technologies and changes in design features, where negative expectations regarding the potential effect on sales are associated with lower innovation expenditures. We also find that the interaction effects of instruments may be driven by specific design features of core policy instruments.

Clearly, this novel empirical research is not free of limitations, and should be seen as a first step to analyzing the impact of instrument mixes on eco-innovation. First, choosing the German *Energiewende* for such an exploratory study makes it possible to draw lessons from one of the most advanced cases of low-carbon transition. The focus on one country and one sector obviously implies that our results may not be readily transferable to other contexts. Second, while operationalizing instrument design features proved feasible within an innovation survey, and the correlations found between innovation and the policy variables build upon and support earlier qualitative findings, we also recognize the caveats inherent to survey-based research such as recall bias, social desirability bias and common method bias. In particular, cross-sectional analyses are limited to correlations rather than causal inference. Third, our operationalization of the measurement of perceptions of the instrument mix should be regarded as a first attempt only.

Our exploratory empirical study on how instrument design and instrument interaction affect eco-innovation also suggests areas for future research (see also Rogge and Schleich 2018; Rogge and Dütschke 2018). Future empirical research on the impact of instrument mixes and eco-innovation could try to establish causality rather than correlations. Specifically, a periodic innovation survey among low-carbon technology manufacturers could eventually allow panel analyses. Such a survey could take a broader systems perspective and also cover providers of complementary or enabling technologies, such as storage or grid technologies. These surveys could also be implemented across several countries, thus providing for a better understanding of policy mix aspects such as the relevance of instrument design and interaction. For example, additional insights into the links between policy and low-carbon innovation could be generated by comparing findings for countries with a similar industry structure but alternative governance approaches regarding the transition of the energy system. Finally, if the CIS or similar surveys included policy mix questions enabling *cross-sectoral comparisons*, it would be possible to assess the role of instrument mixes for eco-innovation in the more general greening of the economy.

Ultimately, we hope the findings of our explorative study will initiate a critical assessment of how policy and eco-innovation are measured in innovation surveys and beyond. Clearly, further research is needed to help establish new standards in innovation surveys and in the analysis of policy instruments' interactions.

Acknowledgements This chapter was written in the context of Fraunhofer ISI's internal project TransPoSi, which investigates transformative policy processes for system innovation, and the corresponding role of visions, targets and instruments. The data underlying our analysis was collected as part of the GRETCHEN project (2012–2015), funded by the Federal Ministry of Education and Research (BMBF) within its FONA funding initiative "Economics of Climate Change" (Econ-C-026).

References

Barney, J. (2001). Resource-based theories of competitive advantage: A ten-year retrospective on the resource-based view. *Journal of Management, 27*(6), 643–650. https://doi.org/10.1016/S0149-2063(01)00115-5.

Bergek, A., & Berggren, C. (2014). The impact of environmental policy instruments on innovation: A review of energy and automotive industry studies. *Ecological Economics, 106*, 112–123. https://doi.org/10.1016/j.ecolecon.2014.07.016.

BMWi. (2014). *Act on the development of renewable energy sources (Renewable Energy Sources Act – RES Act 2014)*. EEG, Berlin.

BMWi. (2015). *The energy of the future – Fourth "energy transition" monitoring report – Summary: A good piece of work*. Federal Ministry for Economic Affairs and Energy, Berlin.

BMWi. (2016a). *10-point agenda: A clear roadmap for the energy transition*. Federal Ministry for Economic Affairs and Energy. http://www.bmwi.de/Redaktion/EN/Dossier/energy-transition.html. Accessed 27 Jun 2017.

BMWi. (2016b). *Bundesbericht Energieforschung 2016: Forschungsförderung für die Energiewende*. Federal Ministry for Economic Affairs and Energy, Berlin.

BMWi, BMU. (2010). *Energy concept for an environmentally sound, reliable and affordable energy supply, federal ministry of economics and technology*. Federal Ministry for the Environment, Berlin.

Borghesi, S., Cainelli, G., & Mazzanti, M. (2015). Linking emission trading to environmental innovation: Evidence from the Italian manufacturing industry. *Research Policy, 44*(3), 669–683. https://doi.org/10.1016/j.respol.2014.10.014.

Bröcker, M. (2013). *Regierung plant Bremse für den Strompreis*. Rheinische Post.

Bruns, E., Ohlhorst, D., Wenzel, B., & Köppel, J. (2011). *Renewable energies in Germany's electricity market: A biography of the innovation process*. Dordrecht: Springer.

Cantner, U., Graf, H., Herrmann, J., & Kalthaus, M. (2016). Inventor networks in renewable energies: The influence of the policy mix in Germany. *Research Policy, 45*(6), 1165–1184. https://doi.org/10.1016/j.respol.2016.03.005.

Costantini, V., Crespi, F., Martini, C., & Pennacchio, L. (2015). Demand-pull and technology-push public support for eco-innovation: The case of the biofuels sector. *Research Policy, 44*(3), 577–595. https://doi.org/10.1016/j.respol.2014.12.011.

Cunningham, P., Edler, J., Flanagan, K., & Laredo, P. (2013). *Innovation policy mix and instrument interaction: A review*. Nesta working paper series 13/20. NESTA, p. 47. www.nesta.org.uk/wp13-20.

Dechezleprêtre, A., & Glachant, M. (2014). Does foreign environmental policy influence domestic innovation?: Evidence from the wind industry. *Environmental and Resource Economics, 58*(3), 391–413. https://doi.org/10.1007/s10640-013-9705-4.

del Río, P. (2006). The interaction between emissions trading and renewable electricity support schemes: An overview of the literature. *Mitigation and Adaptation Strategies for Global Change, 12*(6), 1363–1390.

del Río, P. (2009). The empirical analysis of the determinants for environmental technological change: A research agenda. *Ecological Economics, 68*(3), 861–878. https://doi.org/10.1016/j.ecolecon.2008.07.004.

del Río, P. (2010). Analysing the interactions between renewable energy promotion and energy efficiency support schemes: The impact of different instruments and design elements. *Energy Policy, 38*(9), 4978–4989. https://doi.org/10.1016/j.enpol.2010.04.003.

del Río, P. (2012). The dynamic efficiency of feed-in tariffs: The impact of different design elements. *Energy Policy, 41*, 139–151. https://doi.org/10.1016/j.enpol.2011.08.029.

del Río, P., & Cerdá, E. (2017). The missing link: The influence of instruments and design features on the interactions between climate and renewable electricity policies. *Energy Research & Social Science, 33*, 49–58. https://doi.org/10.1016/j.erss.2017.09.010.

del Río, P., & Peñasco, C. (2014). Innovation effects of support schemes for renewable electricity. *Universal Journal of Renewable Energy, 2*, 45–66.

del Río, P., Carrillo-Hermosilla, J., Könnölä, T., & Bleda, M. (2015). Resources, capabilities and competences for eco-innovation. *Technological and Economic Development of Economy, 22*(2), 274–292. https://doi.org/10.3846/20294913.2015.1070301.

del Río, P., Peñasco, C., & Romero-Jordán, D. (2016). What drives eco-innovators?: A critical review of the empirical literature based on econometric methods. *Journal of Cleaner Production, 112*, 2158–2170. https://doi.org/10.1016/j.jclepro.2015.09.009.

Demirel, P., & Kesidou, E. (2011). Stimulating different types of eco-innovation in the UK: Government policies and firm motivations. *Ecological Economics, 70*(8), 1546–1557. https://doi.org/10.1016/j.ecolecon.2011.03.019.

Di Stefano, G., Gambardella, A., & Verona, G. (2012). Technology push and demand pull perspectives in innovation studies: Current findings and future research directions. *Research Policy, 41*(8), 1283–1295. https://doi.org/10.1016/j.respol.2012.03.021.

Doblinger, C., Dowling, M., & Helm, R. (2015). An institutional perspective of public policy and network effects in the renewable energy industry: Enablers or disablers of entrepreneurial behaviour and innovation? *Entrepreneurship & Regional Development, 28*(1–2), 126–156. https://doi.org/10.1080/08985626.2015.1109004.

Edler, J., Georghiou, L., Blind, K., & Uyarra, E. (2012). Evaluating the demand side: New challenges for evaluation. *Research Evaluation, 21*(1), 33–47. https://doi.org/10.1093/reseval/rvr002.

Flanagan, K., Uyarra, E., & Laranja, M. (2011). Reconceptualising the 'policy mix' for innovation. *Research Policy, 40*(5), 702–713. https://doi.org/10.1016/j.respol.2011.02.005.

Frondel, M., Horbach, J., & Rennings, K. (2008). What triggers environmental management and innovation? Empirical evidence for Germany. *Ecological Economics, 66*(1), 153–160. https://doi.org/10.1016/j.ecolecon.2007.08.016.

Gallagher, K. S., Grübler, A., Kuhl, L., Nemet, G., & Wilson, C. (2012). The energy technology innovation system. *Annual Review of Environment and Resources, 37*(1), 137–162. https://doi.org/10.1146/annurev-environ-060311-133915.

Gawel, E., Strunz, S., & Lehmann, P. (2014). A public choice view on the climate and energy policy mix in the EU — How do the emissions trading scheme and support for renewable energies interact? *Energy Policy, 64*, 175–182. https://doi.org/10.1016/j.enpol.2013.09.008.

Ghisetti, C., & Pontoni, F. (2015). Investigating policy and R&D effects on environmental innovation: A meta-analysis. *Ecological Economics, 118*, 57–66. https://doi.org/10.1016/j.ecolecon.2015.07.009.

Grau, T. (2014). Responsive feed-in tariff adjustment to dynamic technology development. *Energy Economics, 44*, 36–46. https://doi.org/10.1016/j.eneco.2014.03.015.

Greene, W. H. (2012). *Econometric analysis*. Boston: Pearson.

Guerzoni, M., & Raiteri, E. (2015). Demand-side vs. supply-side technology policies: Hidden treatment and new empirical evidence on the policy mix. *Research Policy, 44*(3), 726–747. https://doi.org/10.1016/j.respol.2014.10.009.

Hašcic, I., Johnstone, N., & Kalamova, M. (2009). Environmental policy flexibility, search and innovation. *Finance a Uver – Czech Journal of Economics and Finance, 59*(5), 426–441.

Helfat, C. E., Finkelstein, S., Mitchell, W., Peteraf, M. A., & Singh, H. (2007). *Dynamic capabilities: Understanding strategic change in organizations*. Oxford: Blackwell.

Hermwille, L. (2016). The role of narratives in socio-technical transitions—Fukushima and the energy regimes of Japan, Germany, and the United Kingdom. *Energy Research & Social Science, 11*, 237–246. https://doi.org/10.1016/j.erss.2015.11.001.

Hoffmann, V. H., Trautmann, T., & Schneider, M. (2008). A taxonomy for regulatory uncertainty-application to the European emission trading scheme. *Environmental Science & Policy, 11*(8), 712–722.

Hoppmann, J., Peters, M., Schneider, M., & Hoffmann, V. H. (2013). The two faces of market support—How deployment policies affect technological exploration and exploitation in the solar photovoltaic industry. *Research Policy, 42*(4), 989–1003. https://doi.org/10.1016/j.respol.2013.01.002.

Hoppmann, J., Huenteler, J., & Girod, B. (2014). Compulsive policy-making—The evolution of the German feed-in tariff system for solar photovoltaic power. *Research Policy, 43*(8), 1422–1441. https://doi.org/10.1016/j.respol.2014.01.014.

Horbach, J. (2008). Determinants of environmental innovation—New evidence from German panel data sources. *Research Policy, 37*(1), 163–173. https://doi.org/10.1016/j.respol.2007.08.006.

Horbach, J., Rammer, C., & Rennings, K. (2012). Determinants of eco-innovations by type of environmental impact—The role of regulatory push/pull, technology push and market pull. *Ecological Economics, 78*, 112–122. https://doi.org/10.1016/j.ecolecon.2012.04.005.

Horbach, J., Oltra, V., & Belin, J. (2013). Determinants and specificities of eco-innovations compared to other innovations—An econometric analysis for the French and German industry based on the community innovation survey. *Industry and Innovation, 20*(6), 523–543. https://doi.org/10.1080/13662716.2013.833375.

Huenteler, J., Schmidt, T. S., Ossenbrink, J., & Hoffmann, V. H. (2016). Technology life-cycles in the energy sector—Technological characteristics and the role of deployment for innovation. *Technological Forecasting & Social Change, 104*, 102–121. https://doi.org/10.1016/j.techfore.2015.09.022.

IEA. (2011). *Interactions of policies for renewable energy and climate*. Paris.
IEA, IRENA. (2017). *Perspectives for the energy transition: Investment needs for a low-carbon energy system*. International Energy Agency, International Renewable Energy Agency, Paris, Bonn.
IRENA. (2013). *Renewable energy innovation policy: Success criteria and strategies*. International Renewable Energy Agency, Bonn.
Jacobsson, S., & Bergek, A. (2004). Transforming the energy sector: The evolution of technological systems in renewable energy technology. *Industrial and Corporate Change, 13*(5), 815–849. https://doi.org/10.1093/icc/dth032.
Jaffe, A. B., Newell, R. G., & Stavins, R. N. (2002). Environmental policy and technological change. *Environmental & Resource Economics, 22*(1–2), 41–69.
Johnstone, N. (Ed.). (2007). *Environmental policy and corporate behaviour*. Cheltenham: Elgar.
Johnstone, N., Haščič, I., & Popp, D. (2010). Renewable energy policies and technological innovation: Evidence based on patent counts. *Environmental and Resource Economics, 45*(1), 133–155. https://doi.org/10.1007/s10640-009-9309-1.
Kammerer, D. (2009). The effects of customer benefit and regulation on environmental product innovation. *Ecological Economics, 68*(8–9), 2285–2295. https://doi.org/10.1016/j.ecolecon.2009.02.016.
Kemp, R. (1997). *Environmental policy and technical change*. Cheltenham: Edward Elgar.
Kemp, R., & Pearson, P. (2007). *MEI D15 – Final report MEI project about measuring eco innovation: Deliverable 15*. UM-MERIT, Maastricht.
Kemp, R., & Pontoglio, S. (2011). The innovation effects of environmental policy instruments—A typical case of the blind men and the elephant? *Ecological Economics, 72*, 28–36. https://doi.org/10.1016/j.ecolecon.2011.09.014.
Kesidou, E., & Demirel, P. (2012). On the drivers of eco-innovations: Empirical evidence from the UK. *Research Policy, 41*(5), 862–870. https://doi.org/10.1016/j.respol.2012.01.005.
Lauber, V., & Jacobsson, S. (2016). The politics and economics of constructing, contesting and restricting socio-political space for renewables – The German renewable energy act. *Environmental Innovation and Societal Transitions, 18*, 147–163. https://doi.org/10.1016/j.eist.2015.06.005.
Lehmann, P. (2010). *Using a policy mix to combat climate change – An economic evaluation of policies in the German electricity sector*. PhD thesis, Universität Halle-Wittenberg.
Matthes, F. C. (2017). Energy transition in Germany: A case study on a policy-driven structural change of the energy system. *Evolutionary and Institutional Economics Review, 14*(1), 141–169. https://doi.org/10.1007/s40844-016-0066-x.
Mowery, D., & Rosenberg, N. (1979). The influence of market demand upon innovation: A critical review of some recent empirical studies. *Research Policy, 8*(2), 102–153.
Nelson, R. R., & Winter, S. G. (1982). *An evolutionary theory of economic change*. Cambridge: Belknapp Press of Harvard University Press.
OECD. (2005). *Oslo manual: Guidelines for collecting and interpreting innovation data*. 3rd ed., Paris.
OECD. (2009). *Eco-innovation in industry: Enabling green growth*. Paris.
OECD. (2011). *Better policies to support eco-innovation*. Paris: OECD Studies on Environmental Innovation.
OECD. (2015). *System innovation: Synthesis report*. Paris.
OECD/IEA/NEA/ITF. (2015). *Aligning policies for a low-carbon economy* (p. 242). Paris: OECD.
Pavitt, K. (1984). Sectoral patterns of technical change: Towards a taxonomy and a theory. *Research Policy, 13*(6), 343–373. https://doi.org/10.1016/0048-7333(84)90018-0.
Pegels, A., & Lütkenhorst, W. (2014). Is Germany's energy transition a case of successful green industrial policy? Contrasting wind and solar PV. *Energy Policy, 74*, 522–534. https://doi.org/10.1016/j.enpol.2014.06.031.

Peters, M., Schneider, M., Griesshaber, T., & Hoffmann, V. H. (2012). The impact of technology-push and demand-pull policies on technical change – Does the locus of policies matter? *Research Policy, 41*(8), 1296–1308. https://doi.org/10.1016/j.respol.2012.02.004.

Quitzow, R. (2015). Dynamics of a policy-driven market: The co-evolution of technological innovation systems for solar photovoltaics in China and Germany. *Environmental Innovation and Societal Transitions, 17*, 126–148. https://doi.org/10.1016/j.eist.2014.12.002.

Quitzow, L., Canzler, W., Grundmann, P., Leibenath, M., Moss, T., & Rave, T. (2016). The German energiewende – What's happening? Introducing the special issue. *Utilities Policy, 41*, 163–171. https://doi.org/10.1016/j.jup.2016.03.002.

Reichardt, K., & Rogge, K. (2016). How the policy mix impacts innovation: Findings from company case studies on offshore wind in Germany. *Environmental Innovation and Societal Transitions, 18*, 62–81. https://doi.org/10.1016/j.eist.2015.08.001.

Rennings, K. (2000). Redefining innovation—Eco-innovation research and the contribution from ecological economics. *Ecological Economics, 32*(2), 319–332. https://doi.org/10.1016/S0921-8009(99)00112-3.

Rennings, K., & Rammer, C. (2011). The impact of regulation-driven environmental innovation on innovation success and firm performance. *Industry and Innovation, 18*(3), 255–283. https://doi.org/10.1080/13662716.2011.561027.

Rogge, K. S. (2016). Reviewing the evidence on the innovation impact of the EU emission trading system. In S. Weishaar (Ed.), *Research handbook on emissions trading* (pp. 161–194). Cheltenham: Edward Elgar.

Rogge, K. S., & Dütschke, E. (2018). What makes them believe in the low-carbon energy transition? Exploring corporate perceptions of the credibility of climate policy mixes. *Environmental Science and Policy, 87*, 74–84. https://doi.org/10.1016/j.envsci.2018.05.009.

Rogge, K. S., & Hoffmann, V. H. (2010). The impact of the EU ETS on the sectoral innovation system for power generation technologies. *Energy Policy, 38*(12), 7639–7652. https://doi.org/10.1016/j.enpol.2010.07.047.

Rogge, K. S., & Reichardt, K. (2016). Policy mixes for sustainability transitions: An extended concept and framework for analysis. *Research Policy, 45*(8), 1620–1635. https://doi.org/10.1016/j.respol.2016.04.004.

Rogge, K. S., & Schleich, J. (2017). *Do policy mix characteristics matter for low-carbon innovation? A survey-based exploration for renewable power generation technologies in Germany*. SPRU working paper series (SWPS) 19. University of Sussex, Brighton.

Rogge, K. S., & Schleich, J. (2018). Do policy mix characteristics matter for low-carbon innovation? A survey-based exploration of renewable power generation technologies in Germany. *Research Policy* (forthcoming).

Rogge, K. S., Kern, F., & Howlett, M. (2017). Conceptual and empirical advances in analysing policy mixes for energy transitions. *Energy Research & Social Science, 33*, 1–10. https://doi.org/10.1016/j.erss.2017.09.025.

Schleich, J., Walz, R., & Ragwitz, M. (2017). Effects of policies on patenting in wind-power technologies. *Energy Policy, 108*, 684–695. https://doi.org/10.1016/j.enpol.2017.06.043.

Schmidt, T. S., Schneider, M., Rogge, K. S., Schuetz, M. J. A., & Hoffmann, V. H. (2012). The effects of climate policy on the rate and direction of innovation: A survey of the EU ETS and the electricity sector. *Environmental Innovation and Societal Transitions, 2*, 23–48. https://doi.org/10.1016/j.eist.2011.12.002.

Sorrell, S., & Sijm, J. (2003). Carbon trading in the policy mix. *Oxford Review of Economic Policy, 19*(3), 420–437.

Spyridaki, N.-A., & Flamos, A. (2014). A paper trail of evaluation approaches to energy and climate policy interactions. *Renewable and Sustainable Energy Reviews, 40*, 1090–1107. https://doi.org/10.1016/j.rser.2014.08.001.

Strunz, S. (2014). The German energy transition as a regime shift. *Ecological Economics, 100*, 150–158. https://doi.org/10.1016/j.ecolecon.2014.01.019.

Taylor, M. (2008). Beyond technology-push and demand-pull: Lessons from California's solar policy. *Energy Economics, 30*(6), 2829–2854. https://doi.org/10.1016/j.eneco.2008.06.004.

Teece, D. J., Pisano, G., & Shuen, A. (1997). Dynamic capabilities and strategic management. *Strategic Management Journal, 18*(7), 509–533. https://doi.org/10.1002/(SICI)1097-0266(199708)18:7<509:AID-SMJ882>3.0.CO;2-Z.

Veugelers, R. (2012). Which policy instruments to induce clean innovating? *Research Policy, 41*(10), 1770–1778. https://doi.org/10.1016/j.respol.2012.06.012.

Vollebergh, H. (2007). *Impacts of environmental policy instruments on technological change*. OECD, Paris. p. 34. Accessed 29 Jun 2014.

Weber, K. M., & Rohracher, H. (2012). Legitimizing research, technology and innovation policies for transformative change: Combining insights from innovation systems and multi-level perspective in a comprehensive 'failures' framework. *Research Policy, 41*(6), 1037–1047. https://doi.org/10.1016/j.respol.2011.10.015.

Wernerfelt, B. (1984). A resource-based view of the firm. *Strategic Management Journal, 5*(2), 171–180.

Wieczorek, A. J., & Hekkert, M. P. (2012). Systemic instruments for systemic innovation problems: A framework for policy makers and innovation scholars. *Science and Public Policy, 39*(1), 74–87. https://doi.org/10.1093/scipol/scr008.

Chapter 12
Corporate Social Responsibility in the Fashion Industry: How Eco-Innovations Can Lead to a (More) Sustainable Business Model in the Fashion Industry

Thomas Cleff, Gwen van Driel, Lisa-Marie Mildner, and Nadine Walter

12.1 Introduction

The clothing industry is one of the world's leading industries and one of the largest economic sectors. However, the clothing industry is also one of the most polluting and human-unfriendly businesses, with various negative social and ecological effects. These effects include child labor, poor working conditions, excessive water consumption, and accelerated accumulation of waste (Rossum 2012). The increased awareness of these major global issues has resulted in some parts of the fashion industry having a higher demand for environmentally friendly products, thereby forcing manufacturers to reconsider their production, portfolio and communication strategies in order to make their products greener (Launois 2008).

In Sect. 12.3, this study uncovers the most important insights into the challenges encountered by fashion companies that want to incorporate eco-innovation to market a justified, sustainable fashion brand. In Sect. 12.4, we examine the opportunities and risks and highlight the major changes that fashion companies should make to become more sustainable. To support the main objective, an interview questionnaire was designed to compare sustainability experts' perceptions with current fashion brand attitudes towards social responsibility. The semi-structured interviews were conducted over a 3-week period in August 2017 with ten participants. The results of the interviews with three experts from fashion companies and seven sustainability consultants are described in Sect. 12.5.

T. Cleff (✉) · L.-M. Mildner · N. Walter
Pforzheim University, Pforzheim, Germany
e-mail: thomas.cleff@hs-pforzheim.de; nadine.walter@hs-pforzheim.de

G. van Driel
Tias School for Business and Society, Tilburg University, Utrecht, The Netherlands

J. Horbach, C. Reif (eds.), *New Developments in Eco-Innovation Research*, Sustainability and Innovation, https://doi.org/10.1007/978-3-319-93019-0_12

The results indicate that even though the current fashion industry is taking steps to become more sustainable, this is a slow process. Preserving the fashion industry is a vicious cycle, as not every stakeholder cooperates with change. Among many others, the first challenge is for governments, businesses, and consumers to become more aware of the importance of sustainability to break the current pattern in which "fast fashion" dominates "slow fashion" and to speed up developments.

Section 12.6 includes recommendations to the fashion industry for eco-innovation in its processes and products. In conjunction with the literature research, it has been found that existing companies and new entrants must integrate eco-innovation into their business model in order to eliminate negative effects on social, ecological, and economic issues. Moreover, eco-innovation is an indispensable requirement, as the clothing industry is one of the world's leading industries and the second largest polluter in the world.

12.2 Corporate Social Responsibility

The idea of corporate social responsibility (CSR) suggests that a company has not only economic or legal obligations but also responsibilities that extend beyond these obligations (McGuire 1963). CSR requires companies to voluntarily consider its impact on other stakeholders other than just shareholders, such as society and the general public. Walton (1967) stresses that social responsibility requires companies to make investment decisions that might not be beneficial to short run profit maximization. These decisions exceed a firm's narrow compliance of the law, but they encompass the firm's effect on the whole social system (Davis and Blomstrom 1966).

CSR has its historical roots in the USA in the 1950s. This origin is due to two reasons. First, corporate goals in the US have traditionally been more focused on profit maximization than in Europe. Second, governmental regulations are higher in Europe than in the USA (Carroll and Buchholtz 2014; Matten and Moon 2008). In the 1950s, Bowen (1953), who is often named the "father of social responsibility," published "Social Responsibilities of the Businessman" and wrote about the responsibilities of managers with respect to society. This view was emphasized by other authors of this time, e.g., Peter Drucker (1956). In the 1960s, this view changed slightly. McGuire (1963) and Davis (1967) see the responsibility not with single managers but with the firm as an institution. Davis states in his "power-responsibility-equation" that an organization with higher levels of power and influence has a larger societal responsibility. During the 1970s, various authors worked on a conceptualization of CSR. One of the most popular frameworks was developed by Carroll (1979), who identifies the four major dimensions of CSR as economical, legal, ethical and philanthropic. In the 1980s, CSR was more integrated into the overall corporate strategy (Drucker 1984). In addition, other related concepts are being developed, such as the "Stakeholder Approach" (Freeman 1984) and the "World Conservation Strategy" of the International Union of Conversation of Nature and Natural Resources (1980). During the 1990s, CSR became a popular worldwide concept. This popularity was largely driven by violations of human rights issues in production facilities in developing countries. These cases led to public protests and a debate about stronger regulations. After 2000, CSR further established itself as a corporate practice. Carroll and Shabana (2010) rightly stated, "CSR is evolving into a core business function which is central to the firm's overall strategy and vital to its success."

Therefore, CSR is part of a company's management concept and strategy (United Nations Industrial Development Organization 2017).

Hence, the concept of CSR consists of the following aspects (Dahlsrud 2006; Marsden 2001):

- **Economic**: CSR entails gaining profits but also performing socially and environmentally responsible acts.
- **Social and environmental**: CSR means that companies limit their negative impacts on society and the environment.
- **Ethical**: CSR dictates that companies have a social obligation towards society and need to comply with ethical values and standards.
- **Voluntariness**: CSR counts on companies' voluntary actions.
- **Stakeholder**: CSR considers stakeholder's interests in addition to shareholders' goals.
- **Strategy**: CSR is part of the strategy and management concept of a firm.

Various CSR instruments have been developed, and worldwide commitment for CSR has increased. This commitment includes three areas.

- **International guidelines and principles**: The two most important international guidelines and principles include the "OECD Guidelines for Multinational Enterprise" (OECD 2011) and the "Tripartite declaration of principles concerning multinational enterprises and social policy" of the ILO (International Labour Organisation—ILO 2017). The OECD guidelines cover various areas, including human rights, employment, environment, bribery, consumer interests, science/technology, competition and taxation. The ILO focuses on employment and labor only but details areas, including forced labor, child labor, equality, security of employment and work safety.

  The previous work of the ILO influenced the OECD principles on employment. These guidelines were signed by all OECD countries, which are committed to promoting and recommending these to firms within their countries.
- **Code of conduct**: A code of conduct is an agreement on behavioral rules. The commitment to these rules is voluntary. The most prominent code of conduct concerning CSR is the UN Global Compact, which contains guidelines on human rights, labor, the environment and anti-corruption. Over 9000 companies and 4000 non-businesses have signed this code so far, including large multinational corporations such as BASF, Daimler, Nestle, Nike and Shell (United Nations Global Compact 2017).
- **Management systems**: A management system contains all organizational measures (policies, processes and procedures) used by an organization to achieve its objectives (Wieland and Grüninger 2000). Various management systems exist

that contain CSR-relevant topics. The most prominent systems are ISO26000, SA8000 and ILO-OSH 2001 (ISO 2017; SAI 2017; ILO 2009). These systems all provide standards on how businesses and organizations can operate in a socially responsible and ethical way. Firms that earn an accreditation prove an organization's compliance to these standards. All three options are possible ways of implementing CSR in a firm's organization.

12.3 Corporate Social Responsibility in the Fashion Industry

The fashion industry is one of the key industries worldwide due to its sheer size and global expansion. The global garment market is valued at \$3.5 trillion USD, and it accounts for 2% of the world's gross domestic product. Women's clothing is valued at \$621 billion USD and men's at \$402 billion USD (Fashion United 2016). Worldwide, 40 million people work in the textile industry. The production of clothing is mostly done in developing Asian countries. In 2015, China (45%), Bangladesh (7%), Vietnam (6%), India (5%), and Cambodia (2%) accounted for 64% of the global clothing exports (World Trade Organization 2016, 108).

Due to multiple negative influences on the environment and society, CSR is an important topic in the fashion industry for a fairly long period of time. Two concerns are prominent in the public eye. The first issue is poor working conditions and low wages in the production countries, and the second is environmental damage due to chemical substances or pollution.

Both topics illustrate the discussion areas of labor conditions and the environment.

The majority of people in the fashion industry are employed in the countries of production. Low wages are the major reason for fashion companies to produce in countries such as Bangladesh; low transportation costs and the speed of communication foster that trend (Starmanns 2013). CSR violations in these countries include poor working conditions, underpayment, long working days, abuse, child labor and the lack of health/social insurance (Martens 2006).

The garment industry impacts the environment on multiple dimensions along its production chain. First, there are large amounts of pesticides or chemicals, and a large amount of water is consumed during cotton production. In addition, the worldwide transportation is responsible for CO_2 emissions that have grown in the recent decades (Hill et al. 2012). The treatment of textiles, such as dyeing, washing, printing, and finishing, results in the discharge of waste water that includes toxic substances (Choi et al. 2011). The increased scale of production worsens the problem. Since "fast fashion" offers fashionable clothing at a low price, demand has increased. Consumers buy more pieces but wear them for a shorter period of time (Remy et al. 2016). Moreover, unsold clothing supplies are another issue. Estimations show that approximately 4% of clothes are not sold due to wrong size, fit or color, which is a problem whose consequences have not yet been fully investigated (Wijnia 2016).

A pre-requisite for successful CSR activities is transparency. CSR guidelines or codes of conduct oblige companies to publish a certain amount of information. "Commercially sensitive information" is now being publicly shared. This information can include the location of production sites, names of suppliers, levels of material waste or working conditions.

As the largest consumer of clothing, the European Union has taken a leading role regarding transparency in the fashion industry. With the "EU Report on the Flagship Initiative on Garment", voted on 3 May 2017, the European Parliament suggests that companies have an obligation to disclose their value chain. On a regular basis, firms will have to publish names, addresses, and contact details of all links within the chain. The Parliament further proposes a due diligence obligation (European Parliament 2017). Thus, clothing companies are required to identify, address, and correct the human rights violations in their supply chain.

Currently, there are no enforceable legal obligations for transparency. As a consequence, companies can select the information they want to publicly share. Thus, firms can stress their positive impact and disguise their negative impact, thereby pretending to be more socially and environmentally friendly than they are. This green washing is fairly common and reduces consumers' confidence in sustainable products (Delmas and Burbano 2011). Paradoxically, companies who do better in terms of transparency are more likely to be caught in a scandal since they are supervised (Reuters 2017).

12.4 Eco-Innovation to Create a (More) Sustainable Business Model in the Fashion Industry

Improving social and environmental conditions requires significant efforts from both the fashion industry and the consumer. The consumer, who is addressed as a conscious citizen, needs to adjust his or her buying behaviors. The fashion industry needs to "loosen" its existing models in term of pricing and propositions towards the customer (retail and end consumer) and consider all steps of the chain and their fair costs (production and trade) (Koppert and Brouw 2017).

12.4.1 Creating New Business Models: Change the Focus to Eco-Innovation to Create a (More) Sustainable Business Model in the Fashion Industry

CSR forces fashion companies to change their traditional business models. Currently, firms from the fashion industry are focusing mainly on the final phase of the value chain, the waste phase. Additive eco-innovations (such as waste disposal, waste recovery, and emissions control) are still dominant. In contrast to this, integrative eco-innovations (e.g., eco-friendly products, process integrated eco-innovation) are less common. For example, only one percent of clothing companies are using organic cotton (Kennedy 2015). However, circular systems would enable firms to maximize the reusability of materials and minimize the value destruction on all steps. This process requires a change of the whole system, including design, production, consumption and reuse (Jonker 2014).

12.4.2 Integrated Eco-Innovation and Green Sourcing

Sourcing affects 40–45% of the cost base of most companies and is considered to be an important tool for change (Turner and Houston 2008). Companies can shape their procurement efforts in a sustainable way. By focusing on suppliers that respect the environment and society, they can follow so-called inclusive sourcing, such as sourcing from suppliers that ensure employment equality for men and women. This process forces companies to introduce supplier minimum standards and conduct supplier audits. Being more involved in the whole process improves the environmental and social impact (Clinton and Whisnant 2014). Moreover, cooperation with other organizations that can lead to alliances and networks in the value chain are an important first step towards circularity. Cooperation emphasizes trust between the actors in the value chain and transparency of the information being exchanged. Both trust and transparency are prerequisites for sustainable sourcing (Witjes 2016). Sourcing also is a good starting point for integrating green initiatives. So-called green sourcing can reduce a company's environmental burden while saving costs and providing better relationships with suppliers and communities. Green sourcing results in the influence of a particular choice, whether it is transportation, materials, energy sources, and/or packaging design (Turner and Houston 2008). Environmental innovators frequently see increased resource efficiency as a component in a comprehensive efficient company strategy. This strategy expresses itself in the fact that they give cost reductions and quality management as innovation goals significantly more frequently than other innovators (Cleff and Rennings 1999a, 2000). "Innovators in integrated environmental protection at the product level pursue the goal of maintaining or increasing their market shares more frequently. In contrast, it is compliance with existing legislation and anticipation of future laws which has especial importance for innovators in integrated environmental protection at the level of (production) processes. This can be explained by the fact that integrated environmental protection at the level of the process generally confers little or no additional benefit on the customer and therefore receives comparatively little reward from the market, while integrated environmental protection at the product level (…) brings added benefit to the customer, explaining the strategic interest" (Cleff and Rennings 1999b, 331).

12.4.3 Transparency Along the Value Chain

The results of different market studies in logistics and supply chain management agree that transparency in the chain is a top priority. Making supply chains transparent empowers fashion companies to identify, assess, reduce, and realize a solution to minimize effects on society and the environment. However, the willingness of chain partners and employees is a critical factor. All stakeholders in the supply chain need to cooperate to achieve full transparency. This only works in a culture of cooperation and not by exerting pressure or imposing sanctions. With regards to technology, systems for full transparency along the value chain already exist. Visibility software based on platforming technology can demonstrate all resources, capacities, stocks, and processes in the supply chain and simplify the information exchange between supply chain partners (Groenendijk 2016).

12.4.4 Eco-Innovation in Sustainable Materials

The transition from a linear supply to a closed loop system also affects the materials. In a closed loop model, materials are continually recycled in the production system. Efforts are made to reduce waste in the production process (Clinton and Whisnant 2014). For instance, old clothes can be recycled, and the old fibres can then be woven into new fabrics. Organizations increasingly highlight and apply this recycle and reuse concept (Rikkert 2013). However, major challenges exist with establishing such systems. Firstly, it is difficult to create take-back schemes on a global basis—and the fashion industry is a global industry. Infrastructure, logistics and regulations vary in different countries. Secondly, it can be expensive and inefficient to create a closed loop system for a single company, since consumers are usually not sorting textiles according to brands. So collected materials are usually a mix of garments from various producers and these contain fiber types or chemical components that are unknown to non-producers. Therefore, collective system with a central organization or with all participating brands' participation might be the only solution (Watson et al. 2018).

12.4.5 Commitment on All Levels of a Company

To be able to truly anchor all sustainable elements in a company, the entire firm must commit. Companies perform significantly better with employees who feel involved in the organization and who feel acknowledged as humans. It is not systems, protocols, or procedures that are the most crucial for success, but the commitment and qualities of the employees. For this reason, sustainability should be stored in employees' routines by clearly repeating and communicating it. Sharing CSR ambitions and being transparent in expectations and goals ensures trust. This trust positively influences employees, which may result in a change of behaviors (Houtum 2015).

To conclude, eco-innovation to ensure the transition to a (more) sustainable business model affects the entire organization, including its value chain, processes, goals and structure, employees, cash flow, balance sheet, and its collaboration. According to Petersen et al. (2017), a number of critical factors are crucial at the beginning of the transition. The entire organization should undertake the transformation and commit. Every employee should invest in the growth of the CSR-focused organization (level, competencies, expertise). Finally, it is important to sustain resources specifically for the transformation. The transformation into a sustainable business model is attainable with full commitment, a business plan with realistic goals, a willingness to invest, and a phased approach.

12.5 Expert Interviews on CSR and Eco-Innovation in the Fashion Industry

12.5.1 Methodology

In this section, we investigate the risks and opportunities for fashion companies to transform into a circular economy and gradually become more transparent. The analysis is achieved by analyzing companies' motives for incorporating CSR into their strategy and how to integrate sustainability and eco-innovation into the business model. This extensive perspective requires an exploratory approach using a semi-structured questionnaire in which current changes within the fashion industry are analyzed. In addition, the analysis includes the changeover and urgency of fast fashion to slow fashion, which encompasses the conversion from non-sustainable to sustainable fashion.

The following questions have served to guide these analyses. (1) "What is the role of sustainability in the past, present, and future?" (2) "Why is sustainability important to the fashion industry?" (3) "What are the opportunities and risks in conducting business on a sustainable basis?" (4) "In what way can fashion companies integrate sustainability into their business model?" These questions support the main research question of, "What challenges face organizations wishing to launch a sustainable fashion brand?"

The research consisted of ten semi-structured interviews with both fashion industry experts and CSR/sustainability consultants. To allow for a comparison between these two groups, a distinction was made with regards to the interview questions.

The purpose of the interviews with three CSR/sustainability managers from fashion companies was to survey the extent to which (a part of) the fashion retail industry is active in CSR and which processes are already being applied. Communication from companies to consumers was taken into account, and the possible problems they could face were considered. In contrast, the interviews with seven consultants went into detail and discussed the overall subject and importance of sustainability within the fashion industry.

Among the interviewees, three participants are employed by a fashion company with multiple (physical) stores, and seven are sustainability consultants (not working under their own fashion brand or owning a single brand with or without a physical store). The interviews were planned and scheduled within the 3 weeks from August 7, 2017, to August 23, 2017. All participants were approached by email, LinkedIn or by phone. Prior to the interview, all interviewees received an overview of the research topic and the purpose of the interview. The theme and focus points were extended in a questionnaire, which is included in Appendix.

12.5.2 *Results*

12.5.2.1 What Is the Role of Sustainability in the Past, Present, and Future?

The experts have indicated that the world has changed with respect to how sustainability is handled today compared to some years ago. They see the growing number of graduated young people as a major factor. "What strikes is that there are increasingly more young people who hold a diploma and are specialized in a particular subject." "Increased knowledge and expertise has led to a better ability to manage developments and being able to renew." Interviewees indicated that paying close attention to young professionals brings a fresh look to sustainability-related issues.

An additional effect is the increased use of technology. As stated by four experts, digitalization enables the media to transmit any form of message instantaneously to the entire world. Thus, there is more transparency in the value chain that shows how companies produce, where they produce and by what means they produce. In addition, disclosure has resulted in great movements within the clothing industry in which the consumer has become aware of a deterioration of the earth and the living and health conditions of poor people in textile-producing countries. Therefore, there is an increased need to implement CSR throughout the entire company. All experts confirmed that, from the company's point of view, there is a remarkable difference between CSR in the past and today. Of the ten experts, five indicated that CSR used to be a "nice accessory" to consider, but today, it is an indispensable opportunity to carry out eco-innovations and sustainable initiatives throughout the organization. Despite companies being aware of the urgency to change to a more sustainable production process, this proves to be difficult. All ten interviewees confirmed that one of the largest challenges is the fact that durable fashion is still in its infancy and is slowly developing. The progress is relatively new, and consumers are reluctant regarding this subject. Eight experts confirmed that companies that want to be sustainable and adapt their business model face a major task. "The challenge is to create a straight alignment within the company, in which all departments pursue the same sustainability goals. If each employee does not pursue the same goals, the entire (internal) organization can never change, and the business model thus cannot be sustained."

In line with the experts' view, it seems that CSR is not yet fully incorporated in the company's DNA at most major fashion chains. The main reason seems to be inconsistency with consumer behavior. Respondents emphasized that people are making different choices as a consumer rather than as a citizen. As a citizen, CSR and sustainability are considered to be important. However, as a consumer, sustainability seems to be less important compared to other purchase influencing factors such as product price. Moreover, the experts stated that consumers became slightly more critical in the past due to negative media attention. Negative media attention has made it possible for companies to slowly accelerate the transition to sustainability.

12.5.2.2 Why Is Sustainability Important to the Fashion Industry?

The following factors were stated by the interviewees to be the most important reasons why the fashion industry must develop sustainability.

1. *To prevent resource depletion*: three out of ten participants confirmed that sustainability is important to companies because there is a great demand from society. Therefore, they see sustainability primarily as self-interest, and not changing could adversely affect the company's continued sales. On the other hand, the remaining seven interviewees indicated that change within the fashion industry is necessary because resources get depleted, and the costs to produce clothing in a very price-competitive sales market increase.
2. *To remain competitive*: for the existence of a company, sustainability is important for survival. The experts emphasized that increasingly more new entrants into the textile market include sustainability in their missions, visions and business models. Therefore, traditional companies perceive a pressure that sustainability is important to be able to stay competitive.
3. *To reduce environmental damage*: regardless of the necessity of eco-innovation, eight experts expressed great concern about the ongoing negative impact of the current fashion industry on the world's environment. However, all interviewees agreed that it is important to respond quickly and launch the phasing out of harmful substances throughout production chains. "Large-scale pollution by the textile industry has always been a problem, but the recent use of persistent and harmful chemicals is a greater, often invisible, threat to the ecosystem and human health." Major brands have great economic influence and are therefore able to take the lead in this phasing-out.
4. *To improve working conditions*: five out of ten interviewees confirmed the increased awareness about pitiful working conditions in textile producing countries. Therefore, more companies need to be transparent and make their processes

clearer. To this end, honesty plays a major role. As indicated by the experts, it is important not to change everything at once. "Small steps should be taken so that each one can be measured and evaluated afterwards." After all, the upcoming developments represent a gradual process.

12.5.2.3 What Are the Opportunities and Risks for Companies in Conducting Sustainable Business?

As a major opportunity, the experts stated that eco-innovation may lead to both positive effects on the profits and benefits for the employees. Sustainable solutions generate direct improvements on the health of employees affected by the environment in which they work. With regards to becoming a more sustainable fashion brand, the interviewees mentioned the following as the most important opportunities and risks.

Opportunities:

- Taking care of society and the environment reinforces the company's reputation, recognition, appreciation, and success.
- Sustainability results in cooperation (shared value), financial returns, innovation, and proud and committed employees.
- Maximizing the reusability of materials and commodities and minimizing value destruction leads to cost savings.
- Value chain transparency helps in identifying factories where labour rights are being violated.
- Strong supplier relationships ensure more consistency and improved traceability.

Risks:

- Companies who do better in terms of transparency are more likely to be caught in a scandal since they are supervised by media.
- Green washing can have negative effects on consumers' confidence in sustainable products.
- Producing sustainable products is more expensive than non-sustainable products.
- Rising wages and more restrictive safety rules in producing countries puts pressure on the shelf life of fast fashion and other discounters.

Seven out of ten interviewees unanimously agree on the mentioned catalogue of opportunities and risks. The remaining three interviewees have stated that it is a major challenge to balance opportunities against the risks; they fear the unknown or think that sustainable initiatives are too costly in the long run. However, increasingly more companies are seeing the importance and benefits of CSR. The experts stated that the main opportunity and motivation for companies to eco-innovate is that it reinforces reputation. A strong reputation attracts employees, customers, and investors, creating a positive image and strong competitiveness, ultimately leading to higher sales. Although disclosure seems difficult for many organizations, it has positive results on the company's internal and external environment. In addition, to stimulate consumer confidence, storytelling also appeared to be a great opportunity of sustainable initiatives. Thus, the experts indicated that an important opportunity is that companies have a real, true reason to tell their story and that they can transfer the business initiatives through messages to consumers. Moreover, apart from the confidence that consumers will have in businesses, this communication increases environmental and social awareness. "Storytelling is therefore indispensable." Transferring messages that contain knowledge, emotions, and feelings makes consumers feel more bound to the brand and they increasingly want to know more.

In another vein, it is currently difficult to address the prices of non-sustainable products. "The margins are much smaller for sustainable fashion and there is currently a worsening price ratio," said one expert. All interviewees confirmed that consumers are to a great extent price-sensitive. Communication is highly difficult, but, according to the experts, transparency can help reduce this problem. Furthermore, as awareness grows and interest in sustainable initiatives from the fashion world increases, sustainable clothing will eventually become cheaper than it is at present. To ensure the competitiveness and profitability of a company, taking sustainable steps is crucial. According to some of the experts, it is important that companies influence sustainable buying behavior. "Companies offer clothing, so they have the most impact. The government should support this, for example in the form of grants and raising awareness."

Finally, all interviewees hardly agreed on the risks of sustainability. However, as claimed by four experts, being completely transparent and making all processes measurable is difficult. It may counteract the economic interests of a company. Developments are slow because there is little demand for sustainability. Furthermore, setting sustainable goals within a company is difficult since no one pursues the same sustainable vision.

12.5.2.4 In What Way Can Fashion Companies Integrate Sustainability into Their Business Model?

Even though all experts confirmed that it is a major challenge to change an existing business model into a more sustainable one, five experts expressed the following first steps that fashion companies should take to integrate sustainability into their business model.

1. Formulate a sustainable vision and a sustainable mission for the company.
2. Choose an appropriate mission-driven strategy.
3. Expand sustainability into activities that affect every part of the business.
4. Create shared value (e.g., save energy and reduce CO_2 emissions).
5. Make supply chains transparent (disclosure).
6. Implement a circular design process that focuses on how material usage can create value instead of saving costs (closing the loop—eliminate waste).

7. Shape partnerships (co-ordination and co-organization: sustainable enterprising is created between companies, rather than within companies).
8. Introduce supplier minimum standards and conduct audits (inclusive sourcing).
9. Be more involved in the process and support suppliers (increasing social impact).

From the point of view of the experts, the first five factors seem to be the most important. However, it was also alleged that the preservation of a traditional business model extended by only small aspects of sustainability is not sufficient to reach long-term sustainability goals.

The experts argued that companies must think in a whole new way and consider a new model as has already occurred in many other industries. Experts claim that starting from a whole new concept is the basis of a sustainable model as opposed to the model of existing companies. Therefore, the experts claimed that organizations should start from the ground up and integrate sustainability into their processes and strategies. However, this is not currently realized practice in today's fashion industry. Only two of the experts clearly state that they "(...) really try to stay out of that industry as much as possible. That means, not making use of existing streams, production and stores, because (they) do not believe one can find a solution within that current system."

12.6 Conclusion

This study has examined the most recent developments within the sustainable fashion industry. It has demonstrated the opportunities, risks, and ways in which companies can launch a sustainable fashion brand. Based on the results, the main research question, "What challenges face organizations wishing to launch a sustainable fashion brand?" can be answered.

Sustainability has a number of complex features that make it a challenge, including the following.

1. It is difficult for companies to be completely transparent and to radically change the traditional processes along the value chain.
2. Sustainability within a company is challenging since not all employees and stakeholders pursue the same sustainability vision and goals.
3. Developments are slow because there is currently insufficient demand for sustainable fashion.

It can be said that the developments in this industry are in a vicious cycle. To break out of the cycle, governments, companies, and consumers have to contribute. The government should provide a legal framework to enhance eco-innovation. Voluntary environmental policy measures may be sufficient for eco-pioneers since those firms are highly committed to sustainable production. However, the majority of firms in the fashion industry still act "passive", so strong regulation measures seem to be necessary (Cleff and Rennings 1999a, 201).

Companies have to create new and sustainable value chains and must make consumers more aware of the importance of sustainable products. "Environmental innovators clearly see increased resource efficiency frequently as a component in a comprehensive company efficiency strategy. This expresses itself in the fact that they give cost reductions and quality management as innovation goals significantly more frequently than other innovators" (Cleff and Rennings 1999b, 335). Last but not least, consumers have to recognize the importance of a sustainable product and adapt their buying behaviors to be in line with their responsibility as a citizen.

Companies can influence customers' behavior with the right marketing campaigns and storytelling that shows that sustainable fashion is equivalent to or better than traditional fashion. Currently, one of the few things consumers know is that sustainably produced fashion is more expensive. A price-sensitive consumer prefers to go for traditional and cheap fashion. This again highlights the importance of transferring sustainable information. As soon as awareness is increased and there is a higher demand for sustainable products from the fashion world, economies of scale will lead to cheaper sustainable fashion products.

In addition, eco-innovation will occur in the future, including the deployment of sustainable (raw) materials, systems requiring less water, production both locally and abroad using circular methods [such as recycling and re-use of materials (cradle-to-cradle)], and employing automation. The effects in the long term remain unknown. The pursuits of sustainable initiatives contribute to improving social and ecological aspects. Most initiatives are recent and difficult to currently measure. Will automation in production countries contribute to improvements for employees there? Will automation cause a higher unemployment rate, resulting in a worsening of gross national income? Will the ecological impact of the fashion industry diminish, or will we in fact disturb the earth even more with new systems, methods, and technologies? Only in the course of time will we recognize the consequences of our activities.

Appendix: Interview Questionnaire

The aim of the expert interviews is to gain valuable insights. The questions contribute to the validity of this research. All the questions asked related to the main question. Within these interviews the following structure has been used.

1. Acquaintance
2. Personal introduction
3. Introduction to thesis subject
4. Interview
5. Discussion
6. Completing

Interview Questionnaire: Fashion Companies

1. How does the company define 'Corporate Social Responsibility'?
2. What different domains/forms of CSR does the company know in general?
3. What form(s) of CSR does the company already apply?
4. What is / are the main reason(s) to apply CSR?
5. Is the company affiliated with organizations that support CSR? If so, which one (s)?
6. What does the connection with these organizations mean to the company?
7. To what extent is CSR policy included in the company's internal processes?
8. Are there measurable goals related to CSR in the company's KPIs? If so, which one?
9. Does the company communicate to customers about CSR? If so, in what way (s)?
10. Is the company familiar with (general) consumer opinion about the use of CSR by retail companies? If so, what is that general opinion according to your company?
11. Does the company experience difficulties communicating with its customers about CSR? If so, what are these difficulties?
12. Are there any plans made by the company to communicate differently/more about CSR in the future? If so, what do they want to change?

Interview Questionnaire: Sustainable Fashion Consultants

Personal

1. Could you please tell me about your background in sustainable fashion and why sustainability has become important to you?
2. What has been your greatest lesson as an expert in sustainability?
3. What do you see as the biggest changes in sustainable fashion from five years ago to today?

Sustainability and CSR

4. Sustainability is trending. It is a hot topic that consists of many definitions. In what way should you describe sustainability best with regards to businesses?
5. Why should fashion companies care about sustainability and what would they get out of it if they consider sustainability to be important?
6. What are the risks for companies to do business on a sustainable basis? (How do you experience this within your own company?)
7. What are the opportunities for companies to do business on a sustainable basis? (How do you experience this within your own company?)

8. What are the challenges to incorporating sustainability within the company? (How do you experience this within your own company?)

Transparency

9. As you may know, Fashion for Good has developed a sewing robot that makes it possible to produce clothes closer to the consumer. This technology makes the production of clothing cheaper, resulting in a reduction in transportation, supplies and CO_2 emissions. In what way do you think this will affect workers in production countries such as Bangladesh? Will there be a future for them after this development/automation?
10. In what way do companies/you control the working conditions in production countries today?
11. How do companies/you build sustainable relationships with producers and suppliers?
12. How do companies/you find the right factories that produce durably?
13. Integrating sustainability means adjusting the business model. What could companies do in terms of sourcing, environmental friendly raw materials, and working conditions? (How and in what way do/did you accomplish this within your own business?)
14. How important is sustainability for the marketing department? Has it become a way to (successfully) distinguish your company from the competition?
15. Why do companies 'act' sustainable only to get a better reputation?
16. Do you believe that these companies will eventually fall because they cannot meet the actual standards? Why?
17. Why do you think there is, in terms of sustainability, such a big 'gap' in what companies claim they do and what they actually do?
18. If you had to advise organizations with regard to launching a sustainable fashion brand, what would you recommend?

Circular Economy

19. In recent years, the concept of a 'circular economy' has risen exponentially. Do you think this is a term trend, or do you really think it is the future? Why?
20. At the end of last year, the Dutch government decided that the Netherlands should be a circular economy by 2050. This means producing with less raw materials and energy, reuse and recycling, and not affecting biodiversity. Do you think that this will continue to be intentional, or will it really work and what is needed?
21. A circular economy means that everyone must cooperate. How do you think we can ever achieve this?
22. Today, we live quite individually. Do you think organizations can also be sceptical? In other words, with regards to transparency, how will companies deal with sharing their own information? (How do you accomplish this?)

23. Who will and who will not survive?

Consumer (Behaviour)

24. Many consumers are willing to invest in fair and environmentally friendly clothes, but in reality, they do this only moderately. What are the obstacles?
25. Is there a way to offer the same prices as non-durable products? Why does this (or not) prove to be difficult?
26. In what way can the marketing department contribute to fulfilling the global intentions? In other words, what can marketers do to influence the behaviour? (Not only the buying behaviour, but the overall behaviour that sustainability is needed).
27. Do you think companies have the responsibility to encourage young people to buy durable clothing items? Why?
28. Do you believe in a future where clothes are leased? Why (not)?

References

Bowen, H. R. (1953). *Social responsibilities of the businessman*. New York: Harper & Brothers.

Carroll, A. B. (1979). A three-dimensional conceptual model of corporate performance. *The Academy of Management Review, 4*(4), 497–505.

Carroll, A. B., & Buchholtz, A. (2014). *Business and society: Ethics, sustainability and stakeholder management* (9th ed.). Stamford: Cengage Learning.

Carroll, A. B., & Shabana, K. M. (2010). The business case for corporate social responsibility: A review of concepts, research and practice. *International Journal of Management Reviews, 12*(1), 85–105.

Choi, J., Cobbing, M., Crawford, T., Erwood, S., Harjono, M., & Hojsík, M., et al. (2011). *De vuile was van de kledingindustrie: De relatie tussen de kledingindustrie en giflozingen in China*. http://www.greenpeace.nl/Global/nederland/report/2011/DirtyLaundry_LR.pdf. Accessed 3 Nov 2017.

Cleff, T., & Rennings, K. (1999a). Determinants of environmental product and process innovation. *European Environment, 9*(5), 191–201.

Cleff, T., & Rennings, K. (1999b). Determinants of environmental innovation – Empirical evidence from the Mannheim innovation panel and a follow-up telephone survey. In P. Klemmer (Ed.), *Innovation and the environment: Case studies on the adaptive behavior in society and the economy* (Vol. 6, pp. 315–335). Berlin: Analytica.

Cleff, T., & Rennings, K. (2000). Determinants of environmental product and process innovation – Evidence from the Mannheim innovation panel and a follow-up telephone survey. In J. Hemmelskamp, K. Rennings, & F. Leone (Eds.), *Innovation-oriented environment regulation: Theoretical approaches and empirical analysis, ZEW economic studies* (Vol. 10, pp. 331–347). Heidelberg: Physica-Verlag.

Clinton, L., & Whisnant, R. (2014). *Model behavior: 20 business model innovations for sustainability*. http://sustainability.com/our-work/reports/model-behavior/. Accessed 3 Nov 2017.

Dahlsrud, A. (2006). How corporate social responsibility is defined: An analysis of 37 definitions. *Corporate Social Responsibility and Environmental Management, 15*(1), 1–13.

Davis, K. (1967). Understanding the social responsibility puzzle: What does the businessman owe society? *Business Horizons, 10*(4), 45–50.

Davis, K., & Blomstrom, R. L. (1966). *Business and its environment*. New York: McGraw-Hill.

Delmas, M., & Burbano, V. (2011). The drivers of greenwashing. *California Management Review, 54*(1), 64–87.

Drucker, P. F. (1956). *The practice of management*. New York: Harper Business.

Drucker, P. F. (1984). The new meaning of corporate social responsibility. *California Management Review, 26*(2), 53–63.

European Parliament. (2017). *Bericht über die EU-Leitinitiative für die Bekleidungsbranche*. http://www.europarl.europa.eu/sides/getDoc.do?pubRef=-//EP//TEXT+REPORT+A8-2017-0080+0+DOC+XML+V0//DE#title2. Accessed 27 Mar 2017.

Fashion United. (2016). *Global fashion industry statistics – International apparel*. https://fashionunited.com/global-fashion-industry-statistics. Accessed 3 Nov 2017.

Freeman, E. R. (1984). *Strategic management: A stakeholder approach*. Cambridge: Cambridge University Press.

Groenendijk, R. (2016). *Transparantie in supply chain: Een niet ingeloste belofte?* http://www.logistiek.nl/supply-chain/blog/2016/6/transparantie-in-supply-chain-een-niet-ingeloste-belofte-101144749. Accessed 3 Nov 2017.

Hill, N., Walker, H., Beevor, J., & James, K. (2012). *Conversion factors for company reporting: methodology paper for emission factors*. 6 July 2012. https://www.gov.uk/government/uploads/system/uploads/attachment_data/file/69568/pb13792-emission-factor-methodology-paper-120706.pdf. Accessed 3 Nov 2017.

Houtum, P. V. (2015). *MVO en het MKB: Een praktijkvoorbeeld*. http://www.mvoloont.com/sites/default/files/uploads/MVO%20een%20praktijkvoorbeeld.pdf. Accessed 3 Nov 2017.

ILO. (2009). *Guidelines on occupational safety and health management systems*. http://www.ilo.org/wcmsp5/groups/public/---ed_protect/---protrav/---safework/documents/normativeinstrument/wcms_107727.pdf. Accessed 3 Nov 2017.

ILO. (2017). *Tripartite declaration of principles concerning multinational enterprises and social policy*. http://www.ilo.org/wcmsp5/groups/public/---ed_emp/---emp_ent/---multi/documents/publication/wcms_094386.pdf. Accessed 3 Nov 2017.

International Union for Conservation of Nature and Natural Resources. (1980). *World conservation strategy – Living resource conservation for sustainable development*. https://portals.iucn.org/library/efiles/documents/WCS-004.pdf. Accessed 3 Nov 2017.

ISO. (2017). *ISO 26000 – Social responsibility*. https://www.iso.org/iso-26000-social-responsibility.html. Accessed 3 Nov 2017.

Jonker, J. (2014). *Nieuwe business modellen*. Nijmegen: Academic Service.

Kennedy, S. (2015). *A cut above: Which big brands are using more organic cotton?* https://goodonyou.eco/big-brands-are-using-more-organic-cotton/. Accessed 28 Feb 2018.

Koppert, P., & Brouw, H. O. (2017). *Roadmap circulair textiel*. May 2017. https://www.afvalcirculair.nl/publish/pages/105419/roadmap_circulair_textiel_juni_2017.pdf. Accessed 3 Nov 2017.

Launois, A. (2008). *Consumers go for green products, mintel*. 9 April 2008. https://www.bakeryandsnacks.com/Article/2008/04/10/Consumers-go-for-green-products-Mintel. Accessed 3 Nov 2017.

Marsden, C. (2001). *The role of public authorities in corporate social responsibility*. http://www.alter.be/socialresponsibility/people/marchri/en/displayPerson. Accessed 3 Nov 2017.

Martens, J. (2006). *Die öffentliche Armut der Entwicklungsländer, DGB-Bildungswerk, Global Policy Forum Europe, Terre des Hommes: Düsseldorf, Bonn and Osnabrück*. http://www.younicef.de/fileadmin/Medien/PDF/oeffentliche_armut_01.pdf. Accessed 3 Nov 2017.

Matten, D., & Moon, J. (2008). "Implicit" and "explicit" CSR: A conceptual framework for a comparative understanding of corporate social responsibility. *The Academy of Management Review, 33*(2), 404–424.

McGuire, J. W. (1963). *Business and society*. New York: McGraw-Hill.

OECD. (2011). *Guidelines for multinational enterprise*. http://www.oecd.org/daf/inv/mne/48004323.pdf. Accessed 3 Nov 2017.

Petersen, T., Quandt, J. H., & Schmidt, M. (2017). *Führung in Verantwortung*. Wiesbaden: Springer Gabler.

Remy, N., Speelman, E., & Swart, S. (2016). *Style that's sustainable: a new fast-fashion formula.* https://www.mckinsey.com/business-functions/sustainability-and-resource-productivity/our-insights/style-thats-sustainable-a-new-fast-fashion-formula. Accessed 3 Nov 2017.

Reuters. (2017). *Textielindustrie ontdekt Myanmar: Lage lonen, lange dagen en kinderarbeid.* 6 Feb 2017. https://nos.nl/artikel/2156864-textielindustrie-ontdekt-myanmar-lage-lonen-lange-dagen-en-kinderarbeid.html. Accessed 3 Nov 2017.

Rikkert, S. (2013). *Maatschappelijk Verantwoord Ondernemen 2.0.* April 2013. http://www.ncdo.nl/sites/default/files/Globaliseringsreeks%20deel%206%20onderzoek.pdf. Accessed 3 Nov 2017.

Rossum, M. V. (2012). *Goedkope mode bestaat niet.* 14 Dec 2012. https://www.nrc.nl/nieuws/2012/12/14/naar-een-groenere-mode-doorbreek-de-cirkel-van-kopen-kopen-kopen-a1482901. Accessed 3 Nov 2017.

SAI. (2017). *SA8000 Standard.* http://www.sa-intl.org/index.cfm?fuseaction=Page.ViewPage&pageId=1711. Accessed 3 Nov 2017.

Starmanns, M. (2013). Wie viel Unternehmensverantwortung fordern Sozialstandardinitiativen? BSCI, ETI, FLA, FWF und SA 8000 im Vergleich. In G. Burckhardt (Ed.), *Corporate social responsibility – Mythen und maßnahmen: Unternehmen verantwortungsvoll führen, regulierungslücken schließen* (2nd ed., pp. 97–101). Wiesbaden: Springer Gabler.

Turner, M., & Houston, P. (2008). *Sourcing als uitgangspunt van een succesvolle groene strategie, management executive.* http://www.managementexecutive.nl/artikel/3475/Sourcing-als-uitgangspunt-van-een-succesvolle-groene-strategie. Accessed 3 Nov 2017.

United Nations Global Compact. (2017). *UN global compact progress report.* https://www.unglobalcompact.org/docs/publications/UN%20Impact%20Brochure_Concept-FINAL.pdf. Accessed 3 Nov 2017.

United Nations Industrial Development Organization. (2017). *What is CSR?* http://www.unido.org/csr/o72054.html. Accessed 3 Nov 2017.

Walton, C. C. (1967). *Corporate social responsibilities.* Belmont: Wadsworth.

Watson, D., Elander, M., Gylling, A., Andersson, T., & Heikkilä, P. (2018). *Stimulating textile-to-textile recycling.* Copenhagen: Nordic Council of Ministers.

Wieland, J., & Grüninger, S. (2000). EthikManagementSysteme und ihre Auditierung. In T. Bausch, A. Kleinfeld, & H. Steinmann (Eds.), *Unternehmensethik in der Wirtschaftspraxis.* Munich: Rainer Hampp Verlag.

Wijnia, G. (2016). *Mapping obsolete inventory in the Dutch apparel industry.* https://mvonederland.nl/sites/default/files/media/Thesis%20-%20Mapping%20obsolete%20inventory%20in%20the%20Dutch%20clothing%20industry.pdf. Accessed 3 Nov 2017.

Witjes, S. (2016). *Samenwerking voor een circulaire economie: De verbinding tussen het duurzaam inkoopproces en duurzamere business modellen.* Utrecht: Copernicus Instituut voor Duurzame Ontwikkeling.

World Trade Organization. (2016). *World statistical review 2016.* https://www.wto.org/english/res_e/statis_e/wts2016_e/wts2016_e.pdf. Accessed 3 Nov 2017.

Chapter 13
Towards a Dynamic Understanding of Innovation Systems: An Integrated TIS-MLP Approach for Wind Turbines

Rainer Walz

13.1 Introduction

Meeting global challenges requires both a higher level of innovations and changes in the direction of innovations. Rennings (2000) grasped the differences between normal innovations and environmental innovations early on, and emphasized that these differences also lead to a double externality problem, because very often the ecological benefits of environmental innovations are not included in the market prices. Thus, environmental policy becomes a key driver in creating demand for environmentally-friendly solutions, and at the same time acts as demand-led innovation policy. However, the challenge remains of how to foster the development of an innovation system that supports these societal goals. A thorough analysis of the dynamics of the system would benefit such an endeavor. Therefore, there have recently been numerous applications of the heuristic of technological innovation systems (TIS), many of them in the field of sustainability technologies. Furthermore, various studies have been performed which look into niche developments and regime shifts from a multi-level perspective (MLP). Each of these approaches has merits and limitations in contributing to a dynamic analysis of transitions to sustainability. Both TIS and MLP have also come under criticism. The former is seen as too narrow because it does not account for wider aspects; the latter as not specific enough; and both are criticized for not being actor-specific.

This paper looks into the feasibility of a combined TIS-MLP approach which also relates to political economy issues, and whether this could form the basis for analyzing the dynamics driving the development of the innovation system. In order to ensure its compatibility with real-world problems, the concept is applied

R. Walz (✉)
Competence Center Sustainability & Infrastructure Systems, Fraunhofer Institute for Systems and Innovation Research ISI, Karlsruhe, Germany
e-mail: rainer.walz@isi.fraunhofer.de

J. Horbach, C. Reif (eds.), *New Developments in Eco-Innovation Research*, Sustainability and Innovation, https://doi.org/10.1007/978-3-319-93019-0_13

to wind energy development in China. Thus, the paper augments more general conceptual advances with an evidence-based case study and extends conceptual analysis by including empirical results.

The paper is structured as follows: Sect. 13.2 starts with background information on the state-of-the-art of TIS and MLP analysis, respectively. It continues with outlining the concept for how to base the analysis of innovation system development on the dynamics which can be derived from the TIS and MLP approaches. Section 13.3 describes the case study and summarizes the state and main steps in the historical development of China's wind energy innovation system. Section 13.4 applies the concept of an integrated TIS-MLP approach to the Chinese wind turbine industry. The development of the innovation system and how it is influenced by both internal dynamics and landscape and regime factors are analyzed in a dynamic setting for three phases. For each phase, a diagram shows the dynamics and feedback loops. Finally, Sect. 13.5 summarizes the lessons learned and presents the overall conclusions.

13.2 Conceptual Background

13.2.1 Technological Innovation Systems

The heuristic of systems of innovation has been developed for national, sectoral and technological systems (see, e.g. Lundvall et al. 2002; Edquist 2005; Malerba 2005; Carlsson et al. 2002). The innovation system concept also has great potential for analyzing sustainability-oriented innovation systems. Innovations in such systems are typically more influenced by public needs and public discourse than "traditional" sectoral or technological innovation systems. Regulation must address environmental externalities, and the long-time horizons of sunk costs in infrastructure supported by traditional economic sector regulation poses a triple regulatory challenge (Walz 2007).

It has been suggested that a technological innovation system can be analyzed by looking at how the different functions it is supposed to carry out are fulfilled (Hekkert et al. 2007; Bergek et al. 2008a, b, c; Hekkert and Negro 2009; Suurs and Hekkert 2009). Abstracting from differences in wording, the following categories of an innovation system's functions can be distinguished:

- *Knowledge generation* (F1)
- *Knowledge diffusion (F2)* by exchanging information in networks, but also along the value chain (including supplier-user interactions)
- *Guidance of search (F3)*, directing R&D and the search for new solutions with respect to technology and market
- *Entrepreneurial experimentation (F4)* leading to diversity and a variety of solutions in order to allow a sufficiently large stock of technologies enabling the selection process to result in a dominant design

- *Facilitation of market formation (F5)*, which enables learning in the market and scale effects
- *Legitimization (F6)* of a new technology, which is closely connected with recognizing a growth potential for the technology and the ability to counteract political resistance and to push for political support
- *Resource mobilization (F7)*, which is especially important for new technologies associated with a higher risk of failure.

These functions are not disjunctive. Bergek et al. (2008a) point out that the mechanisms and interactions of the actors in an innovation system, and the feedback loops between the different functions need to be taken into account to properly understand the innovation process. These feedback mechanisms can induce an increase in innovations, but can also block further development (Bergek et al. 2008b; Hekkert and Negro 2009). It is within these internal dynamic relationships that the development of an innovation system takes place. Thus, the feedback mechanisms between functions create the internal dynamics.

So far, the majority of applications of technological systems of innovation to green innovations have been qualitative case studies in the renewable energy field. Some of them have been performed for emerging economies (e.g. Mohamad, 2011; Lema and Lema 2012; Walz and Delgado 2012; Lema et al. 2015). Typically, such case studies are based on desktop research, interviews and questionnaires. They analyze the components of a TIS and their interrelationships, research the level of activity with regard to the different functions, and derive the underlying pattern of the innovation system.

The empirical evidence suggests that policy has had a strong impact on innovations in renewable energy technologies for power generation. Both public R&D spending as well as policies which induce domestic demand increase these innovation activities. Likewise, policy factors such as introducing targets for renewable energy, increasing the diffusion of innovations, and providing stable policy support lead to higher innovation output. Insofar, the results of TIS case studies can be seen in line with the results of econometric studies (see, e.g. Horbach et al. 2012, or Schleich et al. 2017). On the other hand, the TIS case studies also use variables which are difficult to quantify. Furthermore, they link the different variables in feedback loops. This makes it impossible to distinguish between independent and dependent variables, and the causal relationships become much more complex compared to typical econometric studies, which quantify the relationship between independent and dependent variables.

However, the innovation systems approach has also been criticized for being too inward-looking, and not taking a wider systems perspective into account. Thus, aspects such as embedment of the innovation system into societal development, and competition with existing technologies are seen as not being represented enough (Markard et al. 2012; Weber and Rohracher 2012).

13.2.2 Multi-Level Perspective

The innovation process is embedded in institutions, knowledge production, and socioeconomic development. Thus, innovation follows certain paths which can even lead to path dependencies and problems with moving towards new technological solutions. Innovations require organizational adaptations and the co-evolution of institutions supporting the further development of the technologies. Dosi (1982) explains the existence of path dependency of innovation processes, which has been taken up in the climate change literature by Unruh (2000) under the heading of carbon lock-in. At the beginning of a radical innovation, the selection processes towards a dominant design are important, as is the availability of diverse solutions to select from. In later phases, market formation and feedbacks between users and producers become more important, and the co-evolution of technologies and institutions supports further incremental innovations. However, the co-evolution of technology and its surrounding institutions can also lead to path dependency. A new technology not only has to compete with a traditional one, but with an entire system consisting of a traditional technology together with the institutions that have co-evolved around it.

The notion of path dependency and co-evolution also shows up in the multi-level perspective, which is advocated by scholars such as Geels and Schot (2007), Smith et al. (2010), or Geels (2011). This approach distinguishes landscape, regime and niche. The landscape represents the broader socioeconomic system; the regime consists of the established technological paradigm. A radical alternative has to grow in a niche together with its own social network before it is able to compete with the established paradigm.

The notion of co-evolution in the tradition of evolutionary scholars such as Dosi or Nelson is visible at various points of the multi-level perspective. This can be horizontal co-evolution within the regime between the established paradigm and institutions. Furthermore, selection processes lead to an adaptation of strategies or routines of companies oriented towards the paradigm. Co-evolution can also take place vertically, e.g. between the paradigm and the regime. Another form of vertical interaction is the competition between a new and the established paradigm, with the latter using its surrounding institutions to combat the new one. However, it might also be that the landscape benefits the growth of the niche.

According to MLP, niches gain momentum if a dominant design, powerful actors and networks emerge. The niche grows, and starts to become an important economic component. It can be closely associated with empowerment, which Smith and Raven (2012) advocate as a specific function of niches as protective spaces.

The MLP approach has been criticized in the past for being too functional, and not putting enough emphasis on power and actor aspects (Smith et al. 2005; Geels and Schot 2007; Smith and Raven 2012). Furthermore, it has been suggested that transition research needs to take space into consideration (Markard et al. 2012), and there are calls to integrate MLP with the economics of geography.

Energy innovations share the double externality problem described by Rennings (2000). In addition to regulating the protection of knowledge and R&D, energy innovations also face the externality of environmental costs. There is not much demand for green energy innovations, unless some form of environmental regulation levels the playing field between new and older, environmentally more harmful innovations. Thus, demand is highly policy driven, and policies such as standards, emissions trading systems, feed-in-tariffs or quota systems are simultaneously both environmental and demand-led innovation policies. Furthermore, changes on the landscape level such as increasing environmental awareness, changing perceptions of human-environment relationships, or the development of a political system giving green issues higher priority all affect green energy innovation technologies. Thus, the different regimes and niches of green energy innovations are all affected by the same specific changes on the landscape level.

Renewable power generation technologies belong to infrastructure-related regimes. The specificities within this regime lead to similar selection environments. Electricity technologies and related technologies share the following specificities (see Markard 2010; Lema et al. 2015):

- Asset durability: a lot of these technologies are characterized by very long lifetimes (e.g. power stations, investments in related infrastructure such as electricity or water supply networks, roads and rail). Thus, the high asset durability limits the opportunity for reinvestments. Furthermore, the investments in infrastructure-related technologies tend to be very capital-intensive (Markard 2010). Thus, it would be very costly to substitute them before they have reached their end-of-life. Both factors foster "technical path dependency" and technological lock-in.
- Technical systemness of physical networks: if the technologies are physically connected with each other via a grid, technical systemness (Markard 2010) increases the path dependency. Problems with integrating renewable electricity supply, for example, can arise from a grid structure which is optimized for the existing carbon-intensive power system. If the grid structure is not suitable, even large investments in low-carbon electricity supply will not necessarily increase the market share of low-carbon alternatives unless they are supported by vast investments in a new grid structure. Thus, the specific features of technical systemness lead to a comparatively high path dependency.
- Cultural significance: access to energy, water and transportation are all related to basic needs. This becomes apparent, for example, in their prominence among the future global challenges.
- Monopolistic bottleneck: despite the calls for deregulation and liberalization, it is still acknowledged that monopolistic bottlenecks characterized by both sunk costs and natural monopoly cost functions should be regulated. Clearly, infrastructure systems based on physical networks such as electricity/gas, water supply and sewage treatment, or railways include such monopolistic bottlenecks. Even potentially competitive stages generally require access to these monopolistic bottlenecks. This also holds for power produced by independent power producers, e.g. the operators of renewable energy, or railway operators. However, how the economic sector is regulated also influences the speed and direction of

related technology innovations. From the point of view of innovation, these infrastructure sectors pose a third regulatory challenge (Walz 2007).
- Actor structure and political economy: infrastructure innovation systems are characterized by a specific structure of actors. The incumbents driving the existing regime, such as public utilities or multinational energy companies, are typically very powerful and sometimes influence government. Many of the actors driving the niches, however, are small and medium enterprises, and often newcomers. However, in addition to this actor constellation—which can also be found in other innovation systems—there are also community-based groups and NGO-type actors who are among the key proponents for eco-innovation niches. This reflects the characteristic of infrastructure systems as a social need, which cannot be left to individual market-based decisions alone. To sum up the argument, important actors in infrastructure innovation systems differ from the typical actors in other innovation systems. Thus, it can be expected that their behavior also differs. Furthermore, the regime-niche constellation can be characterized as an arena with a very uneven power structure: on the one hand, the large companies which profit from existing lock-ins and are sometimes directly linked to government, versus the drivers of eco-innovations which are very often not part of the established innovation system and possess neither capital reserves nor experience in upscaling innovations, on the other hand.

Geels and Schot (2007) proposed that, depending on the state of development and the timing of transformations taking place, the interplay between niche and regime can give rise to different transition pathways. The specificities of infrastructure technologies imply that the fossil fuel-based regime is rather strong. This has led Walz and Köhler (2014) to expect that a transition pathway, which Geels and Schot (2007) have called "technological substitution", will emerge more often in the case of renewable energies: radical innovations which have developed in niches remain stuck because the regime is stable and entrenched. Only after strong disruptive changes in the landscape will the regime be challenged. Strong growth of the niche due to policy measures might prove to be too expensive, which again reduces the legitimacy of further growth of the niche. In such instances, the narrative of transition typically points towards future cost reductions of the niche technologies due to learning (Smith and Raven 2012). The link to niche growth in other countries can strengthen such a narrative: export successes in radical new technologies are an important argument to counterbalance the critique of rising economic costs. If the niche technologies promise to reduce or even to phase out monopolistic bottlenecks, this can also add to bolstering the transition narrative.

13.2.3 Conceptual Basis for the Case Study

The previous sections have shown that both the TIS and MLP approaches offer good starting points to analyze the dynamics of an energy transition. The following aspects form the conceptual core upon which mental models are based for analyzing the dynamics of innovation systems:

- First, authors such as e.g. Bergek et al. (2008a) or Hekkert and Negro (2009) see the development of innovation systems as influenced by virtuous or vicious circles among the different innovation functions. Thus, the feedbacks between these functions make it possible to account for the internal dynamics of innovation systems (Smith and Raven 2012).
- Secondly, the MLP approach sees the development of a sustainable niche as influenced by the interaction of a niche with landscape and regime, which puts the development of a specific technology into a wider perspective of transition pathways and regime shifts (Geels and Schot 2007; Geels 2011). Thus, drawing on the MLP approach makes it possible to take the external dynamics into account.
- Third, both approaches have been criticized for neglecting advocacy, political economy, and also spatial dimensions, that are important for the interaction of innovation systems on a global scale (Markard et al. 2012; Smith and Raven 2012). However, the elements of a socio-political environment (Geels 2014) are also used to make the approach more actor-specific. Thus, aspects of political economy could be used to translate the dynamics between niche and regime into the logic of actor behavior within a TIS.
- Fourth, a combination of TIS and MLP can be achieved by interpreting a specific renewable energy TIS as one niche among others within a broader, fossil fuel-dominated energy system. This niche draws on the common systemic relationship between renewable energy niches and the fossil fuel-based regime described above.

Walz and Köhler (2014) see a sustainability transition as characterized by the development of various niches which share a common systemic relationship with a regime (see Sect. 13.2.2). For electricity supply technologies, this regime is based on fossil fuels and large central nuclear power stations, around which institutions have co-evolved. Green energy technologies such as renewable energies form niches, which address the core of socio-technical regimes. The energy technologies share common features which justify their distinction from other technologies.

In the case of the electricity system transition, various renewable energy technologies each form a niche (Fig. 13.1). They face a common regime characterized by centralized, fossil fuel-based power stations, around which a complex web of institutions, complementary technologies, and markets has co-evolved, which together perpetuate carbon lock-in. It is assumed that regime-niche interaction follows a disruptive transition path. There is an internal dynamic within each niche described, e.g. in various TIS case studies. There are also landscape influences on both the regime and the niches. Furthermore, one niche might be directly or indirectly influenced by developments in the niche in another country, or by the internal development of other renewable energy niches. However, the internal dynamics of the latter are neglected in order to reduce complexity of the analysis.

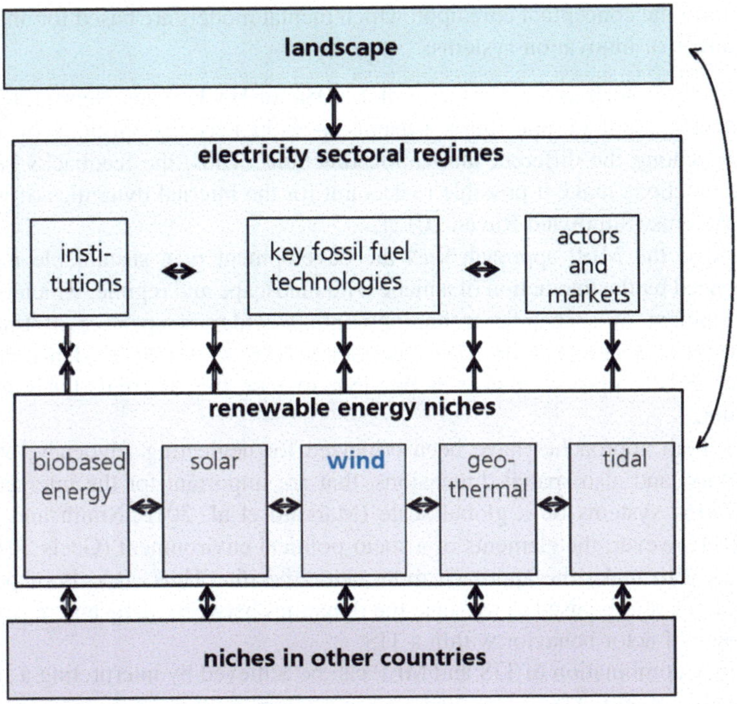

Fig. 13.1 Level of aggregation of technological innovation system within the multi-level perspective. Source: adapted from Walz and Köhler (2014)

The concept of using an integrated TIS-MLP approach is elaborated for the case of wind turbine innovations in China. Applying the concept to a specific case helps to test the feasibility of combining the elements described above. Wind energy has been chosen because it is one of the most thoroughly analyzed sustainable TIS, and is a good example of successful TIS development.

13.3 Status and History of Wind Energy in China

The development of wind energy in China is widely seen as a success story for the diffusion of renewable energy in an emerging economy. The development of the Chinese wind energy industry and the analysis of the innovation system have been studied intensively in recent years (e.g. Walz and Delgado 2012; Klagge et al. 2012; Zhao et al. 2012; Ru et al. 2012; Wang et al. 2012; Zhang et al. 2013; Zheng et al. 2013; Gosens and Lu 2013, 2014; Dai et al. 2014; Schmitz and Lema 2015; Koch-Weser and Meick 2015; Gandenberger 2017; Gandenberger and Strauch 2017; Binz et al. 2017). Figure 13.2 shows the impressive development of wind energy in China. China's share of globally accumulated installed capacity rose from a mere 2% in

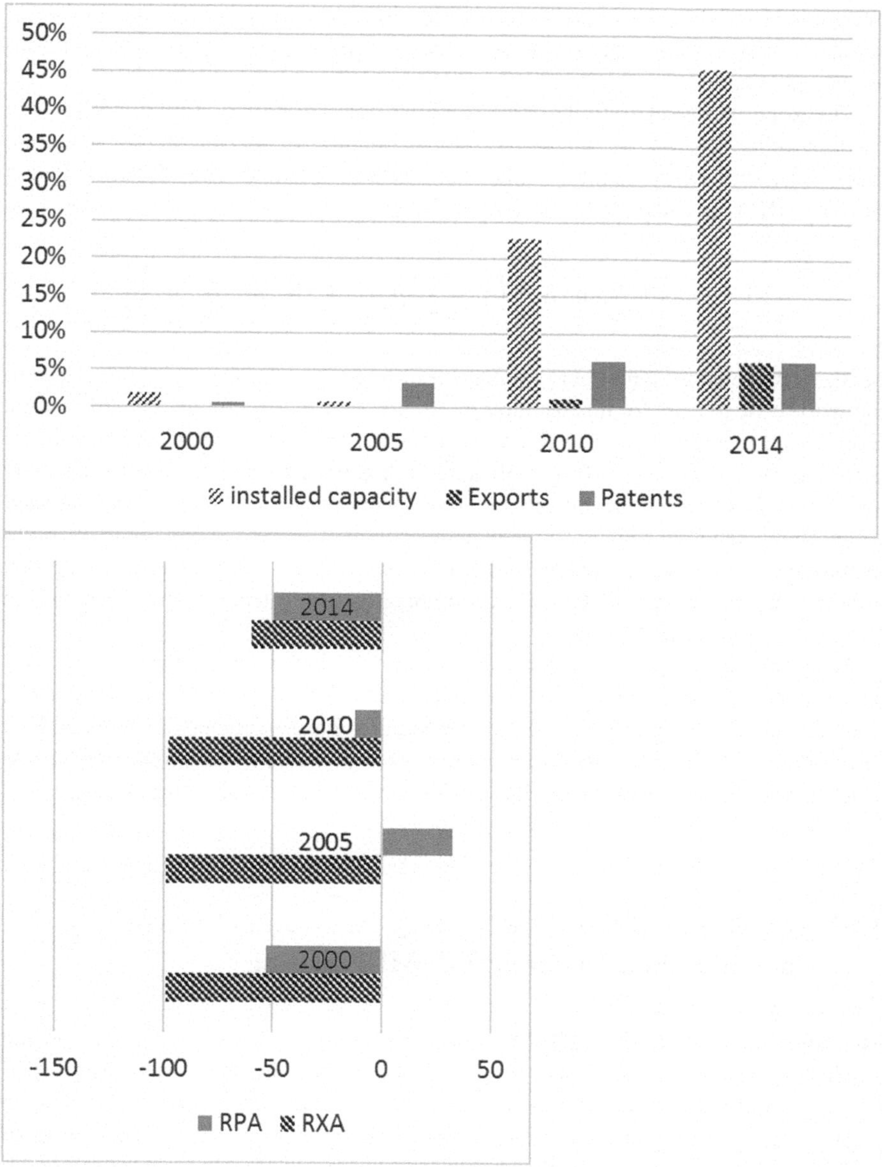

Fig. 13.2 World shares and specialization of China's wind power industry. Source: calculations of Fraunhofer ISI

2000 to over 45% in 2014, and this growth is still continuing. Clearly, China has become the biggest market for wind turbines. Patents have started to rise too, and China has reached a global share of 6%. However, Chinese wind turbine exports are still significantly lower than those in other industrial fields, indicating that its wind turbine industry has not yet reached full international competitiveness. This is underlined by the gap between China and the world leaders with regard to the size of installed wind turbines, and the widespread perception that China's wind turbines are not top quality.

The Relative Patent Activity (RPA) and the Relative Export Activity (RXA) are used in order to look at whether or not a country is specializing in a certain technology. For every country i and every technology field j, the Relative Patent Activity (RPA) is calculated according to:

$$\mathrm{RPAij} = 100^{*}\tanh\ \ln\left[\left(p_{ij}/\sum_{i} p_{ij}\right)/\left(\sum_{j} p_{ij}/\sum_{ij} p_{ij}\right)\right]$$

i.e. the RPA relates the number of patents p for a given technology j in a country i to the worldwide patents for this technology. This ratio is then compared with the same ratio for all technologies. The RXA is calculated in a similar way and substitutes patents (p) by exports (x). All specialization indicators are normalized between +100 and −100. Positive values indicate an above average specialization in the analyzed technology; negative values show that the country is specializing more in other technologies. The data on patent and export specialization further corroborate that wind energy is not one of China's economically strong technologies, despite the success story described above.

Nevertheless, in sum, the development of Chinese wind energy is seen as a success story. Table 13.1 gives an overview of how markets, turbine sizes, manufacturing, policy targets and policy measures have developed over time. This development has been ascribed to numerous policy interventions. Figure 13.3 illustrates how capacity development has evolved in parallel to policy measures.

13.4 Evolution of the Wind Energy Innovation System in China in a Dynamic TIS-MLP Setting

The application of the TIS-MLP approach explained in Sect. 13.2 draws on the results of the papers on China mentioned above. Among the various studies of Chinese wind industry development, especially Walz and Delgado (2012), Klagge et al. (2012), Gosens and Lu (2013, 2014) and Binz et al. (2017) were used to allocate the various effects to the different functions of the Chinese wind energy innovation system. The development of the wind energy innovation system in China is characterized by three different phases, which reflect the different context and early follower strategy of Chinese producers. The following sections describe each phase.

Table 13.1 History of the Chinese wind energy industry and policy

	Until early 2000s	Early 2000s until late 2000s	Since late 2000s
Market size	• 1985: 55 kW • 2000: 352 MW	Doubling each year, 26 GW by 2009	2015: 130 GW 700 MW exports in 2007
Turbine size	• early 1990s: focus on 20 kW+ • early 2000s: focus on 600 kW +	Development towards 3 MW plants	2016: development of 10 MW plants; goal of 30 MW turbines by 2015
Companies	Small number of key manufacturers, e.g. Goldwind	Development of key players, some of them diversifying from heavy machinery	Goldwind extends lead, consolidation of markets, exit of some smaller companies
Policy target	1990s: 300–400 by 2000 In 2001, 500 MW target by 2006	2007: 5 GW target by 2010 2007: long-term development plan calls for 100 GW installed capacity by 2020	2020 target for 3–5 equipment manufacturers to be at advanced technological level internationally Stronger focus on grid integration; grid parity by 2020
Policies	"Ride the Wind" Program 40% local content requirements Government subsidies Government procurement	2003–2007: concessions tenders 2006: renewable energy law leading to feed-in-tariffs in 2009 2004–2007: local content requirement 70% 2008/2009/cancellation of local content requirement	Since 2012: export credits from China Development Bank Downward adjustment of feed-in-tariffs Encouragement of more distributed utilization to relieve grid stress

Source: (Walz and Delgado 2012; Binz et al. 2017)

13.4.1 Formative Phase of the Chinese Wind Energy Innovation System

The formative phase of the Chinese wind energy industry lasted from the late 1980s to the early 2000s. Important landscape factors were the overriding Chinese policy goals to catch up technologically and to increase energy security. The success stories of wind energy in other countries, which also resulted in manufacturing companies emerging as major suppliers on international markets, provided the guidance of search (F3). China's general policy approach was a landscape factor which was applied to wind energy: government programs aimed at the transfer of knowledge. These programs led to an increase in resource availability (F7) and increase in the absorption of foreign knowledge. The government also initiated a small-scale diffusion program (market formation F5) of wind energy ("Ride the Wind program"). Turbine imports from abroad supported the further diffusion of (foreign) knowledge (F2). The increased level of knowledge diffusion triggered domestic entrepreneurial experimentation (F4).

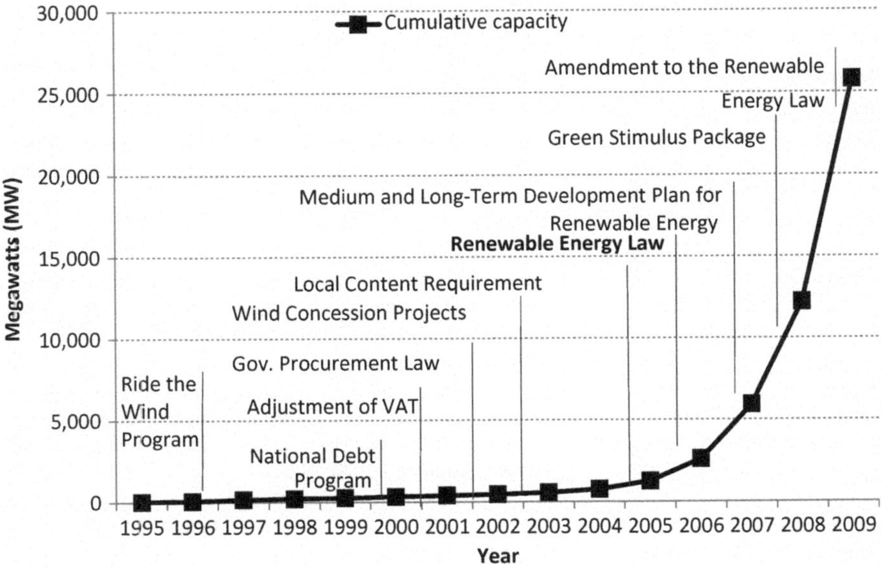

Fig. 13.3 World shares and specialization of China's wind power industry. Source: Walz and Delgado (2012)

Despite all these activities, the link between entrepreneurial experimentation and knowledge generation remained rather weak. No positive feedback loop was established during this phase, which would have accelerated the formation of the innovation system (Fig. 13.4).

13.4.2 Take-Off Phase of the Chinese Wind Energy Innovation System

The take-off phase lasted from the early 2000s to the late 2000s and is characterized by the growth of the niche in a protective space. Diffusion policies were continued and led to a surge in installed capacity (F5). The central government, which is a landscape factor in the Chinese system, used its strong role to bring the utilities that form the basis of the existing regime in line with an expansion of wind energy. This was facilitated by the de facto renewable portfolio standards, which resulted in wind-based projects at large state-owned utilities. Thus, the regime was involved in implementing government policy. However, landscape factors played an additional role. The industrial-based growth paradigm, which is deeply rooted in the Chinese economic model, led to supportive policies such as local content requirements for installed turbines. Thus, diffusion policies were augmented by an industrial policy featuring protective measures. With regard to the importance for the innovation functions, this changed the role of foreign competitors for the innovation system

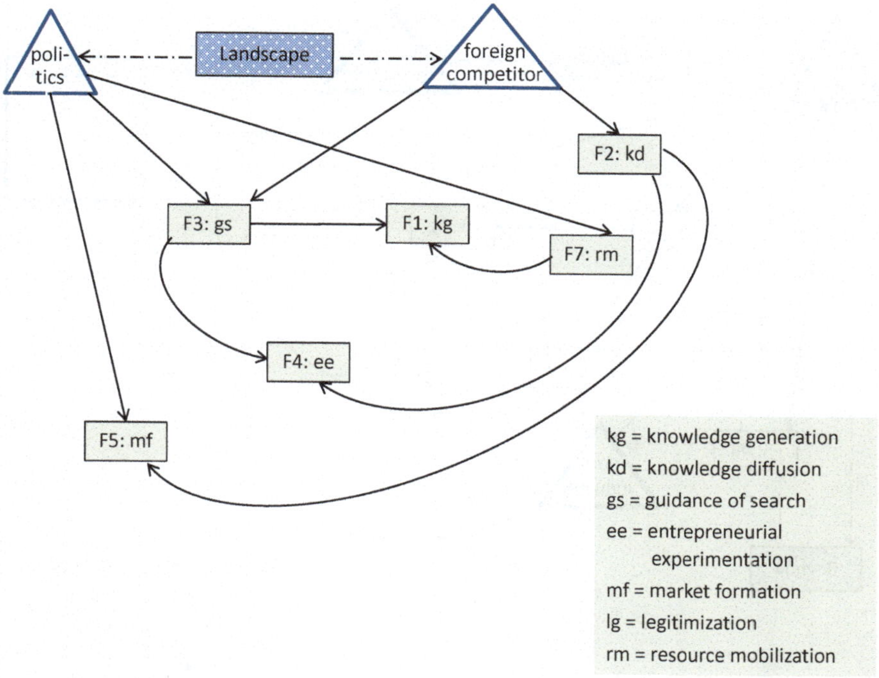

Fig. 13.4 Feedback loops and actors involved in the first development phase of the Chinese wind energy innovation system

functions. From being the major supplier of wind turbines, they moved more towards the joint production of turbines.

The strong impulse for market formation plus the local content requirements also strongly supported the growth of a domestic wind turbine industry. This had several effects on the functions of the innovation system. First, it increased legitimacy (F6), because a domestic industry is the most visible sign of a successful industrial strategy. Second, market formation (F5) was further fostered, as the supply by domestic producers was a much better fit for the political economy of supporting diffusion. Third, these developments led to learning in the market, providing additional guidance of search (F3) and leading to increased entrepreneurial experimentation (F4). Increased profits from deployment and market prospects increased resource availability, which was also used to increase domestically produced new knowledge (knowledge generation (F1), as indicated by rising patents during this time). The increased domestic knowledge started to diffuse, which closed the positive feedback loop and led to accelerated innovation system development. This trend towards the build-up of domestic capabilities can also be seen in the increasing importance of joint ventures instead of licensing as the mode of technology transfer from abroad.

From a systems dynamic perspective, the development of positive feedback loops in the take-off-phase is of the uttermost importance. These feedback loops were

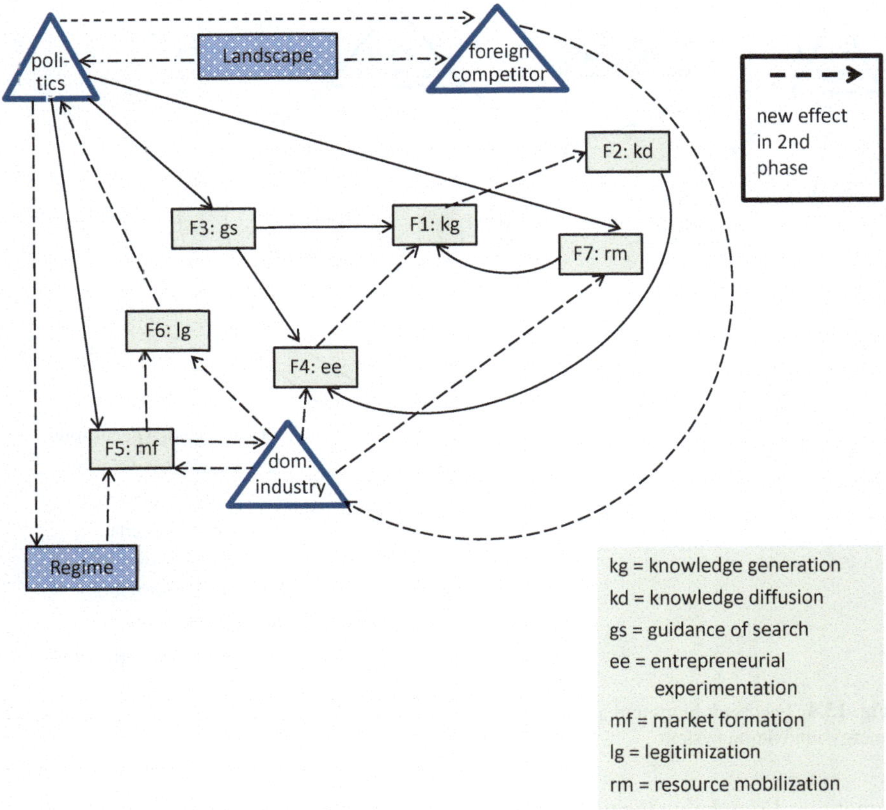

Fig. 13.5 Feedback loops and actors involved in the take-off phase of the Chinese wind energy innovation system

established around the entrepreneurial and knowledge generation functions (upper right part of Fig. 13.5), but also around market formation and legitimacy functions (left part of Fig. 13.5). Finally, these feedback cycles then became interlinked, which further boosted the dynamics. The most important drivers behind this development were the strong role of the Chinese government together with the industrial growth paradigm and industrial policies. Without these landscape factors, the Chinese system would not have been able to develop as quickly as it did (Fig. 13.5).

13.4.3 Mature Phase of the Chinese Wind Energy Innovation System

In the third phase, which started in the late 2000s, acceleration of market formation (F5) and diffusion of knowledge led to a decrease in the costs of newly installed wind turbines, which fell dramatically after 2007. However, the incentive system put the focus on installing capacity (MW), not on feeding electricity to the grid. Perhaps it also suited the interests of the coal-dominated regime not to reduce the importance of the traditional coal-based power plant system (interplay with regime). This development is in line with the surprisingly large volume of installed capacity not linked to the grid, and is also thought to be responsible for the Chinese manufacturers' strategy of not placing higher emphasis on increasing the quality of the turbines (as there was no incentive for doing so). Thus, Chinese turbines became more cost competitive per kW installed compared to foreign ones, and rapidly won market shares in China because quality—demonstrated in higher levels of kWh per installed KW—was not rewarded very strongly on the home market. This also allowed the Chinese government to remove the local content requirement. The success in the domestic build-up of the industry, supported by designating renewable energy one of the emerging strategic technologies in the Five Year Plan, further increased legitimacy (F6) and strengthened the positive feedback cycles even more. Situational context factors such as financial packages in the aftermath of the financial crisis and financing via CDM also benefited green technologies such as wind turbines. The same applies to the growing concern about rising levels of local air pollution (situative context factor in combination with landscape factor). In virtuous cycles, these developments made domestic companies stronger.

The problems with integrating wind turbines into the grid were tackled by correcting the incentives. Feed-in-tariffs provided new guidance of search, and an increase in the quality of Chinese wind turbines can be expected. However, the time lag until the establishment of a first positive feedback cycle and the guidance of search towards low-cost installation instead of low-cost generation have also been responsible for China not yet realizing higher exports of its technology. Furthermore, with increasing renewable electricity supplied to the grid, fluctuating supply and overall grid capacity become serious challenges for grid management. Thus, there is growing pressure on the transmission and distribution operators to increase their ability to integrate wind power. The same problem also leads to signals to reduce the speed of wind power expansion in areas subject to significant grid stress. Thus, there might also be the first signs of negative feedback loops arising from the expansion of the market (Fig. 13.6).

13.5 Lessons Learnt and Next Steps

Explaining the dynamics of innovation system development by integrating the internal dynamics of TIS with the external dynamics of MLP seems to be a very promising approach. The example of wind energy in China shows feedback loops which explain the internal dynamics between the innovation functions. Landscape is important for starting positive cycles; this demonstrates the importance of external dynamics. In China, it will be decisive whether or not wind energy reaches higher

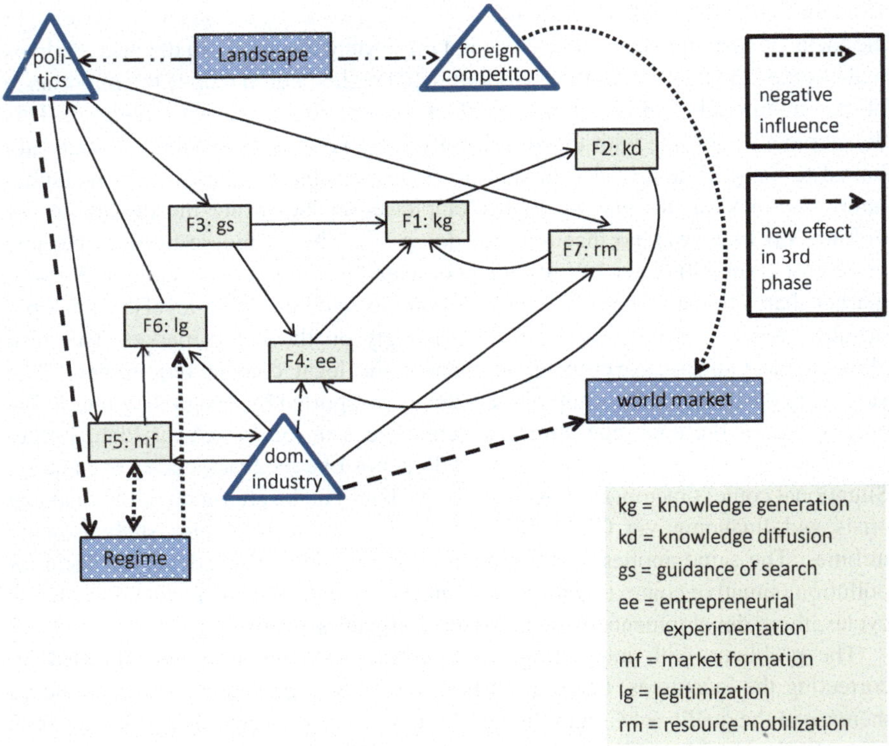

Fig. 13.6 Feedback loops and actors involved in the mature phase of the Chinese wind energy innovation system

levels of quality and of expansion which weaken the regime that is still based on central fossil fuel power stations. Situative context factors and landscape influence play an important role in deciding which way the dynamics will continue. Aspects of advocacy and political economy play an important role, and are connected with the effects of globalizing value chains. In China, this can be seen by the importance accorded to developing domestic production capacities supported by local content requirement and building on the absorption of foreign knowledge. The successful build-up of large wind turbine manufacturing companies has provided jobs, hopes of increasing exports, and enhanced the legitimacy of wind energy in comparison to traditional power stations. The importance of these arguments is rooted in a general paradigm of economic strategy that can be assigned to the landscape level. Thus, political economy considerations link aspects of landscape, and of regime-niche interaction to specific innovation system functions. This can be interpreted as the political economy acting as a link to connect the MLP and TIS approaches.

The analysis of the Chinese dynamics of innovation systems also points to some peculiarities when compared to developments in OECD countries:

- The build-up of capacities to absorb knowledge in China makes the diffusion of knowledge (F2) from foreign countries very important in the beginning.
- So far, no strong negative feedback loop has developed for China; thus, the system is likely to expand further; the effects of rising policy costs due to the rapid expansion of wind energy are not very visible and not as pronounced in China.
- There seems to be no pressure from landscape factors on siting in China; perhaps reflecting different cultural and political landscape factors.

The results also offer interesting feedback concerning the question of whether countries that are catching up differ systematically from traditional OECD-countries with regard to MLP mechanisms. The niche-regime interaction role of the regime in China seems to be less antagonistic than in some OECD-countries. In the terminology of Geels and Schot (2007), this is perhaps an indication that renewable energy in China follows a reconfiguration pathway rather than a technological substitution process pathway (see Sect. 13.2.2). By the same token, the lower policy costs in China might be explained as indicating a second mover advantage. Thus, countries catching up might be less locked into a fossil fuel-based regime. From a general point of view, this raises the perspective that an integrated TIS-MLP approach might also contribute to explaining the leapfrogging potential for catching-up countries.

Finally, the results also give rise to interesting methodological perspectives. The internal and external dynamics lead to various feedback loops which drive the system's development. Analyzing these dynamics becomes highly complex rather quickly. Thus, additional tools to support such an analysis might be helpful. The use of system dynamics modeling (see, e.g. Sterman 2001; Walrave and Raven 2016), which portrays dynamic processes as a combination of stock and flow problems, and which can be used to combine quantitative and more qualitative aspects within a systemic, consistent modeling framework, might be an interesting option to support a dynamic understanding of innovation system development.

Acknowledgements This chapter draws on research performed within the SINCERE project. The financial contribution of the German DFG is acknowledged.

References

Bergek, A., Hekkert, M., & Jacobsson, S. (2008a). Functions in innovation systems: A framework for analysing energy system dynamics and identifying system building activities by entrepreneurs and policy makers. In T. Foxon, J. Köhler, & C. Oughton (Eds.), *Innovations in low-carbon economy* (pp. 79–111). Cheltenham: Edward Elgar.

Bergek, A., Jacobsson, S., & Sandén, B. (2008b). 'Legitimation' and 'development of positive external economies': Two key processes in the formation phase of technological innovation systems. *Technology Analysis & Strategic Management, 20*(5), 575–592.

Bergek, A., Jacobsson, S., Carlsson, B., Lindmark, S., & Rickne, A. (2008c). Analyzing the functional dynamics of technological innovation systems: A scheme of analysis. *Research Policy, 37*, 407–429.

Binz, C., Gosens, J., Hansen, T., & Hansen, U. E. (2017). Toward technology-sensitive catching-up policies: Insights from renewable energy in China. *World Development, 96*, 418–437.

Carlsson, B., et al. (2002). Innovation systems: Analytical and methodological issues. *Research Policy, 31*, 233–245.

Dai, Y., Zhou, Y., Xia, D., Ding, M., & Xue, L. (2014). *Innovation paths in the Chinese wind power industry*. Draft discussion paper of GDI, Bonn.

Dosi, G. (1982). Technological paradigms and technological trajectories. *Research Policy, 11*, 147–162.

Edquist, C. (2005). Systems of innovation: Perspectives and challenges. In J. Fagerberg et al. (Eds.), *The Oxford handbook of innovation* (pp. 181–208). Oxford: Oxford University Press.

Gandenberger, C. (2017). *Giant and Dwarf – China's two faces in wind energy innovation*. Working paper sustainability and innovation no. S 07/2017, Fraunhofer ISI, Karlsruhe.

Gandenberger, C., & Strauch, M. (2017). Wind energy technology as opportunity for catching-up? A comparison of the TIS in Brazil and China. *Innovation and Development, 8*(1), 1–22.

Geels, F. (2011). The multi-level perspective on sustainability transitions: Responses to seven criticisms. *Environmental Innovation and Societal Transitions, 1*(1), 24–40.

Geels, F. (2014). Reconceptualising the co-evolution of firms-in-industries and their environments: Developing an inter-disciplinary triple embeddedness framework. *Research Policy, 43*, 261–277.

Geels, F., & Schot, J. (2007). Typology of sociotechnical transition pathways. *Research Policy, 36*, 399–417.

Gosens, J., & Lu, Y. (2013). From lagging to leading? Technological innovation systems in emerging economies and the case of Chinese wind power. *Energy Policy, 60*, 234–250.

Gosens, J., & Lu, Y. (2014). Prospects for global market expansion of China's wind turbine manufacturing industry. *Energy Policy, 67*, 301–318.

Hekkert, M. P., & Negro, S. O. (2009). Functions of innovation systems as a framework to understand sustainable technological change: Empirical evidence for earlier claims. *Technological Forecasting and Social Change, 76*(4), 584–594.

Hekkert, M., et al. (2007). Functions of an innovation system: A new approach for analysing technological change. *Technological Forecasting and Social Change, 74*, 413–432.

Horbach, J., Rammer, C., & Rennings, K. (2012). Determinants of eco-innovations by type of environmental impact – The role of regulatory push/pull, technology push and market pull. *Ecological Economics, 78*, 112–122.

Klagge, B., Liu, Z., & Campos Silva, P. (2012). Constructing China's wind energy innovation system. *Energy Policy, 50*, 370–382.

Koch-Weser, I., & Meick, E. (2015). *China's wind and solar sectors: Trends in deployment, manufacturing, and energy policy*. Staff Report, U.S.-China Economic and Security Review Commission, March 9th 2015, Washington.

Lema, R., & Lema, A. (2012). Technology transfer? The rise of China and India in green technology sectors. *Innovation and Development, 2*(1), 23–44.

Lema, R., Iizuka, M., & Walz, R. (2015). Low carbon innovation and development. *Innovation and Development, 5*(2), 173–187. https://doi.org/10.1080/2157930X.2015.1065096.

Lundvall, B. A., et al. (2002). National systems of production, innovation, and competence building. *Research Policy, 32*, 213–231.

Malerba, F. (2005). Sectoral systems: How and why innovation differ across sectors. In J. Fagerberg et al. (Eds.), *The Oxford handbook of innovation* (pp. 308–406). Oxford: Oxford University Press.

Markard, J. (2010). *Transformation of infrastructures: Sector characteristics and implications for fundamental change*. Eawag – Swiss Federal Institute of Aquatic Science and Technology, Dübendorf.

Markard, J., Raven, R., & Truffer, B. (2012). Sustainability transitions: An emerging field of research and its prospects. *Research Policy, 41*, 955–967.

Mohamad, Z. (2011). The emergence of fuel cell technology and challenges to latecomer countries: Insights from Singapore and Malaysia. *International Journal of Technology and Globalisation, 5*(3), 306–326.

Rennings, K. (2000). Redefining innovation – Eco-innovation research and the contribution from ecological economics. *Ecological Economics, 32*, 319–332.

Ru, P., Zhi, Q., Zhang, F., Zhong, X., Li, J., & Su, J. (2012). Behind the development of technology: The transition of innovation modes in China's wind turbine manufacturing industry. *Energy Policy, 43*, 58–69.

Schleich, J., Walz, R., & Ragwitz, M. (2017). Effects of policies on patenting in wind-power technologies. *Energy Policy, 108*, 684–695.

Schmitz, H., & Lema, R. (2015). The global green economy: Competition or cooperation between Europe and China? In J. Fagerberg, S. Laestadius, & B. R. Martin (Eds.), *The triple challenge for Europe economic development, climate change, and governance* (1st ed.). Oxford: Oxford University Press.

Smith, A., & Raven, R. (2012). What is protective space? Reconsidering niches in transitions to sustainability. *Research Policy, 41*, 1025–1036.

Smith, A., Stirling, A., & Berkhout, F. (2005). The governance of sustainable socio-technical transitions. *Research Policy, 34*, 1491–1510.

Smith, A., Voß, J. P., & Grin, J. (2010). Innovation studies and sustainability transitions: The allure of the multi-level perspective and its challenges. *Research Policy, 39*, 435–448.

Sterman, J. (2001). *Business dynamics: Systems thinking for a complex world.* Boston: Irwin.

Suurs, R. A. A., & Hekkert, M. (2009). Cumulative causation in the formation of a technological innovation system: The case of biofuels in the Netherlands. *TFSC, 76*, 1003–1020.

Unruh, G. C. (2000). Understanding carbon lock-in. *Energy Policy, 28*(12), 817–830.

Walrave, B., & Raven, R. (2016). Modelling the dynamics of technological innovation systems. *Research Policy, 45*(9), 1833–1844.

Walz, R. (2007). The role of regulation for sustainable infrastructure innovations: The case of wind energy. *International Journal of Public Policy, 2*(1/2), 57–88.

Walz, R., & Köhler, J. (2014). Using lead market factors to assess the potential for a sustainability transition. *Environmental Innovation and Societal Transitions, 10*, 20–41.

Walz, R., & Nowak Delgado, J. (2012). Innovation in sustainability technologies in newly industrializing countries – Results from a case study on wind energy. *Innovation and Development, 2*(1), 87–109.

Wang, Z., Qin, H., & Lewis, J. (2012). China's wind power industry: Policy support, technological achievements, and emerging challenges. *Energy Policy, 51*, 80–88.

Weber, M., & Rohracher, H. (2012). Legitimizing research, technology and innovation policies for transformative change: Combining insights from innovation systems and multi-level perspective in a comprehensive 'failures' framework. *Research Policy, 41*, 1037–1047.

Zhang, S., Andrews-Speed, P., & Zhao, X. (2013). Political and institutional analysis of the successes and failures of China's wind power policy. *Energy Policy, 56*, 331–340.

Zhao, X., Wang, F., & Wang, M. (2012). Large-scale utilization of wind power in China: Obstacles of conflict between market and planning. *Energy Policy, 48*, 222–232.

Zheng, H., Wang, J., Byrne, J., & Kurdgelashvili, L. (2013). Review of wind power tariff policies in China. *Energy Policy, 53*, 41–50.

The manufacturer's authorised representative in the EU is Springer Nature Customer Service Centre GmbH, Europaplatz 3, 69115 Heidelberg, Germany. If you have any concerns regarding our products, please contact ProductSafety@springernature.com

Printed and bound by CPI Group (UK) Ltd, Croydon, CR0 4YY
05/10/2026
02228122-0009